MOLECULAR BIOLOGY

LABFAX

I: Recombinant DNA

The LABFAX series

Series Editors:

B.D. HAMES Department of Biochemistry and Molecular Biology, University of Leeds, Leeds LS2 9JT, UK

D. RICKWOOD Department of Biology, University of Essex, Wivenhoe Park, Colchester CO4 3SQ, UK

MOLECULAR BIOLOGY LABFAX
CELL BIOLOGY LABFAX
CELL CULTURE LABFAX
BIOCHEMISTRY LABFAX
VIROLOGY LABFAX
PLANT MOLECULAR BIOLOGY LABFAX
IMMUNOCHEMISTRY LABFAX
CELLULAR IMMUNOLOGY LABFAX
ENZYMOLOGY LABFAX
PROTEINS LABFAX

Forthcoming titles
MOLECULAR BIOLOGY LABFAX II: GENE ANALYSIS LABFAX

MOLECULAR BIOLOGY

LABFAX

I: Recombinant DNA

EDITED BY

T.A. BROWN

Department of Biomolecular Sciences
UMIST
Manchester M60 1QD
UK

ACADEMIC PRESS
SAN DIEGO LONDON BOSTON
NEW YORK SYDNEY TOKYO TORONTO

This book is printed on acid-free paper.

Academic Press
525 B Street, Suite 1900, San Diego, California 92101-4495, USA
http://www.apnet.com

Academic Press Limited
24–28 Oval Road, London NW1 7DX, UK
http://www.hbuk.co.uk/ap/

ISBN 0-12-136055-5

Library of Congress Cataloging-in-Publication Data

Brown, T. A. (Terence A.)
Molecular biology labfax / by Terry Brown.—2nd ed.
v. <1>: cm. — (Labfax series)
Contents: 1. Recombinant DNA
Includes bibliographical references and index.
ISBN 0–12–136055–5
1. Molecular biology—Handbooks, manuals, etc. 2. Molecular biology—Tables. I. Title. II. Series.
QH506.B77 1998
572.8—dc21 97–45533
CIP

A catalogue record for this book is available from the British Library

Typeset by Phoenix Photosetting, Chatham
Printed in Great Britain by The University Press, Cambridge.

98 99 00 01 02 03 CU 9 8 7 6 5 4 3 2 1

PREFACE TO VOLUME I OF THE SECOND EDITION

In the eight years since publication of *Molecular Biology Labfax* there has been a vast proliferation of molecular biology techniques. To reflect this growth the second edition therefore differs from the first in being split into *Volume 1: Recombinant DNA* and *Volume 2: Gene Analysis*. The *Recombinant DNA* volume covers the general information that is relevant to most if not all molecular biology experiments. All five chapters – on bacteria and bacteriophages, restriction and methylation, DNA and RNA modifying enzymes, genomes, and cloning vectors – are direct descendents of chapters in the first edition and most are organized along the same lines. The main difference from the first edition, other than the increased information content, is that I have tried to present the data in a more accessible manner, avoiding many of the design errors that I made in the first edition where, to be honest, I tried to squeeze a litre into a 500-ml flask. I hope that the result is a more user friendly book.

I would like to thank Rich Roberts and Yasukazu Nakamura for allowing me to reproduce information from the Web sites that they maintain on, respectively, restriction enzymes and codon usage, and Michael McClelland for letting me use his compilation of data on DNA methylation. I am also grateful to the various suppliers of molecular biology reagents who have made available vector maps and other information: these are acknowledged in full at the relevant places in thc book.

In the Preface to the first edition I stated that you need to be slightly unbalanced to compile a databook. I will not comment on the state of mind of someone who decides to repeat the experience. Once again, my wife Keri is the main reason that I have come through unscathed.

T.A. Brown

PREFACE TO VOLUME I OF THE FIRST EDITION

There is nothing new under the sun and so it would be foolhardy to suggest that *Molecular Biology Labfax* is an entirely new departure in scientific publishing. It is, however, different from the existing cloning manuals in that it is designed as a companion rather than a guide for molecular biology research. *Molecular Biology Labfax* does not contain procedures or methodology but instead is a detailed compendium of the essential information – on genotypes, reagents, enzymes, reaction conditions, cloning vectors and suchlike – that is needed to plan and carry out molecular biology research. Some of this information is already available in cloning manuals, catalogues and possibly on pieces of paper kept somewhere safe, but tracking down exactly what you need to know takes time and can be a frustrating experience. With molecular biology becoming an increasingly sophisticated science, an acute need has arisen for a databook to complement the traditional cloning manuals. *Molecular Biology Labfax* is intended to meet this need.

To be useful, the coverage of *Molecular Biology Labfax* has to be right. The scope of the book is of necessity a compromise between a desire to include everything and a need to keep within a reasonable size limit. The reader will expect to find extensive details of *Escherichia coli* genotypes and genetic markers, restriction enzymes, DNA and RNA modifying enzymes, chemicals and reagents, cloning vectors, restriction fragment patterns and suchlike. These topics are covered in as comprehensive a way as possible, so for instance in Chapter 4 all the known restriction enzymes are described along with reaction conditions for all the commercially available ones. A few items that might be expected are not included on the basis that they are of specialist interest, an example being cloning vectors for eukaryotes. Topics such as these will be covered by future editions in the Labfax series. Although experimental protocols are not given, certain key information is provided for subjects such as the growth of *E. coli* strains, use of radiochemicals, electrophoresis of nucleic acids and hybridization analysis. These topics are of widespread importance in molecular biology procedures and so warrant sections of their own. Readers using other standard techniques will find their needs met by the data presented throughout the book. For instance, the DNA sequencer will find data on dideoxynucleotides and Maxam–Gilbert reagents (Chapter 2), radionucleotides and detection methods (Chapter 3), enzymes for chain termination sequencing (Chapter 5), M13 cloning vectors (Chapter 6), and electrophoresis systems (Chapter 8), as well as details of the genetic code and codon usages for interpretation of sequence information (Chapter 7).

A second essential requirement is accuracy. Wherever possible I have double-checked items that I have had doubts about, going back to the original publications if necessary. In a few cases the literature contains annoying contradictions that I have been unable to resolve, with *E. coli* genotypes providing some of the biggest headaches. The relevant entries carry a footnote or other warning to alert the reader and I welcome enlightenment if anyone knows any of the answers.

Without the help of a number of people *Molecular Biology Labfax* would never have been completed. I am very grateful to Rich Roberts, Toshimichi Ikemura and Michael McClelland for their contributions, as well as GIBCO–BRL, Pharmacia, Promega, USB, Clontech, Strategene and FMC for providing artwork for the cloning vectors and electrophoresis sections. I would like to give a general thank-you to the various colleagues and friends who helped me out with points here and there. Half-way through the enterprise it became clear that you need to be slightly unbalanced to compile a databook: I became even more worried on the occasions when I thought I was actually enjoying the experience. Because of this I must thank my wife, Keri, who made sure I survived to tell the tale.

T.A. Brown

CONTENTS

VOLUME II: CONTENTS

ABBREVIATIONS

AMV	avian myeloblastosis virus
ATP	adenosine 5′-triphosphate
BAP	bacterial alkaline phosphatase
Bis	Bis(hydroxymethyl)amino-methane
bp	base pair
BSA	bovine serum albumin
cccDNA	covalently closed circular DNA
cDNA	copy DNA
CDP	cytidine 5′-diphosphate
CEPH	Centre d'Études du Polymorphisme Humain
CHLC	Co-operative Human Linkage Center
CIAP	calf intestine alkaline phosphatase
CIP	calf intestine phosphatase
CMP	cytidine 5′-monophosphate
CoA	coenzyme A
CTP	cytidine 5′-triphosphate
Dam	DNA adenine methylase
Dcm	DNA cytosine methylase
DDBJ	DNA Database of Japan
DMF	dimethylformamide
DMSO	dimethyl sulphoxide
DNA	deoxyribonucleic acid
DNase	deoxyribonuclease
dsDNA	double-stranded DNA
dsRNA	double-stranded RNA
DTT	dithiothreitol
EBI	European Bioinformatics Institute
EDTA	ethylenediaminetetra-acetic acid
EGTA	ethyleneglycobis(aminoethyl) tetra-acetic acid
EMBL	European Molecular Biology Laboratory
EST	expressed sequence tag
exo	exonuclease
F	fertility
GDB	Genome Database
GTP	guanosine 5′-triphosphate
Hfr	high frequency of recombination
IPTG	isopropyl-β-D-thiogalactopyranoside
kb	kilobase
kd	kilodalton
Lac	lactose
LrRNA	large ribosomal RNA
MGD	Mouse Genome Database
M-MuLV	Moloney murine leukaemia virus
mRNA	messenger RNA
NAD	nicotinamide adenine dinucleotide
NCBI	National Center for Biotechnology Information
NDP	nucleotide diphosphate
NMP	nucleotide monophosphate
NTP	nucleotide triphosphate
nt	nucleotides
OMIM	Online Mendelian Inheritance in Man
ONPG	*o*-nitrophenyl-β-D-galactopyranoside
ORF	open reading frame
PEG	polyethylene glycol
PCR	polymerase chain reaction
PIR	Protein Information Resource
RAV2	Rous-associated virus 2
RF	replicative form
RNA	ribonucleic acid
RNase	ribonuclease
rRNA	ribosomal RNA
RT-PCR	reverse transcriptase polymerase chain reaction
SDS	sodium dodecyl sulfate
sp.	species
ssDNA	single-stranded DNA
ssRNA	single-stranded RNA
TDP	thymidine 5′-diphosphate

TIGR	The Institute for Genome Research
T_m	melting temperature
Tris	Tris(hydroxymethyl)aminomethane
tRNA	transfer RNA
UTP	uridine 5′-triphosphate
u.v.	ultraviolet
X-gal	5-bromo-3-indolyl-β-D-galactopyranoside
YAC	yeast artificial chromosome

CHAPTER 1
BACTERIA AND BACTERIOPHAGES

The first section of Chapter 1 provides information on the genotypes of *Escherichia coli* strains used in recombinant DNA experiments, and the second and third sections give genetic data on bacteriophages λ and M13, respectively. The final section describes culture media for *E. coli* and buffers for working with phage λ.

1. *ESCHERICHIA COLI* STRAINS USED IN RECOMBINANT DNA EXPERIMENTS

1.1 Standard format for describing a genotype

Genotypes of *E. coli* strains are described in accordance with a standard nomenclature.

Individual genes

(i) Each mutant locus is described by a three-letter abbreviation.
ara = arabinose utilization

(ii) The capital letter following the locus refers to the individual gene that is mutated.
araD = L-ribulosephosphate 4-epimerase

(iii) Numbers following the gene designation refer to the specific allele involved.
araD139

(iv) A superscript '–' is generally not used as, by convention, only mutated genes are listed in a genotype. A superscript '+' may be used to emphasize a locus or gene that is wild type.
lac^{+} = no mutations in the genes involved in lactose utilization

(v) A superscript 'q' indicates a constitutive mutation.
$lacI^{q}$ = constitutive expression of the gene for the Lac repressor

(vi) An amber mutation is denoted by 'am' following the gene designation.
malBam

(vii) If an antibiotic response is listed in the genotype then a superscript 'r' or 's' is used to denote resistance or sensitivity respectively.
kan^{r} = kanamycin resistant

Deletions

(i) Deletions are denoted by 'Δ' with the deleted gene or genes listed in brackets, possibly followed by an allele designation outside of the brackets.
Δ(*gal-uvrB*)*40* = deletion of the region from *gal* to *uvrB*

Insertions

(i) An insertion is denoted by '::', preceded by the position of the insertion and followed by the inserted DNA.

trpC22::Tn*10* = insertion of Tn*10* into *trpC*

(ii) Alternatively, the map position of the insertion is denoted by a three-letter code. The first letter is always *z*, followed by *a–i* to indicate a 10 min interval, and *a–i* to indicate a 1 min interval.

zhg::Tn*10* = insertion of Tn*10* at 87 min

Fusions

(i) A fusion is denoted in the same way as a deletion, except that the symbol 'Φ' is used.

(ii) A superscript '+' denotes that the fusion involves an operon rather than a single gene.

Φ(*ompF-lacZ*$^+$) = fusion between *ompF* and the *lac* operon

(iii) A prime (′) is used to denote that a fused gene is incomplete. If the prime appears after the gene then the gene is deleted in the 3′ region, if before the gene then the deletion is in the 5′ region.

Φ(*ompC′-lacZ*$^+$) = fusion in which *ompC* is deleted in the 3′ region

Plasmids and phages

(i) A plasmid or lysogenic phage carried by the bacterium is listed at the end of the genotype in brackets, possibly with relevant genetic information about the plasmid or phage.

(pMC9) = carries pMC9 plasmid

(P2) = carries P2 phage

(λ *c*Its857) = carries λ with a temperature-sensitive *c*I mutation

Fertility status

(i) Strains are assumed to be F$^-$ unless the status is given.

(ii) F$^+$ and Hfr strains are denoted by the relevant symbol at the start of the genotype.

(iii) If the strain is F′ then the genes carried by the F plasmid are listed in square brackets. The F′ status is usually placed at the end of the genotype.

F′[*traD36 proAB*$^+$ *lacI*^{q} *lacZΔM15*]

1.2 Genotypes of popular strains

The genotypes of *E. coli* strains used in recombinant DNA experiments are listed in *Table 1*. These genotypes have been checked for accuracy by comparing descriptions from different sources including, wherever possible, the original publications. This has revealed a number of contradictions, though in most cases these can be put down to obvious errors in secondary publications. There are, however, a few important uncertainties that cannot be resolved from the literature, as well as some instances of two strains of different genotypes circulating under the same name. These uncertainties are described in the footnotes to *Table 1*.

You should bear in mind that no genotype can be considered complete and wholly accurate and it is quite possible that a strain contains unrecognized mutations in genes not covered in the genotype. Many of these mutations will be irrelevant to the use of the strain in a recombinant DNA experiment, but others might have an impact. This is one of the reasons why it is worth trying different strains if a particular cloning experiment or other manipulation is not working as planned.

Table 1. Genotypes of *E. coli* strains used in recombinant DNA experiments

Strain	Genotype	References
594	*lac galK2 galT22 rpsL179*	1
1101	F^+ *supE*	2
71/18[a]	*supE thi-1* Δ(*lac-proAB*) F′[*proAB*$^+$ *lacI*^{q} *lacZ*Δ*M15*]	3–6
71/18*mutS*[a]	*supE thi-1* Δ(*lac-proAB*) *mutS*::Tn*10* F′[*proAB*$^+$ *lacI*^{q} *lacZ*Δ*M15*]	3–6
χ1776[b]	*tonA53 dapD8 minA1 supE44* Δ(*gal-uvrB*)*40 minB2 rfb-2 gyrA25 thyA142 oms-2 metC65 oms-1* (*tte-1*) Δ(*bioH-asd*)*29 cycB2 cycA1 hsdR2*	7, 8
ABLE[c, d]	*mcrA mcrCB mcrF mrr hsdS kan*^{r} *lac*(*lacZ* ω^-) F′[*proAB*$^+$ *lacI*^{q} *lacZ*Δ*M15* Tn*10*(*tet*^{r})]	9
AG1[e]	*recA1 endA1 gyrA96 thi-1 hsdR17 supE44 relA1*	10
AM1	*supE44 hsdR17 recA1 endA1 gyrA96 thi-1 relA1* F′[*lacI*^{q} *lacZ*$^+$ *lacAB*$^+$] (pItsCRE3)	
AR58	*sup*^{o} *galK2 galE*::Tn*10* (λcI857 ΔH1 *bio*$^-$ *uvrB kil*$^-$ cIII$^-$) *str*^{r}	11, 12
AR120	*sup*^{o} *galK2 nad*::Tn*10*(*tet*^{r}) (λcI$^+$ *ind*$^+$ pL-*lacZ fusion*) *str*^{r}	11, 12
AS1[f, g]	*supE44 hsdR17 endA1 thi-1* (λcI$^+$)	11, 12
BB4	*supF58 supE44 hsdR514 galK2 galT22 trpR55 metB1 tonA mcrA* Δ(*lac*)*U169* F′[*proAB*$^+$ *lacI*^{q} *lacZ*Δ*M15* Tn*10*(*tet*^{r})]	1
BHB2600	*supE supF* (λCH616)	13
BHB2688	N205 *recA* (λ*imm*434 cIts *b*2 *red*3 *E*am4 *S*am7)/λ	14, 15
BHB2690	N205 *recA* (λ*imm*434 cIts *b*2 *red*3 *D*am15 *S*am7)/λ	14, 15
BJ5183	*endA sbcBC recBC galK met str*^{r} *thi-1 bioT hsdR*	16
BL21[h, i]	*hsdS gal*	17–19
BL21(DE3)[h, i, j]	*hsdS gal* (λcIts857 *ind*1 *S*am7 *nin*5 *lacUV5*-T7 gene 1)	17, 18, 20
BL21(DE3)pLysS[h,i,j]	*hsdS gal* (λcIts857 *ind*1 *S*am7 *nin*5 *lacUV5*-T7 gene 1) (pLysS *cam*^{r})	17, 20
BMH71-18	see 71/18	
BMH71-18*mutS*	see 71/18*mutS*	
BNN93[k]	*supE44 hsdR thi-1 thr-1 leuB6 lacY1 tonA21 mcrA mcrB*	
BNN102[k]	*supE44 hsdR thi-1 thr-1 leuB6 lacY1 tonA21 mcrA mcrB hflA150*[*chr*::Tn*10*(*tet*^{r})]	21, 22
BW313	Hfr *lysA dut ung thi-1 recA spoT1*	23

Table 1. *E. coli* genotypes (continued)

Strain	Genotype	References
BW313(P3)	Hfr *lysA dut ung thi-1 recA spoT1* (P3 amp^r*am* tet^r*am* kan^r)	
C600[k]	*supE44 thi-1 thr-1 leuB6 lacY1 tonA21 mcrA*	16, 21, 22, 24–26
C600*galK*[k]	*supE44 hsdR? thi-1 thr-1 leuB6 lacY1 tonA21 mcrA mcrB? galK*	27
C600*hflA*[k]	*supE44 thi-1 thr-1 leuB6 lacY1 tonA21 mcrA hflA150*[*chr*::Tn*10*(tet^r)]	
CAG597	*rpoH165am zhg*::Tn*10 lacZam trpam phoam supCts malam rpsL*	28
CAG626	*lon lacZam trpam phoam supCts malam rpsL*	29, 30
CAG629	*rpoH165am zhg*::Tn*10 lacZam trpam phoam supCts malam rpsL lon*	
CAG748	*thi leu lacY tonA supE44* Δ(*lac*)*X90 dnaJ259 thr*::Tn*10*(tet^r)	31
CES200	*thr-1 ara-14* Δ(*gpt-proA*)*62 lacY1 tsx33 supE44 galK2 hisG4 rfbD1 rpsL31 kdgD51 xyl-5 mtl-1 argE3 leuB6 hsdR recB21 recC22 sbcB15 sbcC mcrA*	32, 33
CES201	*recA sbcBC recB21 recC22 hsdR*	34
CJ236[l]	*dut1 ung1 thi-1 relA1 mcrA* (pCJ105[cam^r F′])	35
CJ236(P3)[l]	*dut1 ung1 thi-1 relA1 mcrA* (pCJ105[cam^r F′]) (P3 amp^r*am* tet^r*am* kan^r)	
CPLK[c]	*mcrA mcrCB mcrF mrr hsdS* str^r *lac*(*lacZ* ω^-)	
CR34	see C600	
CSH18	*supE thi-1* Δ(*lac-pro*) F′[$proAB^+$ Δ(*lacZ*)*H125*]	36, 37
D1210	*supE44 recA13 ara-14* Δ(*gpt-proA*)*62* $lacI^q$ *galK2 rpsL20 xyl-5 mtl-1* Δ(*mcrCB-hsdSMR-mrr*)	38
D1210HP	*supE44 recA13 ara-14* Δ(*gpt-proA*)*62* $lacI^q$ *galK2 rpsL20 xyl-5 mtl-1* Δ(*mcrCB-hsdSMR-mrr*) (λ*c*Its857 *xis kil*)	39
DB1316	*recD1014 mcrB1 hsdR2 zjj202*::Tn*10* (tet^r)	33, 40
DH1[m]	*supE44 hsdR17 recA1 endA1 gyrA96 thi-1 relA1*	16, 41
DH5	*supE44 hsdR17 recA1 endA1 gyrA96 thi-1 relA1*	16, 41
DH5α	*supE44* Δ(*lac*)*U169 hsdR17 recA1 endA1 gyrA96 thi-1 relA1 deoR* (ϕ80 *lacZ*Δ*M15*)	16
DH5αF′	F′ *supE44* Δ(*lac*)*U169 hsdR17 recA1 endA1 gyrA96 thi-1 relA1 deoR* (ϕ80 *lacZ*Δ*M15*)	16

Table 1. *E. coli* genotypes (continued)

Strain	Genotype	References
DH5αF′IQ	*supE44* Δ(*lac*)*U169 hsdR17 recA1 endA1 gyrA96 thi-1 relA1 deoR* F′[*proAB*$^+$ *lacI*^{q} *lacZ*Δ*M15 zzf*::Tn5(*kan*^{r})] (φ80 *lacZ*Δ*M15*)	
DH5αMCR	*supE44* Δ(*lac*)*U169 hsdR17 recA1 endA1 gyrA96 thi-1 relA1 deoR mcrA* Δ(*mcrCB-hsdSMR-mrr*) (φ80 *lacZ*Δ*M15*)	
DH5α*mcrAB*	*supE44* Δ(*lac*)*U169 hsdR17 recA1 endA1 gyrA96 thi-1 relA1 deoR mcrA* Δ(*mcrCB-hsdSMR-mrr*) (φ80 *lacZ*Δ*M15*)	
DH10B	Δ(*lac*)*X74 deoR recA1 endA1 araD139* Δ(*ara-leu*)*7679 galU galK rpsL nupG mcrA* Δ(*mcrCB-hsdSMR-mrr*) (φ80 *lacZ*Δ*M15*)	
DH10B(P3)	Δ(*lac*)*X74 deoR recA1 endA1 araD139* Δ(*ara-leu*)*7679 galU galK rpsL nupG mcrA* Δ(*mcrCB-hsdSMR-mrr*) (φ80 *lacZ*Δ*M15*) (P3 *amp*^{r}*am tet*^{r}*am kan*^{r})	
DH11S	Δ(*lac-proAB*) Δ(*recA*)*1398 deoR rpsL srl thi mcrA* Δ(*mcrCB-hsdSMR-mrr*) F′[*proAB*$^+$ *lacI*^{q} *lacZ*Δ*M15*] (φ80 *lacZ*Δ*M15*)	
DH12S	*araD139* Δ(*ara-leu*)*7679* Δ(*lac*)*X74 galU galK rpsL deoR nupG recA1 mcrA* Δ(*mcrCB-hsdSMR-mrr*) F′[*proAB*$^+$ *lacI*^{q} *lacZ*Δ*M15*] (φ80 *lacZ*Δ*M15*)	
DH20	*supE44 hsdR17 recA1 endA1 gyrA96 thi-1 relA1* F′[*proAB*$^+$ *lacI*^{q} *lacZ*$^+$]	16
DH21	*supE44 hsdR17 recA1 endA1 gyrA96 thi-1 relA1* F′[*proAB*$^+$ *lacI*^{q} *lacZ*$^+$]	16
DK1	*hsdR2 araD139* Δ(*ara-leu*)*7679* Δ(*lac*)*X74 galU galK rpsL mcrA mcrB1* Δ(*srl-recA*)*306*	11
DL538	*hsdR mcrA mcrB supE44 recD1009 sbcC201*	40, 42
DM1	*dam zac*::Tn9(*cam*^{r}) *dcm mcrB hsdR gal1 gal2 ara lac thr leu ton*^{r} *tsx*^{r} *su*^{o}	
DP50*supF*^{h,n}	*supE44 supF58 hsdS3 dapD8 lacY1* Δ(*gal-uvrB*)*47 gyrA29 tonA53* Δ(*thyA*)*57 mcrA*	43
ED8654	*supE44 supF58 hsdR514 metB1 lacY1 or lac-3 galK2 galT22 trpR55 mcrA*	44, 45
ED8767^o	*supE44 supF58 hsdS3 recA56 galK2 galT22 metB1 mcrA mcrB*	45
ER1370	*tonA2* Δ(*lacZ*)*r1 supE44 trp31 his-1 argG6 rpsL104 xyl-7 mtl-2 metB1 serB28*	25
ER1378	*tonA2* Δ(*lacZ*)*r1 supE44 trp31 his-1 argG6 rpsL104 xyl-7 mtl-2 metB1 mcrB1 hsdR2*	25
ER1381	*tonA2* Δ(*lacZ*)*r1 supE44 trp31 his-1 argG6 rpsL104 xyl-7 mtl-2 metB1 hsdR2*	25

Table 1. *E. coli* genotypes (continued)

Strain	Genotype	References
ER1398[g]	*supE44 hsdR2 endA1 thi-1 mcrB1*	25
ER1414	*IN(rrnD-rrnE)1 mcrB1 hsdR2 zjj202*::Tn*10* (*tet*^{r}) *serB28*	46
ER1451	*supE44 endA1 hsdR2 gyrA96 relA1 thi-1 mcrA mcrB1* Δ(*lac-proAB*) F′[*traD36 proAB*$^+$ *lacI*^{q} *lacZ*Δ*M15*]	25
ER1458[p]	Δ(*lac*)*U169 lon-100 araD139 rpsL supF mcrA trpC22*::Tn*10*(*tet*^{r}) *hsdR2 mcrB1 serB28*	46, 47
ER1562[g]	*supE44 hsdR2 endA1 thi-1 mcrA1272*::Tn*10* (*tet*^{r}) *mcrB1*	
ER1563[g]	*supE44 hsdR17 endA1 thi-1 mcrA1272*::Tn*10* (*tet*^{r})	
ER1564	*tonA2* Δ(*lacZ*)*r1 supE44 trp31 his-1 argG6 rpsL104 xyl-7 mtl-2 metB1 mcrA1272*::Tn*10* (*tet*^{r}) *hsdR2*	46
ER1565	*tonA2* Δ(*lacZ*)*r1 supE44 trp31 his-1 argG6 rpsL104 xyl-7 mtl-2 metB1 mcrA1272*::Tn*10* (*tet*^{r}) *hsdR2 mcrB1*	46
ER1578[q]	Δ(*lac*)*U169 lon-100 araD139 rpsL supF mcrA zjj202*::Tn*10*(*tet*^{r}) *hsdR2 mcrB1 serB28* (pMC9)	
ER1647	*tonA2* Δ(*lacZ*)*r1 supE44 mcrA1272*::Tn*10*(*tet*^{r}) *trp-31 his-1 rpsL104 xyl-7 mtl-2 metB1 recD1014* Δ(*mcrCB-hsdSMR-mrr*)*102*::Tn*10*(*tet*^{r})	40, 48
ER1648	*tonA2* Δ(*lacZ*)*r1 supE44 mcrA1272*::Tn*10*(*tet*^{r}) *trp-31 his-1 rpsL104 xyl-7 mtl-2 metB1* Δ(*mcrCB-hsdSMR-mrr*)*102*::Tn*10*(*tet*^{r})	40, 48
ER1793	*tonA2* Δ(*lacZ*)*r1 supE44 trp31 mcrA his-1 rpsL104 xyl-7 mtl-2 metB1* Δ(*mcrCB-hsdSMR-mrr*)*114*::IS*10*	49
ER1821[g]	*mcrA endA1 supE44 thi-1* Δ(*mcrCB-hsdSMR-mrr*)*114*::IS*10*	
ER2267[g]	*mcrA endA1 supE44 thi-1* Δ(*mcrCB-hsdSMR-mrr*)*114*::IS*10* Δ(*lac*)*U169 recA1* F′[*proAB*$^+$ *lacI*^{q} *lacZ*Δ*M15 zzf*::mini-Tn*10*(*kan*^{r})]	
ER2507[h]	*leuB6 supE44 ara-14 galK2 lacY1 rpsL20 xyl-5* Δ(*mcrCB-hsdSMR-mrr*) *mtl-1* Δ(*malB*) Δ(*argF-lac*)*U169 zjc*::Tn*5*(*kan*^{r}) *tonA2*	
ER2508[h]	*leuB6 supE44 ara-14 galK2 lacY1 rpsL20 xyl-5* Δ(*mcrCB-hsdSMR-mrr*) *mtl-1* Δ(*malB*) Δ(*argF-lac*)*U169 zjc*::Tn*5*(*kan*^{r}) *tonA2 lon*::miniTn*10*(*tet*^{r})	29
ES1301*mutS*	*lacZ53 thyA36 rha-5 metB1 deoC IN*(*rrnD-rrnE*) *mutS201*::Tn*5*	
GM48	*thr leu thi lacY galK galT ara tonA tsx dam dcm supE44*	3
GM2163	*ara-14 leuB6 thi-1 tonA31 lacY1 tsx-78 galK2 galT22 supE44 hisG4 rpsL136 xyl-5 mtl-1 dam13*::Tn*9*(*cam*^{r}) *dcm-6 mcrB1 hsdR2 mcrA*	25, 50

Table 1. *E. coli* genotypes (continued)

Strain	Genotype	References
GM2929	*ara-14 leuB6 thi-1 tonA31 lacY1 tsx-78 galK2 galT22 supE44 hisG4 rpsL136 xyl-5 mtl-1 dam13*::Tn*9*(*cam*^{r}) *dcm-6 mcrB1 hsdR2 mcrA recF13*	50
HB101[h, r]	Δ(*gpt-proA*)*62 leuB6 supE44 ara-14 galK2 lacY1 rpsL20 xyl-5* Δ(*mcrCB-hsdSMR-mrr*) *mtl-1 recA13*	16, 25, 51, 52
HMS174	*recA1 hsdR rif*^{r}	18, 53
INVαF′	F′ *endA1 recA1 hsdR17 supE44 thi-1 gyrA96 deoR relA1* Δ(*lac*)*U169* (ϕ80 *lacZ*Δ*M15*)	
JC7623	*recB1 recC22 sbcB15 mcrA*	54
JC9956	*recA99 thr-1 leu-6 thi-1 lacY1 galK2 ara-14 xyl-5 mtl-1 proA2 his-4 argE3 str-31 tsx-33*	55
JM83[s]	*ara* Δ(*lac-proAB*) *rpsL* (ϕ80 *lacZ*Δ*M15*)	3, 56
JM101	*supE thi-1* Δ(*lac-proAB*) F′[*traD36 proAB*$^+$ *lacI*^{q} *lacZ*Δ*M15*]	3, 25, 57
JM103[t]	*supE thi-1 endA1 sbcBC rpsL* Δ(*lac-proAB*) F′[*traD36 proAB*$^+$ *lacI*^{q} *lacZ*Δ*M15*] (P1)	16, 58
JM103Y	see KK2186	
JM105[u]	*endA sbcBC hsdR4 rpsL thi-1* Δ(*lac-proAB*) F′[*traD36 proAB*$^+$ *lacI*^{q} *lacZ*Δ*M15*]	3
JM106	*supE44 endA1 hsdR17 gyrA96 relA1 thi-1 mcrA* Δ(*lac-proAB*)	3
JM107	*supE44 endA1 hsdR17 gyrA96 relA1 thi-1 mcrA* Δ(*lac-proAB*) F′[*traD36 proAB*$^+$ *lacI*^{q} *lacZ*Δ*M15*]	3, 25
JM108	*recA1 supE44 endA1 hsdR17 gyrA96 relA1 thi-1 mcrA* Δ(*lac-proAB*)	3
JM109	*recA1 supE44 endA1 hsdR17 gyrA96 relA1 thi-1 mcrA* Δ(*lac-proAB*) F′[*traD36 proAB*$^+$ *lacI*^{q} *lacZ*Δ*M15*]	3
JM109(DE3)[j]	*recA1 supE44 endA1 hsdR17 gyrA96 relA1 thi-1* Δ(*lac-proAB*) F′[*traD36 proAB*$^+$ *lacI*^{q} *lacZ*Δ*M15*] (λcIts857 *ind*1 *S*am7 *nin*5 *lacUV5*-T7 gene 1)	59
JM110[v]	*dam dcm supE44 thi-1 leu rpsL lacY galK galT ara tonA thr tsx* Δ(*lac-proAB*) F′[*traD36 proAB*$^+$ *lacI*^{q} *lacZ*Δ*M15*]	3
K802[w]	*supE44 hsdR galK2 galT22 metB1 mcrA mcrB1*	25, 60
K803	see WA803	
KB35	F$^+$ *supE44*	2
KK2186[t]	*supE thi-1 endA1 hsdR4 sbcBC rpsL* Δ(*lac-proAB*) F′[*traD36 proAB*$^+$ *lacI*^{q} *lacZ*Δ*M15*]	61

Table 1. *E. coli* genotypes (continued)

Strain	Genotype	References
KLF41	*leuB6 hisG1 recA1 argG6 metB1 lacY1 gal-6 xyl-7 mtl-2 malA1 rpsL104 tonA tsx supE44* F′141	1
KM392	*hsdR514 supE44 supF58 lacY galK2 galT22 metB1 trp55 mcrA* Δ(*lac*)*U169 proC*::Tn5	11
Kro	*recA*	55
KS1000	*ara* Δ(*lac-proAB*) *nalA arg1am rif thi-1* Δ(*tsp*)::*kan*^{r} *eda-51*::Tn10(*tet*^{r}) F′[*lacI*^{q} *lac pro*]	62
KW251	*supE44 galK2 galT22 metB1 hsdR2 mcrA mcrB1 argA81*::Tn*10 recD1014*	
LE30	*mutD5 rpsL azi galU95*	1
LE292	HfrH *argEam rpoB galT*::[λΔ(*int-F*II)]	1
LE392	*supE44 supF58 hsdR514 galK2 galT22 metB1 trpR55 lacY1* or Δ(*lacIZY*)*6 mcrA*	16, 25, 44, 45
LE392.23	*supE44 supF58 hsdR514 galK2 galT22 metB1 trpR55 lacY1* or Δ(*lacIZY*)*6 mcrA* Δ(*argF-lac*)*U169*	1
LG90	Δ(*lac-proAB*)	63
M5219	*lacZ trpA rpsL* (λ*bio*252 *c*Its857 *H*1)	64, 65
MAL103	Δ(*gpt-proAB-argF-lac*)*XIII rpsL* [Mud1 (*lac, Ap*)] (Mucts62)	1
MB100	Δ(*argF-lac*)*U169 rpsL150 relA1 flbB3501 deoC1 ptsF25 rbsR leuABCD*::Tn*10*	1
MB101	*araCam araD* Δ(*argF-lac*)*U169 trpam malBam rpsL relA thi supF* Φ(*araBA′-lacZ*$^+$)*101* [λ*p*1(209)]	1
MB408	*recF recB21 recC22 sbcB15 hflA hflB hsdR* (*tet*^{r})	66
MBM7007	*araCam araD* Δ(*argF-lac*)*U169 trpam malBam rpsL relA thi*	1
MBM7014	*araCam araD* Δ(*argF-lac*)*U169 trpam malBam rpsL relA thi supF*	1
MBM7014.5	*hsdR2 zjj202*::Tn*10*(*tet*^{r}) *araD139 araCU25am* Δ(*lac*)*U169 mcrB*	25
MBM7060	*araCam araD* Δ(*argF-lac*)*U169 trpam malBam rpsL relA thi supF* (λ*p*1048)	1
MC1000	*araD139* Δ(*araABC-leu*)*7679 galE15 galK16* Δ(*lac*)*X74 rpsL thi-1*	1
MC1061	*araD139* Δ(*araABC-leu*)*7679 galE15 galK16* Δ(*lac*)*X74 rpsL hsdR2 mcrA mcrB*	22, 67, 68
MC1061(P3)	*araD139* Δ(*araABC-leu*)*7679 galE15 galK16* Δ(*lac*)*X74 rpsL hsdR2 mcrA mcrB* (P3 *amp*^{r}*am tet*^{r}*am kan*^{r})	67, 69

Table 1. *E. coli* genotypes (continued)

Strain	Genotype	References
MC4100	*araD139 Δ(argF-lac)U169 rpsL150 relA1 flbB3501 deoC1 ptsF25 rbsR*	1
MH225	*araD139 Δ(argF-lac)U169 rpsL150 relA1 flbB3501 deoC1 ptsF25 rbsR Φ(ompC′-lacZ*$^+$*)10-25* [λ*p*1(209)]	1
MH513	*Δ(argF-lac)U169 rpsL150 relA1 flbB3501 deoC1 ptsF25 rbsR Φ(ompF′-lacZ*$^+$*)16-23* [λ*p*1(209)]	1
MH760	*araD139 Δ(argF-lac)U169 rpsL150 relA1 flbB3501 deoC1 ptsF25 rbsR ompR472*	1
MH1160	*araD139 Δ(argF-lac)U169 rpsL150 relA1 flbB3501 deoC1 ptsF25 rbsR ompR101*	1
MH1471	*araD139 Δ(argF-lac)U169 rpsL150 relA1 flbB3501 deoC1 ptsF25 rbsR envZ473*	1
MH2101	*araD139 Δ(argF-lac)U169 rpsL150 relA1 flbB3501 deoC1 ptsF25 rbsR ompR101 Φ(ompC′-lacZ*$^+$*)10-25* [λ*p*1(209)]	1
MH2472	*araD139 Δ(argF-lac)U169 rpsL150 relA1 flbB3501 deoC1 ptsF25 rbsR ompR472 Φ(ompC′-lacZ*$^+$*)10-25* [λ*p*1(209)]	1
MH5101	*Δ(argF-lac)U169 rpsL150 relA1 flbB3501 deoC1 ptsF25 rbsR ompR101 Φ(ompF′-lacZ*$^+$*)16-23* [λ*p*1(209)]	1
MH5473	*Δ(argF-lac)U169 rpsL150 relA1 flbB3501 deoC1 ptsF25 rbsR envZ473 Φ(ompF′-lacZ*$^+$*)16-23* [λ*p*1(209)]	1
MK30-3	*recA galE strA Δ(lac-proAB)* F′[*proAB*$^+$ *lacI*^{q} *lacZΔM15*]	
MM294[g]	*supE44 hsdR17 endA1 thi-1*	16, 25
MM294*c*I$^+$	see AS1	11,12
MV1184[x]	*ara Δ(lac-proAB) rpsL thi* (φ80 *lacZΔM15*) *Δ(srl-recA)306*::Tn*10*(*tet*^{r}) F′[*traD36 proAB*$^+$ *lacI*^{q} *lacZΔM15*]	56
MV1193	*Δ(lac-proAB) rpsL thi endA sbcB15 hsdR4 Δ(srl-recA)306*::Tn*10*(*tet*^{r}) F′[*traD36 proAB*$^+$ *lacI*^{q} *lacZΔM15*]	70
MZ-1	*galK* 8*att*L BamN_7N_{53} *c*Its857 *H*1 *his ilv bio N*$^+$	71
N99*c*I$^+$	*galK strA IN(rrnD-rrnE)1* (λ*c*I$^+$)	
N3098	*lig7ts supF*	1
N4830	see N4830-1	
N4830-1	*su*^{o} *his ilvA galK8 thi-1 thr1 leuB6 lacY1 tonA1 supE44 rfbD1 mcrA1 Δ(hemF-esp) Δ(bio-uvrB)* [λΔ*B*am *Δ(cro-attR)N*$^+$ *c*I857]	72

Table 1. *E. coli* genotypes (continued)

Strain	Genotype	References
NK5486	*thyA rha strA lacZam*	73
NM477	*mcrA thr-1 leuB6 thi-1 lacY1 supE44 rfbD1 tonA21* Δ(*hsdMS-mcrB*)*5*	40
NM514	*hsdR514 argH galE galX lycB7 strA mcrA*	45
NM519	*hsdR recB21 recC22 sbcA23*	74
NM522	*supE thi-1* Δ(*hsdMS-mcrB*)*5* Δ(*lac-proAB*) F′[*proAB*$^{+}$ *lacI*^{q} *lacZ*Δ*M15*]	75
NM531	*supE supF hsdR trpR lacY recA13 metB gal*	74
NM538	*supF hsdR trpR lacY mcrA*	76
NM539	*supF hsdR lacY mcrA* (P2cox)	76
NM554	*recA13 araD139* Δ(*ara-leu*)*7696 galE15 galK16* Δ(*lac*)*X74 rpsL hsdR2 mcrA mcrB1*	77
NM621	*hsdR mcrA mcrB supE44 recD1009*	42
P2392	*supE44 supF58 hsdR514 galK2 galT22 metB1 trpR55 lacY1* or Δ(*lacIZY*)*6 mcrA* (P2)	78
P2CPLK[c]	*mcrA mcrCB mcrF mrr hsdS str*^{r} *lac*(*lacZ* ω^{-}) (P2)	
P2PLK-17	*mcrA mcrB1 hsdR2 supE44 galK2 galT22 metB1* (P2)	
PLK-17	*mcrA mcrB1 hsdR2 supE44 galK2 galT22 metB1*	79
PLK-A	*mcrA mcrB1 hsdR2 supE44 galK22 galT22 metB1 recA lac*	79
PLK-F′	*mcrA mcrB1 hsdR2 supE44 galK2 galT22 metB1 recA lac* F′[*proAB*$^{+}$ *lacI*^{q} *lacZ*Δ*M15*Tn*10*(*tet*^{r})]	79
pop2136	*endA thi hsdK malT malPQ* (λcIts857)	
PR700[h]	*leuB6 supE44 ara-14 galK2 lacY1 rpsL20 xyl-5* Δ(*mcrCB-hsdSMR-mrr*) *mtl-1* Δ(*malB*) Δ(*argF-lac*)*U169 zjc*::Tn*5*(*kan*^{r})	16, 25, 51, 52
PR745[h]	*leuB6 supE44 ara-14 galK2 lacY1 rpsL20 xyl-5* Δ(*mcrCB-hsdSMR-mrr*) *mtl-1* Δ(*malB*) Δ(*argF-lac*)*U169 zjc*::Tn*5*(*kan*^{r}) *lon*::miniTn*10*(*tet*^{r})	29, 80
Q358	*supE hsdR mcrA* $\phi 80^{r}$	81
Q359	*supE hsdR mcrA* $\phi 80^{r}$ (P2)	81
R594	*galK2 galT22 rpsL179 lac*	82
RB791	W3110 *lacI*^{q}*L8*	83
RR1[h, r]	Δ(*gpt-proA*)*62 leuB6 supE44 ara-14 galK2 lacY1 rpsL20 xyl-5* Δ(*mcrCB-hsdSMR-mrr*) *mtl-1*	16, 25, 51, 52, 84, 85

Table 1. *E. coli* genotypes (continued)

Strain	Genotype	References
RT3	*araD139* Δ(*argF-lac*)*U169 rpsL150 relA1 flbB3501 deoC1 ptsF25 rbsR envZ3*	1
RT203	*araD139* Δ(*argF-lac*)*U169 rpsL150 relA1 flbB3501 deoC1 ptsF25 rbsR envZ3* Φ(*ompC′-lacZ*$^+$)*10-25* [λ*p*1(209)]	1
SCS110	*rpsL thr leu endA thi-1 lacY galK galT ara tonA tsx dam dcm supE44* Δ(*lac-proAB*) F′[*traD36 proAB*$^+$ *lacI*^{q} *lacZ*Δ*M15*]	86
SE3001	*araD139* Δ(*argF-lac*)*U169 rpsL150 relA1 flbB3501 deoC1 ptsF25 rbsR* Δ(*malK-lamB*)*1*	1
SE5000	*araD139* Δ(*argF-lac*)*U169 rpsL150 relA1 flbB3501 deoC1 ptsF25 rbsR recA56*	1
SG263	*araCam araD* Δ(*argF-lac*)*U169 trpam malBam rpsL relA thi supF malPQ*::Tn*10*	1
SG265	Δ(*gpt-proAB-argF-lac*)*XIII ara argEam gyrA rpoB thi-1 supP* (P1*cry*)	1
SG404	*araD139* Δ(*argF-lac*)*U169 rpsL150 relA1 flbB3501 deoC1 ptsF25 rbsR asd* (P1*cam*^{r}) F′141	1
SG480	*araD139* Δ(*argF-lac*)*U169 rpsL150 relA1 flbB3501 deoC1 ptsF25 rbsR* Δ(*malPQ-bioH-ompB*)*61*	1
SG608	*araD139* Δ(*argF-lac*)*U169 rpsL150 relA1 flbB3501 deoC1 ptsF25 rbsR* Φ(*ompC′-lacZ*$^+$)*10-25* [λ*p*RT2.3]	1
SG624	*araD139* Δ(*argF-lac*)*U169 rpsL150 relA1 flbB3501 deoC1 ptsF25 rbsR envZ22* Φ(*ompC′-lacZ*$^+$)*10-25* [λ*p*1(209)]	1
SG626	*araD139* Δ(*argF-lac*)*U169 rpsL150 relA1 flbB3501 deoC1 ptsF25 rbsR aroB* Φ(*ompC′-lacZ*$^+$)*10-25* [λ*p*1(209)]	1
SK1590	*gal thi-1 sbcBC endA hsdR4*	87
SK1592	*thi-1 supE endA sbcBC hsdR4*	3
SK2267	*endA1 hsdR4 supE44 thi-1 lacZ4* or *lac-61 gal-44 ton58 [rfa] recA1 sbcBC*	16
SL10	HfrH *thi-1 sup*^{o} Δ(*lac-proAB*) *galE* Δ(*pgl-bio*)	3
SMR10[c]	(λ*cos*2 Δ*B xis*1 *red*3 *gam*am210 *c*Its857 *nin*5 *S*am7)/λ	88, 89
SOLR[y]	*mcrA* Δ(*mcrCB-hsdSMR-mrr*)*171 sbcC recB recJ uvrC umuC*::Tn*5*(*kan*^{r}) *lac gyrA96 relA1 thi-i endA1* λ^r F′[*proAB*$^+$ *lacI*^{q} *lacZ*Δ*M15*]	90
SRB[y]	*mcrA* Δ(*mcrCB-hsdSMR-mrr*)*171 sbcC recJ uvrC umuC*::Tn*5*(*kan*^{r}) *supE44 lac gyrA96 relA1 thi-i endA1* F′[*proAB*$^+$ *lacI*^{q} *lacZ*Δ*M15*]	91, 92

Table 1. *E. coli* genotypes (continued)

Strain	Genotype	References
SRB(P2)[y]	*mcrA* Δ(*mcrCB-hsdSMR-mrr*)*171 sbcC recJ uvrC umuC*::Tn*5*(*kan*^{r}) *supE44 lac gyrA96 relA1 thi-i endA1* F′[*proAB*$^+$ *lacI*^{q} *lacZ*Δ*M15*] (P2)	91, 92
SURE[y]	*mcrA* Δ(*mcrCB-hsdSMR-mrr*)*171 endA1 supE44 thi-1 gyrA96 relA1 lac recB recJ sbcC umuC*::Tn*5*(*kan*^{r}) *uvrC* F′[*proAB*$^+$ *lacI*^{q} *lacZ*Δ*M15* Tn*10*(*tet*^{r})]	89
SURE2[y]	*mcrA* Δ(*mcrCB-hsdSMR-mrr*)*171 endA1 supE44 thi-1 gyrA96 relA1 lac recB recJ sbcC umuC*::Tn*5*(*kan*^{r}) *uvrC* F′[*proAB*$^+$ *lacI*^{q} *lacZ*Δ*M15* Tn*10*(*tet*^{r}) *Amy cam*^{r}]	89
SV101	*araD139* Δ(*argF-lac*)*U169 rpsL150 relA1 flbB3501 deoC1 ptsF25 rbsR malPQ*::Tn*10*	1
SW101	*araD139* Δ(*argF-lac*)*U169 rpsL150 relA1 flbB3501 deoC1 ptsF25 rbsR* Δ(*araABC-leu*)*7679 zab*::Tn*10*	1
TAP90	*supE44 supF58 hsdR pro leuB6 thi-1 rpsL lacY1 tonA1 recD1903*::mini-*tet*	93
TB1[s]	*hsdR ara* Δ(*lac-proAB*) *rpsL* (ϕ80 *lacZ*Δ*M15*)	
TD1	*araD139* Δ(*argF-lac*)*U169 rpsL150 relA1 flbB3501 deoC1 ptsF25 rbsR recA56 srlC300*::Tn*10*	3
TG1	*supE* Δ(*hsdMS-mcrB*)*5 thi* Δ(*lac-proAB*) F′[*traD36 proAB*$^+$ *lacI*^{q} *lacZ*Δ*M15*]	94
TG2	*supE* Δ(*hsdMS-mcrB*)*5 thi* Δ(*lac-proAB*) Δ(*srl-recA*)*306*::Tn*10*(*tet*^{r}) F′[*traD36 proAB*$^+$ *lacI*^{q} *lacZ*Δ*M15*]	95
TK821	*araD139* Δ(*argF-lac*)*U169 rpsL150 relA1 flbB3501 deoC1 ptsF25 rbsR ompR331*::Tn*10*	1
TK827	Δ(*argF-lac*)*U169 rpsL150 relA1 flbB3501 deoC1 ptsF25 rbsR* Φ(*ompF′-lacZ*$^+$)*16-23 ompR331*::Tn*10* [λ*p*1(209)]	1
TKB1[h, j]	*dcm ompT hsdS gal* (λcIts857 *ind*1 *Sam*7 *nin*5 *lacUV5*-T7 gene 1) (pTK *tet*^{r})	17, 20, 96
TKX1	Δ(*mcrA*)*183* Δ(*mcrCB-hsdSMR-mrr*)*173 endA1 supE44 thi-1 recA1 gyrA96 relA1 lac* F′[*proAB*$^+$ *lacI*^{q} *lacZ*Δ*M15* Tn*5*(*kan*^{r})] (pTK *tet*^{r})	96, 97
TOP10	*mcrA* Δ(*mcrCB-hsdSMR-mrr*) (ϕ80 *lacZ*Δ*M15*) Δ(*lac*)*X74 deoR recA1 araD139* Δ(*ara-leu*)*7679 galU galK rpsL endA1 nupG*	98
TOP10F′	*mcrA* Δ(*mcrCB-hsdSMR-mrr*) (ϕ80 *lacZ*Δ*M15*) Δ(*lac*)*X74 deoR recA1 araD139* Δ(*ara-leu*)*7679 galU galK rpsL endA1 nupG* F′[(*tet*^{r})]	
TOP10F′/P3	*mcrA* Δ(*mcrCB-hsdSMR-mrr*) (ϕ80 *lacZ*Δ*M15*) Δ(*lac*)*X74 deoR recA1 araD139* Δ(*ara-leu*)*7679 galU galK rpsL endA1 nupG* F′[(*tet*^{r})] (P3 *amp*^{r}*am tet*^{r}*am kan*^{r})	

Table 1. *E. coli* genotypes (continued)

Strain	Genotype	References
TOPP1,2	*rif*^{r} F′[*proAB*$^+$ *lacI*^{q} *lacZ*Δ*M15* Tn*10*(*tet*^{r})]	99
TOPP3	*rif*^{r} F′[*proAB*$^+$ *lacI*^{q} *lacZ*Δ*M15* Tn*10*(*tet*^{r}) (*kan*^{r})]	99
UT5600	*ara-14 leuB6 azi-6 lacY1 proC14 tsx-67* Δ(*ompT-fepC*)*266 entA403 trpE38 rfbD1 rpsL109 xyl-5 mtl-1 thi-1*	100
W5449	*hsdR18 supE44 recB21 recC22 sbcBC tonB56 tsx-33 ara14 argE3 galK2 his4 lacY1 leuB6 mtl-1 proA2 rpsL31 xyl-5 trpB9579 thi-1*	16
WA802[w]	*mcrA lacY1* or Δ(*lac*)*6 supE44 galK2 galT22 rfbD1 metB1 mcrB1 hsdR2*	40, 55, 101
WA803	*mcrA lacY1* or Δ(*lac*)*6 supE44 galK2 galT22 rfbD1 metB1 mcrB1 hsdS3*	40, 55, 101
XL1-Blue[y]	*recA1 endA1 gyrA96 thi-1 hsdR17 supE44 relA1 lac* F′[*proAB*$^+$ *lacI*^{q} *lacZ*Δ*M15* Tn*10*(*tet*^{r})]	10
XL1-Blue MR[y]	Δ(*mcrA*)*183* Δ(*mcrCB-hsdSMR-mrr*)*173 endA1 supE44 thi-1 recA1 gyrA96 relA1 lac*	97
XL1-Blue MRA[y]	Δ(*mcrA*)*183* Δ(*mcrCB-hsdSMR-mrr*)*173 endA1 supE44 thi-1 gyrA96 relA1 lac*	102
XL1-Blue MRA(P2)[y]	Δ(*mcrA*)*183* Δ(*mcrCB-hsdSMR-mrr*)*173 endA1 supE44 thi-1 gyrA96 relA1 lac* (P2)	102
XL1-Blue MRF′[y]	Δ(*mcrA*)*183* Δ(*mcrCB-hsdSMR-mrr*)*173 endA1 supE44 thi-1 recA1 gyrA96 relA1 lac* F′[*proAB*$^+$ *lacI*^{q} *lacZ*Δ*M15* Tn*10*(*tet*^{r})]	97
XL1-Blue MRF′ Kan[y]	Δ(*mcrA*)*183* Δ(*mcrCB-hsdSMR-mrr*)*173 endA1 supE44 thi-1 recA1 gyrA96 relA1 lac* F′[*proAB*$^+$ *lacI*^{q} *lacZ*Δ*M15* Tn*5*(*kan*^{r})]	102
XL1-Red	*endA1 gyrA96 thi-1 hsdR17 supE44 relA1 lac mutD5 mutS mutT* Tn*10*(*tet*^{r})	103
XL2-Blue[y]	*recA1 endA1 gyrA96 thi-1 hsdR17 supE44 relA1 lac* F′[*proAB*$^+$ *lacI*^{q} *lacZ*Δ*M15* Tn*10*(*tet*^{r}) *Amy cam*^{r}]	
XL2-Blue MRF′[y]	Δ(*mcrA*)*183* Δ(*mcrCB-hsdSMR-mrr*)*173 endA1 supE44 thi-1 recA1 gyrA96 relA1 lac* F′[*proAB*$^+$ *lacI*^{q} *lacZ*Δ*M15* Tn*10*(*tet*^{r}) *Amy cam*^{r}]	
XL*mutS* Kan^r [y]	Δ(*mcrA*)*183* Δ(*mcrCB-hsdSMR-mrr*)*173 endA1 supE44 thi-1 gyrA96 relA1 lac mutS*::Tn*10*(*tet*^{r}) F′[*proAB*$^+$ *lacI*^{q} *lacZ*Δ*M15* Tn*5*(*kan*^{r})]	104
XL*mutS* Kan^s [y]	Δ(*mcrA*)*183* Δ(*mcrCB-hsdSMR-mrr*)*173 endA1 supE44 thi-1 gyrA96 relA1 lac mutS*::Tn*10*(*tet*^{r}) F′[*proAB*$^+$ *lacI*^{q} *lacZ*Δ*M15* Tn*5*]	104

Table 1. *E. coli* genotypes (continued)

Strain	Genotype	References
XLO[y]	Δ(*mcrA*)*183* Δ(*mcrCB-hsdSMR-mrr*)*173 endA1 thi-1 recA1 gyrA96 relA1 lac* F′[*proAB*$^{+}$ *lacI*^{q} *lacZ*Δ*M15* Tn*10*(*tet*^{r})]	
XLOLR[y]	Δ(*mcrA*)*183* Δ(*mcrCB-hsdSMR-mrr*)*173 endA1 thi-1 recA1 gyrA96 relA1 lac* λ^{r} F′[*proAB*$^{+}$ *lacI*^{q} *lacZ*Δ*M15* Tn*10*(*tet*^{r})]	
XPORT	Δ(*mcrA*)*183* Δ(*mcrCB-hsdSMR-mrr*)*173 endA1 supE44 thi-1 recA1 gyrA96 relA1 lac* F′[*proAB*$^{+}$ *lacI*^{q} *lacZ*Δ*M15*]	
XS101	*recA1 hsdR rpoB331* F′[*kan*^{r}]	105
XS127[z]	*gyrA thi rpoB331* Δ(*lac-proAB*) *argE* F′[*traD36 proAB*$^{+}$ *lacI*^{q} *lacZ*Δ*M15*]	105
XS127(P3)[z]	*gyrA thi rpoB331* Δ(*lac-proAB*) *argE* F′[*traD36 proAB*$^{+}$ *lacI*^{q} *lacZ*Δ*M15*] (P3 *amp*^{r}*am tet*^{r}*am kan*^{r})	
Y1088[q]	Δ(*lac*)*U169 supE supF hsdR metB trpR tonA21 mcrA proC*::Tn*5*(*kan*^{r}) (pMC9)	22, 25, 106
Y1089[q]	*araD139* Δ(*lac*)*U169 lon-100 rpsL hflA150*::Tn*10*(*tet*^{r}) (pMC9)	21, 22
Y1089r^{-} [q]	*mcrB araD139* Δ(*lac*)*U169 lon-100 rpsL hflA150*::Tn*10*(*tet*^{r}) (pMC9)	21, 22
Y1090[q]	*mcrA araD139* Δ(*lac*)*U169 lon-100 rpsL supF trpC22*::Tn*10*(*tet*^{r}) (pMC9)	22, 25, 106
Y1090r^{-} [q, aa]	*mcrA hsdR araD139* Δ(*lac*)*U169 lon-100 rpsL supF trpC22*::Tn*10*(*tet*^{r}) (pMC9)	21, 22, 27
Y1091	Δ(*lac*)*U169 proA* Δ(*lon*) *araD139 strA supF hsdR hsdM trpC22*::Tn*10*(*tet*^{r})	
YK537	*supE44 hsdR hsdM recA1 phoA8 leuB6 thi lacY rpsL20 galK2 ara-14 xyl-5 mtl-1*	107

[a]71/18 is also called BMH 71-18.

[b]The nutritional requirements and detergent sensitivity of χ1776 means that it can be used for high-containment experiments. It has a high transformation frequency.

[c]Derived from *E. coli* C.

[d]ABLE C and K have reduced DNA polymerase I activities and hence support only low copy numbers of cloning vectors with ColE1 or pMB1 replicons. ABLE strains may therefore be successful for propagation of clones that are 'toxic' to conventional host strains (9). ABLE C and ABLE K differ only in their relative abilities to propagate cloning vectors.

[e]AG1 carries an uncharacterized mutation that improves transformation efficiency.

[f]Also called MM294*c*I^{+}.

[g]Some sources suggest that these strains might also be *relA1 rfbD1 spoT1*. Ref. 95 states that MM294, the parent strain, is *pro*$^{-}$, but this is contrary to the original publications (16, 25).

[h]Derived from *E. coli* B.

[i]There is conflicting information about the genotypes of BL21 and its derivatives. The genotype shown agrees with ref. 95. According to various other sources the strains might also be *ompT*, *lon*, *dcm* and/or *gal*$^{+}$.

Table 1. *E. coli* genotypes (continued)

[j]The λDE3 lysogen carries the gene for T7 RNA polymerase.
[k]The genotypes of these strains are confusing. The strains C600 and BNN93 are considered the same by some sources, though as shown here C600 is, in fact, *hsdR*$^{+}$ *mcrB*$^{+}$, whereas BNN93 is *hsdR mcrB*. New England Biolabs state that some C600 strains are *mcrA* and others are *mcrA*$^{+}$. C600*hflA* and BNN102 (again considered the same by some sources) are the *hflA* versions of C600 and BNN93, respectively. The *hsdR mcrB* status of C600*galK* is uncertain. In addition, New England Biolabs describes C600 as *rfbD1* and BNN93 as *rfb*$^{+}$. C600 is the same as CR34.
[l]According to New England Biolabs these strains are also *spoT1*.
[m]According to New England Biolabs DH1 might also be *spoT1 rfbD1*.
[n]DP50*supF* is sometimes called simply DP50. The strain was originally used for high-containment experiments with λ vectors. It has now been superseded by strains that are easier to grow.
[o]According to New England Biolabs ED8767 is *rec*$^{+}$ and *lac-3* or *lacY1*.
[p]New England Biolabs describes the insertion in ER1458 as *zjj202::*Tn*10(tet*^{r}*)*, which is not the same position as *trpC22*. Some sources describe this strain as *ser*$^{+}$.
[q]pMC9 is pBR322 carrying *lacI*^{q}.
[r]HB101 and RR1 are isogenic except that HB101 is *recA* and RR1 is *recA*$^{+}$. Some sources describe these strains as *leu*$^{+}$ and/or *thi-1*.
[s]JM83 and TB1 are isogenic except that JM83 is *hsd*$^{+}$ and TB1 is *hsdR*.
[t]JM103 is now considered to have lost *hsdR4* and become P1 lysogenic (108). KK2186 (= JM103Y) is genetically identical to the original JM103.
[u]Ref. 95 states *supE44* but this is not given in the original publication (3).
[v]Ref. 95 states *hsdR17* but this is not given in the original publication (3). New England Biolabs describes this strain as *tsx*$^{+}$.
[w]New England Biolabs considers K802 and WA802 to be the same, but only list a genotype for WA802. This genotype is not the same as that for the authentic K802. The authentic K802 genotype is given here and the New England Biolabs description is given for WA802.
[x]Some stocks of MV1184 do not carry the F′ episome.
[y]These strains carry an uncharacterized mutation that makes plaques and colonies more intensely blue when plated on X-gal or equivalent.
[z]Some sources describe these strains as *supE44*.
[aa]Y1090r^{-} is also called Y1090*hsdR*. Stratagene describes this strain as *mcrB* rather than *mcrA*.

1.3 Genes relevant to recombinant DNA experiments

Complete listings of relevant genes and deletions
All of the genes mentioned in the strain genotypes (*Table 1*) are listed in *Table 2* along with their phenotypes. Alternative gene symbols are given where these have appeared with any frequency in the recombinant DNA literature. *Table 3* gives more information on the relevance of the important mutations and deletions.

The deletions appearing in the strain genotypes are described in more detail in *Table 4*. When writing a genotype it is conventional to give only the start and end points of a deletion, e.g. *Δ(argF-lac)*. This disguises the fact that most deletions remove more than just two genes and have a more substantial effect on the properties of the strain than is apparent from a cursory glance at the written genotype. *Table 4* shows the genes contained within each deleted region, but is of necessity incomplete as it based on the *E. coli* genetic map (110) which does not include all of the genes being revealed by the *E. coli* sequencing project.

Genes involved in DNA restriction and/or modification
Four sets of *E. coli* genes are involved in DNA restriction and/or modification:

(i) The EcoK system (46, 118, 119), specified by the *hsdMRS* genes, recognizes the sequence 5′-AACNNNNNNGTT-3′. The modification component is coded by *hsdM*, the restriction component by *hsdR*, and the site recognition subunit by *hsdS*. An *hsdM* strain is methylation deficient (m_K^-), an hsdR strain is restriction deficient (r_K^-) and an *hsdS* strain lacks both properties ($r_K^- m_K^-$). The modification component protects the host DNA by methylating the second A in each strand of the target sequence. DNA cloned in a *hsdM* host will be restricted if subsequently transferred to a *hsdR*$^+$ host.

(ii) The McrA and McrBC restriction systems (118, 120–122) cleave DNA at methylated cytosines contained in the target sequences 5′-CG-3′ for McrA and 5′-PuC-3′ for McrBC. DNA from some sources (including human DNA) may be methylated at these sites and will therefore be cloned inefficiently in strains expressing McrA and/or McrBC.

(iii) The Mrr restriction system (118, 121, 123) cleaves DNA at methylated adenines, although the precise recognition sequence is not known. DNA from some sources may be methylated at Mrr recognition sites and will therefore be cloned inefficiently in a *mrr*$^+$ strain.

(iv) The DNA adenine methylase (Dam) and DNA cytosine methylase (Dcm) add methyl groups to DNA (118). Dam modifies the N^6 position of the adenine present in the target sequence 5′-GATC-3′ (124) and Dcm modifies the C^5 positions of the cytosines in 5′-CCAGG-3′ and 5′-CCTGG-3′ (125, 126). These modifications may affect *in vitro* restriction of cloned DNA by methylating residues essential to the target sequence for certain restriction enzymes (see Chapter 2, Section 3). In addition, DNA cloned in a *dam* strain is subject to a relatively high level of ssDNA breakage, and plasmids purified from a *dam*$^+$ strain show low transformation frequencies when cloned into a *dam* strain (127).

Table 5 shows the EcoK, McrA, McrBC, Mrr, Dam and Dcm status for important *E. coli* strains. Only a limited number of strains have been tested and not all genes have been assessed in all strains, hence there are a few gaps in the table. The most reliable source of updated information is the New England Biolabs catalogue.

Suppressor mutations

Some cloning vectors contain nonsense mutations in essential genes as a means of preventing the spread of recombinant molecules in natural populations of bacteria. These vectors replicate only in host *E. coli* strains carrying suppressor mutations that enable the nonsense codon to be read through. Most suppressor mutations lie in tRNA genes and result in altered codon–anticodon recognition so that an amino acid is inserted at the nonsense mutation. Details of relevant suppressor mutations are given in *Table 6*.

Table 2. Phenotypes of *E. coli* loci and genes relevant to recombinant DNA experiments (109–112)

Locus	Gene	Phenotype	Synonym
amp		Ampicillin sensitivity or resistance	
ara		Arabinose utilization	
	araA	L-arabinose isomerase (EC 5.3.1.4)	
	araB	ribulokinase (EC 2.7.1.16)	
	araC	regulatory gene (activator–repressor protein)	
	araD	L-ribulosephosphate 4-epimerase (EC 5.1.3.4)	
arg		Arginine biosynthesis	
	argA	*N*-acetylglutamate synthase (EC 2.3.1.1)	*argI*
	argE	acetylornithine deacetylase (EC 3.5.1.16)	
	argF	ornithine carbamoyltransferase (EC 2.1.3.3)	
	argG	arginosuccinate synthetase (EC 6.3.4.5)	
	argH	arginosuccinate lyase (EC 4.3.2.1)	
aro		Aromatic amino acid biosynthesis	
	aroB	dehydroquinate synthase (EC 4.6.1.3)	
asd		Aspartate semialdehyde dehydrogenase (EC 1.2.1.11)	
azi		Azide sensitivity or resistance	*secA*
bio		Biotin biosynthesis	
	bioH	block before pimeloyl CoA	
	bioT	not in the literature; misprint in the original reference?	
cam		Chloramphenicol sensitivity or resistance	
chl		Chlorate resistance	
	chlD	molybdenum uptake	
cyc	*cycA*	Transport of D-alanine, D-serine and glycine	
dam		DNA adenine methylase[a]	
dap		Diaminopimelate biosynthesis	
	dapD	tetrahydrodipicolinate *N*-succinyltransferase	
dcm		DNA cytosine methylase[a]	
deo		Deoxyribose biosynthesis	
	deoC	deoxyribose–phosphate aldolase (EC 4.1.2.4)	
	deoR	regulatory gene for *deo* operon[b]	
dna	*dnaJ*	DNA biosynthesis[b]	
dut		Deoxyuridinetriphosphatase (EC 3.6.1.23)[b]	
eda		2-Keto-3-deoxygluconate 6-phosphate aldolase (EC 4.1.2.14)	
end	*endA*	DNA-specific endonuclease I[b]	
ent		Enterochelin synthesis	
	entA	2,3-dihydro-2,3-dihydroxybenzoate dehydrogenase	

Table 2. Phenotypes of loci and genes (continued)

Locus	Gene	Phenotype	Synonym
env		Cell envelope	
	envZ	regulation of outer membrane protein biosynthesis	
esp		Efficient packaging of phage T1	
fep		Enterochelin uptake	
	fepC	cytoplasmic membrane protein	
flb	*flbB*	Flagellar synthesis	*flhD*
gal		Galactose utilization	
	galE	UTP-galactose 4-epimerase	
	galK	galactokinase (EC 2.7.1.6)	
	galT	galactose-1-phosphate uridylyltransferase (EC 2.7.7.12)	
	galU	glucose-1-phosphate uridylyltransferase (EC 2.7.7.12)	
gln		Glutamine biosynthesis and activation	
	glnV	glutamine tRNA-2	*supE*
gpt		Guanine–hypoxanthine phosphoribosyltransferase (EC 2.4.2.17)	
gyr		DNA gyrase	
	gyrA	sensitivity or resistance to nalidixic acid	*nalA*
hem		Hemin utilization	
	hemF	coproporphyrinogen III oxidase (EC 1.3.3.3)	
hfl	*hflA*	High frequency of lysogeny by phage λ[b]	*hflC*, *hflK*
	hflB	*c*II protein level	
his		Histidine biosynthesis	
	hisG	ATP phosphoribosyltransferase (EC 2.4.2.17)	
hsd		Host-specific restriction and/or modification[a]	
	hsdM	DNA methylase M	
	hsdR	endonuclease R	
	hsdS	specificity determinant for *hsdM* and *hsdR*	
ilv		Isoleucine and valine biosynthesis	
kan		Kanamycin sensitivity or resistance	
lac		Lactose utilization[b]	
	lacI	repressor protein	
	lacY	galactoside permease	
	lacZ	β-D-galactosidase (EC 3.2.1.23)	
lam	*lamB*	Phage λ receptor protein	*malB*
leu		Leucine biosynthesis	
	leuA	α-isopropylmalate synthase (EC 4.1.3.12)	
	leuB	β-isopropylmalate dehydrogenase (EC 1.1.1.85)	
	leuC	α-isopropylmalate isomerase subunit	
	leuD	α-isopropylmalate isomerase subunit	

Table 2. Phenotypes of loci and genes (continued)

Locus	Gene	Phenotype	Synonym
lig		DNA ligase	
lon		ATP-dependent protease La[b]	
lys		Lysine biosynthesis	
	lysA	diaminopimelate decarboxylase (EC 4.1.1.20)	
mal		Maltose utilization	
	malA	maltose uptake	*malPQT*
	malB	maltose catabolism	*malEFGK*
	malE	periplasmic binding protein	
	malF	cytoplasmic membrane protein	
	malG	active transport of maltose and maltodextrins	
	malK	maltose permeation	
	malP	maltodextrin phosphorylase (EC 2.4.1.1)	
	malQ	amylomaltase (EC 2.4.1.25)	
	malT	positive regulatory gene for *mal* regulon	
mcr		Restriction of DNA at methylcytosine residues[a]	
	mcrA		e14
	mcrBC		
met		Methionine biosynthesis	
	metB	cystathionine γ-synthase (EC 4.1.99.9)	
	metC	cystathionine γ-lyase (EC 4.4.1.1)	
min	*minB*	Formation of minicells containing no DNA	
mrr		Restriction of DNA at methyladenine residues[a]	
mtl		Mannitol utilization	
mut		High mutation rate	
	mutD	DNA polymerase III ε subunit	*dnaQ*
	mutS	methyl-directed mismatch repair	
	mutT	A.T→C.G transversions	
nad		NAD biosynthesis	
nup	*nupG*	Nucleoside transport	
omp		Outer membrane proteins	
	ompC	outer membrane protein 1b	
	ompF	outer membrane protein 1a	
	ompR	positive regulatory gene for *ompC* and *ompF*	*ompB*
	ompT	outer membrane protein 3b	
oms		Osmotic sensitivity	
pgl		6′-Phosphogluconolactonase (EC 3.1.1.31)	
pho		Phosphate utilization	
	phoA	alkaline phosphatase (EC 3.1.3.1)	
pnp		Polynucleotide phosphorylase (EC 2.7.7.8)[b]	

Table 2. Phenotypes of loci and genes (continued)

Locus	Gene	Phenotype	Synonym
pro		Proline biosynthesis[b]	
	proA	γ-glutamyl phosphate reductase (EC 1.2.1.41)	
	proB	γ-glutamyl kinase (EC 2.7.2.11)	
	proC	pyrroline-5-carboxylate reductase (EC 1.5.1.2)	
pts		Phosphotransferase system	
	ptsF	fructose phosphotransferase enzyme II	*fruA*
rbs		Ribose utilization	
	rbsR	regulatory gene	
rec		General recombination and radiation repair[b]	
	recA	general recombination, DNA repair, phage λ induction	*zab*
	recB	exonuclease V (EC 3.1.11.5)	
	recC	exonuclease V (EC 3.1.11.5)	
	recD	exonuclease V (EC 3.1.11.5) α subunit	
	recF	recombination and DNA repair	
	recJ	recombination and DNA repair	
rel		Regulation of RNA synthesis	
	relA	ATP:GTP 3′-pyrophosphotransferase[b]	
rfa		Lipopolysaccharide core biosynthesis (rough colony morphology)	
rfb	*rfbD*	TDP–rhamnose synthetase (rough colony morphology)	
rha		Rhamnose utilization	
rif		Rifampicin sensitivity or resistance	
rna		Ribonuclease I	
rpo		RNA polymerase (EC 2.7.7.6)	
	rpoA	α subunit	
	rpoB	β subunit	
	rpoH	σ^{32} subunit[b]	*htpR*
rps		Small ribosomal protein	
	rpsL	protein S12	*strA*
rrn		rRNA	
	rrnD	rRNA operon at 72 minutes	
	rrnE	rRNA operon at 90 minutes	
sbc		Suppressor for *recBC*[b]	
	sbcA	involved in RecE pathway	
	sbcB	exonuclease I, involved in RecF pathway	
	sbcC	involved in RecF pathway	
ser		Serine biosynthesis	
	serB	phosphoserine phosphatase (EC 3.1.3.3)	
spo	*spoT*	Guanosine 3′,5′-bis(diphosphate) 3′-pyrophosphatase	
srl		Sorbitol utilization	
	srlA	D-glucitol-specific enzyme of phosphotransferase	
	srlC	regulatory gene	

Table 2. Phenotypes of loci and genes (continued)

Locus	Gene	Phenotype	Synonym
str		Streptomycin senstivity or resistance	*rpsL*
sup		Suppressor[c]	
	supC	tyrosine tRNA-1	*tyrT*
	supE	glutamine tRNA-2	*glnV*
	supF	tyrosine tRNA-1	*tyrT*
	supP	leucine tRNA-5	*leuX*
tet		Tetracycline sensitivity or resistance	
thi		Thiamine biosynthesis	
thr		Threonine biosynthesis	
thy		Thymine biosynthesis	
	thyA	thymidylate synthase (EC 2.1.1.45)	
ton		Resistance to phages T1 and ϕ80 and to colicins	
	tonA	outer membrane receptor	*fhuA*
	tonB	uptake protein	
tra		Conjugal transfer of F plasmids	
	traD	suppresses conjugal transfer[b]	
trp		Tryptophan biosynthesis	
	trpA	tryptophan synthase (EC 4.2.1.20) A protein	
	trpB	tryptophan synthase (EC 4.2.1.20) B protein	
	trpC	*N*-(5-phosphoribosyl)anthranilate isomerase indole-3-glycerolphosphate synthetase	
	trpR	aporepressor protein	
tsx		Phage T6 and colicin K resistance	
tyr		Tyrosine biosynthesis and activation	
	tyrT	tyrosine tRNA-1	*supC*, *supF*
umu	*umuC*	Sensitivity to u.v.[b]	
ung		Uracil-DNA-glycosylase[b]	
uvr		Repair of u.v. damage to DNA	
	uvrB	excision nuclease	
	uvrC	excision nuclease[b]	
xyl		Xylose utilization	

[a]For further information see text and *Table 5*.
[b]For further information see *Table 3*.
[c]For further information see text and *Table 6*.

Table 3. Relevance of important *E. coli* mutations to recombinant DNA experiments

Mutation	Relevance
dam	See text.
dcm	See text.
deoR	Facilitates transformation by large plasmids.
dnaJ	Inactivates a chaperonin, improving the yield of some proteins expressed from cloned genes.
dut ung	Allows uracil to be incorporated into cloned DNA, required by some mutagenesis techniques.
endA1	Inactivates an endonuclease, improving plasmid yields especially from minipreps.
hflA	Results in a higher frequency of lysogenization by phage λ vectors that carry a normal *c*I repressor gene.
hsd	See text.
Δ(lac-proAB)	Removes the *lac* operon and surrounding region, including the two genes involved in proline biosynthesis. Most *Δ(lac-proAB)* strains carry the deleted genes on an F′ episome. When maintained on minimal media the F′[*proAB*] genes provide positive selection for cells that retain the F′ status.
lacI^q	Most strains used with vectors employing Lac selection are *lacI^q*. This mutation results in over-production of the *lac* repressor (113), minimizing the low level of *lac* expression that occurs in uninduced wild-type cells.
lacZΔM15	This deletion removes the amino-terminal α peptide (amino acids 11-41) of β-galactosidase. Most cloning vectors that employ Lac selection carry a gene that codes for the α peptide and rescues the *lacZΔM15* mutation by α complementation (114).
lon	Inactivates a protease, increasing the stability of fusion proteins but resulting in a mucoid phenotype. Some plasmids are unstable in *lon* strains.
mcr	See text.
mrr	See text.
pnp	Results in increased stability of eukaryotic mRNA sequences, improving the expression of some eukaryotic genes cloned in *E. coli*. Some plasmids are unstable in *pnp* strains.
recA	A *recA* strain is recombination deficient and has a reduced ability to multimerize plasmids and rearrange cloned sequences (3, 115). However, *recA* strains grow relatively slowly and have low transformation frequencies. Strains that are used to propagate *red gam* λ vectors must be *recA$^+$*.
recBC sbcBC	Strains that are *recBC sbcB* or *recBC sbcBC* display decreased rearrangement of cloned palindromic sequences. However, these strains are relatively slow growing (though better than *recBC* mutants: 54, 116) and have low transformation frequencies. Some plasmids are unstable in *recBC sbcB* and *recBC sbcBC* strains.

Table 3. Important mutations (continued)

Mutation	Relevance
recD	Inefficient plasmid replication but allows growth of λ clones containing palindromic sequences.
recF, *recJ*	Reduced recombination between plasmids.
relA	Uncouples RNA synthesis from protein synthesis, allowing overproduction of RNA from cloned sequences.
rpoH	The unmutated gene codes for a heat shock σ protein component of RNA polymerase. Strains that are *rpoH* show reduced expression of heat-induced proteases, giving improved yields of protein from cloned genes at elevated culture temperature.
sbcA	Results in improved growth of *recB* strains (117).
sbcB, *sbcC*	See *recBC sbcBC*.
sup	See text.
traD	Reduced transmission of F plasmids, providing a degree of biological containment.
tsp	Improved yield of secreted proteins, and of proteins obtained from cell extracts.
umuC	Minimizes rearrangements of cloned palindromic and Z-DNA sequences (89).
uvrC	Minimizes rearrangements of cloned palindromic and Z-DNA sequences (89).
ung	See *dut ung*.

Table 4. Properties of *E. coli* deletions relevant to recombinant DNA experiments

Deletion	Map position (minutes)[a]	Genes in the deletion	Phenotypes
Δ(*ara-leu*)	1.4–1.8	*ara*	Arabinose utilization
		dadB	D-Amino acid dehydrogenase
		leu	Leucine biosynthesis
Δ(*argF-lac*) [= Δ(*lac*)*U169*]	6.5–8.0	*argF*	Arginine biosynthesis
		katC	Catalase activity
		mvrA	Methyl viologen resistance
		strC	Low-level streptomycin resistance
		ilvU	Isoleucine-valine biosynthesis
		trmB	tRNA methyltransferase
		dvl	Sensitivity to SDS and toluidine blue plus light
		betABT	Betaine biosynthesis
		codAB	Cytosine transport and deamination
		fecABD	Citrate-dependent iron transport
		cynS	Cyanase
		lac	Lactose utilization

Table 4. *E. coli* deletions (continued)

Deletion	Map position (minutes)[a]	Genes in the deletion	Phenotypes
Δ(*bioH-asd*)	75.0–75.7	*bioH*	Biotin biosynthesis
		pck	Phosphoenolpyruvate carboxykinase
		gntT	Gluconate utilization
		malA	Maltose uptake
		glpRGED	Glycerol utilization
		glgACB	Glycogen biosynthesis
		asd	Aspartate semialdehyde dehydrogenase
Δ(*bio-uvrB*)	17.4–17.6	*bio*	Biotin biosynthesis
		uvrB	Repair of u.v. damage to DNA
Δ(*gal-uvrB*)	17.0–17.6	*gal*	Galactose utilization
		hemF	Hemin utilization
		chlDJ	Chlorate resistance
		pgl	6′-Phosphogluconolactonase
		esp	Efficient packaging of phage T1
		attλ	Integration site for phages λ, 82 and 434
		bioABFCD	Biotin biosynthesis
		uvrB	Repair of u.v. damage to DNA
Δ(*gpt-proA*)	5.7–5.9	*gpt*	Guanine–hypoxanthine phosphoribosyltransferase
		phoE	Phosphate transport protein
		proA	Proline biosynthesis
Δ(*gpt-proAB-argF-lac*)	5.7–8.0	*gpt*	Guanine–hypoxanthine phosphoribosyltransferase
		phoE	Phosphate transport protein
		proAB	Proline biosynthesis
		thrW	Threonine biosynthesis
		hns	Histone-like DNA-binding protein
		att253	Phage 253 integration site
		attP22	Phage P22 integration site
		cxm	Methylglyoxal biosynthesis
		argF	Arginine biosynthesis
		katC	Catalase activity
		mvrA	Methyl viologen resistance
		strC	Low-level streptomycin resistance
		ilvU	Isoleucine–valine biosynthesis
		trmB	tRNA methyltransferase
		dvl	Sensitivity to SDS and toluidine blue plus light
		betABT	Betaine biosynthesis
		codAB	Cytosine transport and deamination

Table 4. *E. coli* deletions (continued)

Deletion	Map position (minutes)[a]	Genes in the deletion	Phenotypes
		fecABD	Citrate-dependent iron transport
		cynS	Cyanase
		lac	Lactose utilization
Δ(*hemF-esp*)	17.1–17.3	*hemF*	Hemin utilization
		chlDJ	Chlorate resistance
		pgl	6′-Phosphogluconolactonase
		esp	Efficient packaging of phage T1
Δ(*hsdMS-mcrB*)	98.5–98.4	*hsdMS*	Host-specific modification
		mcrBC	Restriction of DNA at 5-methyl cytosine residues
Δ(*lac-proAB*)	8.0–5.9	*lac*	Lactose utilization
		cynS	Cyanase
		fecABD	Citrate-dependent iron transport
		codAB	Cytosine transport and deamination
		betABT	Betaine biosynthesis
		dvl	Sensitivity to SDS and toluidine blue plus light
		trmB	tRNA methyltransferase
		ilvU	Isoleucine–valine biosynthesis
		strC	Low-level streptomycin resistance
		mvrA	Methyl viologen resistance
		katC	Catalase activity
		argF	Arginine biosynthesis
		cxm	Methylglyoxal biosynthesis
		attP22	Phage P22 integration site
		att253	Phage 253 integration site
		hns	Histone-like DNA-binding protein
		thrW	Threonine biosynthesis
		proAB	Proline biosynthesis
Δ(*malK-lamB*)	91.5	*malB*[b]	Maltose catabolism
Δ(*malPQ-bioH-ompB*)	75.3–74.9	*malPQ*	Maltose uptake
		gntT	Gluconate utilization
		pck	Phosphoenolpyruvate carboxy-kinase
		bioH	Biotin biosynthesis
		ompB	Regulatory genes (*ompR*, *envZ*) for production of outer membrane proteins
Δ(*mcrCB-hsdSMR-mrr*)	98.4–98.6	*mcrCB*	Restriction of DNA at 5-methyl cytosine residues
		hsdSMR	Host specific restriction and modification
		mrr	Restriction of DNA at methyl adenine residues

Table 4. *E. coli* deletions (continued)

Deletion	Map position (minutes)[a]	Genes in the deletion	Phenotypes
Δ(*ompT-fepC*)	13.2–13.4	*ompT*	Outer membrane protein
		envY	Envelope protein
		entD	Enterochelin synthesis
		fepA	Enterochelin and colicin receptor
		fes	Enterochelin esterase
		entF	Enterochelin synthesis
		fepEC	Enterochelin uptake
Δ(*pgl-bio*)	17.2 - 17.4	*pgl*	6′-Phosphogluconolactonase
		esp	Efficient packaging of phage T1
		attλ	integration site for phages λ, 82 and 434
		bio	Biotin biosynthesis
Δ(*srl-recA*)	58.3	*srl*	Sorbitol utilization
		recA	Recombination, repair, phage λ induction

[a]Taken from ref. 110.

[b]Deletion is entirely within the *malB* segment of the maltose operon.

Table 5. Restriction and modification properties of important *E. coli* strains

	Phenotype[a]					
Strain	**EcoK**[b]	**McrA**	**McrBC**	**Mrr**	**Dam**	**Dcm**
ABLE	$r_K^- m_K^-$	–	–	–	+	+
BB4	$r_K^- m_K^+$	–			+	+
BNN93	$r_K^- m_K^+$	–	–		+	+
BNN102	$r_K^- m_K^+$	–	–		+	+
C600	$r_K^+ m_K^+$	–[c]	+	+	+	+
C600*galK*	$r_K^- m_K^+$?[c]	–	?[c]	+	+	+
CES200	$r_K^- m_K^+$	–	+		+	+
CJ236	$r_K^+ m_K^+$	–	+		+	+
CPLK	$r_K^- m_K^-$	–	–	–	+	+
D1210	$r_K^- m_K^-$		–	–	+	+
DB1316	$r_K^- m_K^+$	+	–	+	+	+
DH5αMCR	$r_K^- m_K^-$	–	–	–	+	+
DH5α*mcrAB*	$r_K^- m_K^-$	–	–	–	+	+
DH10B	$r_K^- m_K^-$	–	–	–	+	+
DH11S	$r_K^- m_K^-$	–	–	–	+	+
DH12S	$r_K^- m_K^-$	–	–	–	+	+
DK1	$r_K^- m_K^+$	–	–		+	+
DL538	$r_K^- m_K^+$	–	–		+	+
DM1	$r_K^- m_K^+$		–		–	+
DP50*supF*	$r_B^- m_B^-$	–			+	+
ED8654	$r_K^- m_K^+$	–	+		+	+
ED8787	$r_K^- m_K^-$	–	–		+	+
ER1398	$r_K^- m_K^+$	+	–	+	+	+

Table 5. Restriction and modification properties (continued)

Strain	Phenotype[a]					
	EcoK[b]	McrA	McrBC	Mrr	Dam	Dcm
ER1414	$r_K^- m_K^+$	+	−	+	+	+
ER1451	$r_K^- m_K^+$	−	−		+	+
ER1458	$r_K^- m_K^+$	−	−		+	+
ER1562	$r_K^- m_K^+$	−	−	+	+	+
ER1563	$r_K^- m_K^+$	−	+	+	+	+
ER1564	$r_K^- m_K^+$	−	+	+	+	+
ER1565	$r_K^- m_K^+$	−	−	+	+	+
ER1578	$r_K^- m_K^+$	−	−		+	+
ER1647	$r_K^- m_K^-$	−	−	−	+	+
ER1648	$r_K^- m_K^-$	−	−	−	+	+
ER1793	$r_K^- m_K^-$	−	−	−	+	+
ER1821	$r_K^- m_K^-$	−	−	−	+	+
ER2267	$r_K^- m_K^-$	−	−	−	+	+
ER2507	$r_B^- m_B^-$		−	−	+	+
ER2508	$r_B^- m_B^-$		−	−	+	+
GM48	$r_K^+ m_K^+$				−	−
GM2163	$r_K^- m_K^+$	−	−	+	+	+
GM2929	$r_K^- m_K^+$	−	−		+	+
HB101	$r_B^- m_B^-$	+	−	−	+	+
JC7623	$r_K^+ m_K^+$	−	+	+	+	+
JM106	$r_K^- m_K^+$	−			+	+
JM107	$r_K^- m_K^+$	−	+	+	+	+
JM108	$r_K^- m_K^+$	−			+	+
JM109	$r_K^- m_K^+$	−	+		+	+
JM110[d]	$r_K^+ m_K^+$				−	−
K802	$r_K^- m_K^+$	−	−	+	+	+
KM392	$r_K^- m_K^+$	−			+	+
KW251	$r_K^- m_K^+$	−	−		+	+
LE392	$r_K^- m_K^+$	−	+		+	+
LE392.23	$r_K^- m_K^+$	−			+	+
MBM7014.5	$r_K^- m_K^+$		−		+	+
MC1061	$r_K^- m_K^+$	−	−	+	+	+
N4830-1	$r_K^+ m_K^+$	−			+	+
NM477	$r_K^- m_K^-$	−	−		+	+
NM514	$r_K^- m_K^+$	−	+		+	+
NM522	$r_K^- m_K^-$		−		+	+
NM538	$r_K^- m_K^+$	−	+		+	+
NM539	$r_K^- m_K^+$	−	+		+	+
NM554	$r_K^- m_K^+$	−	−		+	+
NM621	$r_K^- m_K^+$	−	−		+	+
P2392	$r_K^- m_K^+$	−			+	+
P2CPLK	$r_K^- m_K^-$	−	−	−	+	+
P2PLK-17	$r_K^- m_K^+$	−	−		+	+
PLK-17	$r_K^- m_K^+$	−	−		+	+
PLK-A	$r_K^- m_K^+$	−	−		+	+
PR700	$r_B^- m_B^-$		−	−	+	+

Table 5. Restriction and modification properties (continued)

Strain	Phenotype[a]					
	EcoK[b]	McrA	McrBC	Mrr	Dam	Dcm
PR745	$r_B^- m_B^-$		–	–	+	+
Q358	$r_K^- m_K^+$	–	+		+	+
Q359	$r_K^- m_K^+$	–	+		+	+
RR1	$r_B^- m_B^-$	+	–	–	+	+
SCS110	$r_K^+ m_K^+$				+	+
SOLR	$r_K^- m_K^-$	–	–	–	+	+
SRB	$r_K^- m_K^-$	–	–	–	+	+
SURE	$r_K^- m_K^-$	–	–	–	+	+
SURE2	$r_K^- m_K^-$	–	–	–	+	+
TG1	$r_K^- m_K^-$		–		+	+
TG2	$r_K^- m_K^-$		–		+	+
TKB1	$r_B^- m_B^-$				+	–
TKX1	$r_K^- m_K^-$	–	–	–	+	+
TOP10	$r_K^- m_K^-$	–	–	–	+	+
WA802	$r_K^- m_K^+$	–	–	+	+	+
WA803	$r_K^- m_K^-$	–	–	+	+	+
XL1-Blue MR	$r_K^- m_K^-$	–	–	–	+	+
XL1-Blue MRA	$r_K^- m_K^-$	–	–	–	+	+
XL1-Blue MRF′	$r_K^- m_K^-$	–	–	–	+	+
XL2-Blue MRF′	$r_K^- m_K^-$	–	–	–	+	+
XL*mutS*	$r_K^- m_K^-$	–	–	–	+	+
XLO	$r_K^- m_K^-$	–	–	–	+	+
XLOLR	$r_K^- m_K^-$	–	–	–	+	+
XPORT	$r_K^- m_K^-$	–	–	–	+	+
Y1088	$r_K^- m_K^+$	–	+		+	+
Y1089r⁻	$r_K^+ m_K^+$		–		+	+
Y1090	$r_K^+ m_K^+$	–	+		+	+
Y1090r⁻	$r_K^- m_K^+$	–			+	+

[a]The absence of an entry indicates that status is uncertain.
[b]The r_B m_B phenotype is given for those strains derived from *E. coli* B.
[c]See footnote k to *Table 1*.
[d]The EcoK status of JM110 is confusing: see footnote v to *Table 1*.

Table 6. Properties of *E. coli sup* mutations

Mutation	Codons suppressed	Amino acid inserted	Gene product	Synonyms
supB	UAA, UAG	Glutamine	Glutamine tRNA-1	*glnU*, su_B
supC	UAA, UAG	Tyrosine	Tyrosine tRNA-1	*tyrT*, *Su-4*, su_C
supD	UAG	Serine	Serine tRNA-2	*serU*, *Su-1*, su_I
supE	UAG	Glutamine	Glutamine tRNA-2	*glnV*, *Su-2*, su_{II}
supF	UAG	Tyrosine	Tyrosine tRNA-1	*tyrT*, *Su-3*, su_{III}
supG	UAA, UAG	Lysine	Lysine tRNA	*lysT*, *Su-5*

Table 6. *E. coli sup* mutations (continued)

Mutation	Codons suppressed	Amino acid inserted	Gene product	Synonyms
supH	UAG	Serine	Serine tRNA-2	*serU*
supK	UGA	Various	Protein release factor 2	*prfB*
supL	UAA, UAG	Lysine	Lysine tRNA	*lysT*, su_{β}
supM	UAA, UAG	Tyrosine	Tyrosine tRNA-2	*tyrU*
supN	UAA, UAG	Lysine	Lysine tRNA	*lysV*
supO	UAA, UAG	Tyrosine	Tyrosine tRNA-1	*tyrT*
supP	UAG	Leucine	Leucine tRNA-5	*leuX*, *Su-6*

2. BACTERIOPHAGE λ

λ is a typical example of a head-and tail phage. The DNA is contained in the polyhedral head structure and the tail serves to attach the phage to the bacterial surface and inject the DNA into the cell. The wild-type λ DNA molecule is 48.5 kb and has been completely sequenced (128). A feature of the genetic map (*Figure 1*) is that genes related in terms of function are clustered together on the genome. The recognized genes with their map positions and functions are listed in *Table 7*. There are several excellent reviews of λ biology (e.g. 131, 132).

Table 7. Locations and functions of phage λ genes (129).

Gene	Base pair coordinates[a]	Function
*Nu*1	191–733	DNA packaging and cohesive end formation
A	711–2633	DNA packaging and cohesive end formation
W	2633–2836	Modification of DNA-filled heads
B	2836–4434	Capsid structural component
C	4418–5734	Capsid structural component
*Nu*3	5132–5734	Transient core for capsid assembly
D	5747–6076	Major component of phage head
E	6135–7157	Major component of capsid
FI	7202–7597	DNA packaging and maturation
FII	7612–7962	Structural component of filled head
Z	7977–8552	Structural component of tail
U	8552–8944	Structural component of tail
V	8955–9692	Major component of tail tube
G	9711–10130	Assembly of tail initiator
T	10115–10546	Structural component of tail
H	10542–13100	Structural component of tail
M	13100–13426	Structural component of tail
L	13429–14124	Structural component of tail initiator
K	14276–14872	Tail assembly
I	14773–15441	Tail initiator assembly
J	15505–18900	Structural component of tail
lom	18965–19582	Membrane protein
ORF-401	19650–20852	No product identified

Figure 1. The genetic map of phage λ. Details of the genes and uncharacterized ORFs (open reading frames) are given in *Table 7*. Figure redrawn with permission from ref. 95.

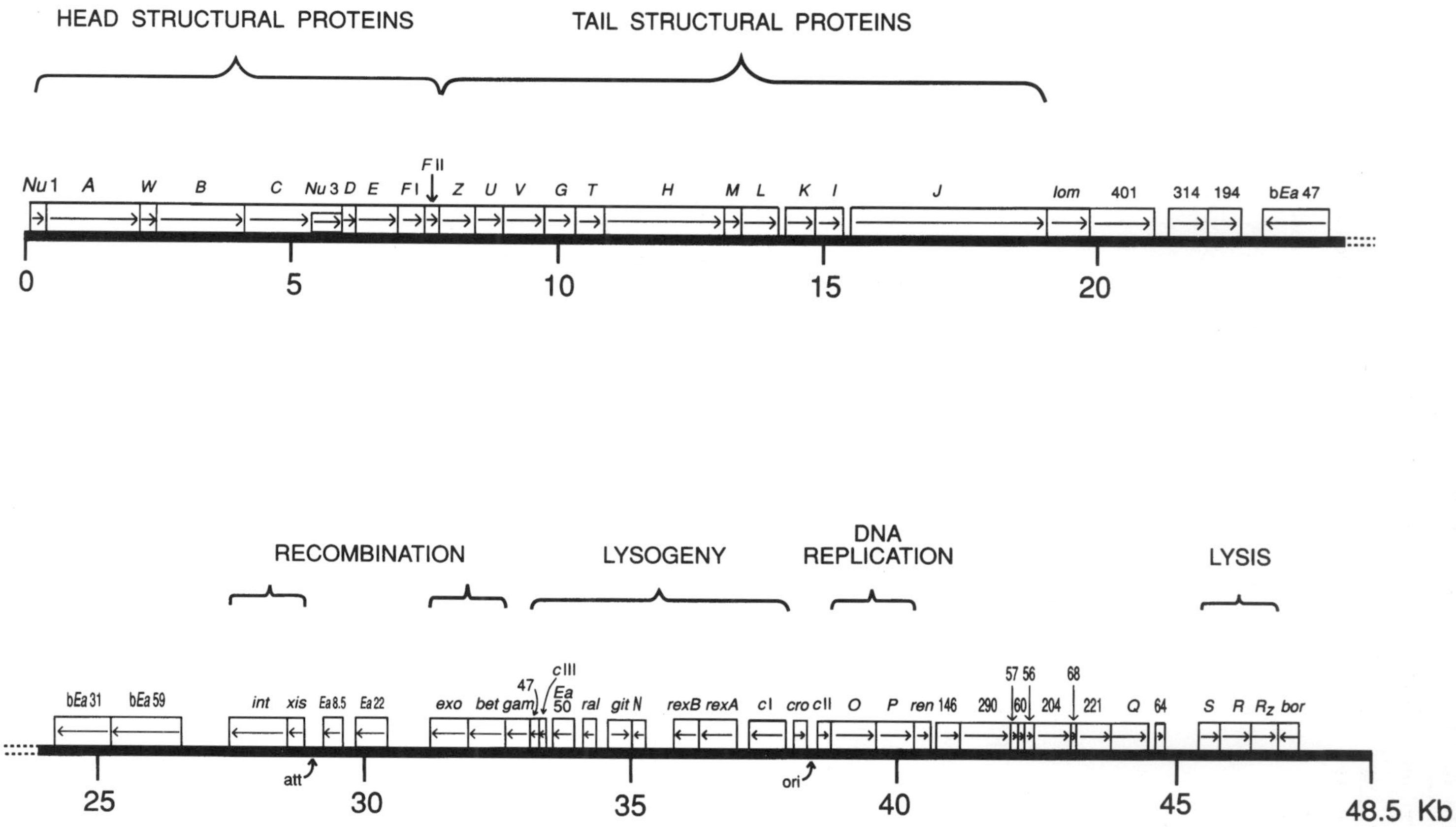

Table 7. Phage λ genes (continued)

Gene	Base pair coordinates[a]	Function
ORF-314	21029–21970	No product identified
ORF-194	21973–22554	No product identified
b*Ea*47	23918–22689	Unknown
b*Ea*31	25399–24512	Unknown
b*Ea*59	26973–25399	Unknown
int	28882–27815	Integrative and excisive recombination
xis	29078–28863	Excisive recombination
*Ea*8.5	29655–29377	Unknown
*Ea*22	30395–29850	Unknown
exo	32028–31351	General recombination: λ exonuclease
bet	32810–32028	General recombination: β-protein
gam	33232–32819	Regulation of DNA replication
ORF-47	33330–33190	Probably *kil*–inhibits host division
*c*III	33463–33302	Establishment of lysogeny
*Ea*10	33904–33539	ssDNA binding protein
ral	34287–34090	Alleviates *Eco*K restriction system
git	34497–35035	Unknown
N	35360–35040	Positive regulation of early development
rexB	36259–35828	Cell growth regulation during lysogeny
rexA	37114–36278	Cell growth regulation during lysogeny
*c*I	37940–37230	Maintenance of lysogeny
cro	38041–38238	Regulation of late stage of lytic cycle
*c*II	38360–38650	Establishment of lysogeny
O	38686–39582	Initiation of early DNA replication
P	39586–40280	Initiation of early DNA replication
ren	40280–40567	Unknown
ORF-146	40644–41081	No product identified
ORF-290	41081–41950	No product identified
ORF-57	41950–42120	No product identified
ORF-60	42090–42269	No product identified
ORF-56	42269–42436	No product identified
ORF-204	42429–43040	No product identified
ORF-68	43040–43243	No product identified
ORF-221	43224–43886	No product identified
Q	43886 44506	Positive regulation of lysis
ORF-64	44621–44812	No product identified
S	45186–45506	Cell lysis
R	45493–45966	Cell lysis
Rz	45966–46423	Cell lysis
bor	46752–46462	Survival of host in animal serum[b]

[a]The sequence coordinates are based on the complete sequence of λ *c*I *ind*1ts857 *S*7 (128) but with minor changes as detailed in ref. 119. The first number is the coordinate of the first nucleotide of the initiation codon and the second number is the coordinate of the third nucleotide of the last codon (not the termination triplet). The DNA strand on which the gene lies can therefore be identified: e.g. gene *B* (2836–4434) lies on the 'plus' strand, and gene *int* (28882–27815) lies on the 'minus' strand.
[b]See ref. 130.

3. BACTERIOPHAGE M13

M13 is a filamentous phage (other examples include f1 and fd) with a ssDNA genome of 6407 nucleotides that takes on a dsDNA form inside the host cell. This circular replicative form (RF) has a copy number of 20–40 and directs synthesis and release of approximately 300 new phage particles per cell generation. The maximum phage titre in an infected culture is ~5×10^{12} ml^{-1} (~150 mg ml^{-1}). M13 phage is male specific because the cell surface receptor is at the tip of the pilus formed only on male cells. Because of this requirement the host *E. coli* for a cloning experiment with an M13 vector must be F^+, Hfr or, as is usually the case, F′. The genetic map of M13 is shown in *Figure 2* and gene functions are listed in *Table 8*. Useful reviews of M13 include refs 133–135.

Table 8. Functions of phage M13 genes

Gene	Function
1	Membrane protein, required for assembly
2	Endonuclease/topoisomerase, required for DNA replication
3	Minor coat protein, recognizes cell surface receptor
4	Membrane protein, required for assembly
5	Regulation of DNA replication and ssDNA synthesis
6	Minor coat protein
7	Unknown role in assembly
8	Major coat protein
9	Unknown role in assembly
X	N-terminal fragment of protein 2, unknown role in DNA replication

Figure 2. Genetic map of phage M13. Details of the genes are given in *Table 8*.

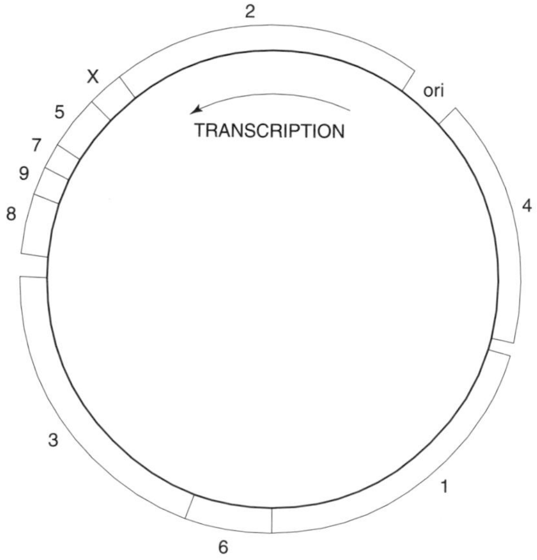

4. CULTURE MEDIA, SUPPLEMENTS AND PHAGE BUFFERS

Many different media have been used for growth of *E. coli*. *Tables 9* and *10* list the most important complex and minimal media, respectively. The complex media are, to a great extent, interchangeable, but specific media may be recommended for particular applications. For example, the NZ series of media are often used for preparation of high-titre phage λ stocks as, although rich, they lack sugars so the LamB phage receptors are not induced and phage particles are not taken out of solution by adsorption to cell debris.

Recipes for soft agar overlays for plaque growth are given in *Table 11*; again these are interchangeable, although phage yields vary according to the medium components.

Bacteria can be stored for several years as stab cultures in the agar media described in *Table 12*, or frozen in any standard medium to which glycerol has been added. Phage stocks are stored at 4°C.

Of the three indicator media described in *Table 13*, the most important in molecular biology is X-gal agar. There are a number of direct alternatives to X-gal (e.g. Indigal, Bluo-gal), used in exactly the same way, which give more intensely blue colorations with lac^+ colonies. ONPG (*o*-nitrophenyl-β-D-galactopyranoside) can also be used instead of X-gal, but the resulting yellow colour is much less distinct.

Buffers for storage and dilution of phage λ are given in *Table 14* and stock and working concentrations of antibiotics are listed in *Table 15*.

Table 9. Complex media for *E. coli* culture

Medium	Recipe (for 1 litre)[a,b]
BBL broth	10 g BBL trypticase peptone, 5 g NaCl
χ1776 broth	25 g bacto-tryptone, 7.5 g bacto-yeast extract, 20 ml 1 M Tris-HCl (pH 7.5). Autoclave, cool, then add 5 ml 1 M $MgCl_2$, 10 ml 1% diaminopimelic acid, 10 ml 0.4% thymidine, 25 ml 20% glucose.
dYT (2YT)	16 g bacto-tryptone, 10 g bacto-yeast extract, 5 g NaCl
λ broth	10 g bacto-tryptone, 2.5 g NaCl
LB medium	10 g bacto-tryptone, 5 g bacto-yeast extract, 10 g NaCl
N-broth	25 g bacto-nutrient broth
NZ[c]	10 g NZ amine A, 5 g bacto-yeast extract, 2 g $MgSO_4{\cdot}7H_2O$
NZC[c]	10 g NZ amine A, 5 g NaCl, 2 g $MgSO_4{\cdot}7H_2O$. Autoclave, cool, add 5 ml 20% casamino acids.
NZYC[c]	10 g NZ amine A, 5 g bacto-yeast extract, 5 g NaCl, 2 g $MgSO_4{\cdot}7H_2O$. Autoclave, cool, add 5 ml 20% casamino acids.
SOC	2 g bacto-tryptone, 0.5 g bacto-yeast extract, 0.6 g NaCl, 0.2 g KCl. Autoclave, cool, then add 10 ml 1 M $MgSO_4$, 10 ml 1 M $MgCl_2$, 20 ml 1 M glucose.
Superbroth	32 g bacto-tryptone, 20 g bacto-yeast extract, 5 g NaCl, 5 ml 1 M NaOH

Table 9. Complex media (continued)

Medium	Recipe (for 1 litre)[a,b]
Terrific broth	12 g bacto-tryptone, 24 g bacto-yeast extract, 4 ml glycerol. Autoclave, cool, then add 100 ml 0.17 M KH_2PO_4, +0.72 M K_2HPO_4.
Tryptone broth	10 g bacto-tryptone, 5 g NaCl

[a]The pH should be checked and, if necessary, adjusted to 7.0–7.2 with NaOH. Media should be sterilized by autoclaving at 121°C, 15 lb in^{-2}, for 20 min. Supplements to be added after autoclaving should be sterilized by filtration.
[b]For agar media add 15 g bacto-agar.
[c]In the NZ series of media the amount of $MgSO_4 \cdot 7H_2O$ can be reduced to 1 g l^{-1}. NZ amine A is the Type A casein hydrolysate.

Table 10. Minimal media for *E. coli* culture

Medium	Recipe (for 1 litre)[a,b]
Davis minimal	7 g KH_2PO_4, 2 g K_2HPO_4, 1 g $(NH_4)_2SO_4$, 0.5 g Na citrate·$2H_2O$. Autoclave, cool, then add 0.1 ml 1 M $MgSO_4$, 5 ml 20% glucose.
M9 minimal	6 g Na_2HPO_4, 3 g KH_2PO_4, 0.5 g NaCl, 1 g NH_4Cl. Adjust the pH to 7.4, autoclave, cool, then add 2 ml 1 M $MgSO_4$, +0.1 M $CaCl_2$, 10 ml 20% glucose.
M63 minimal	3 g KH_2PO_4, 7 g K_2HPO_4, 2 g $(NH_4)_2SO_4$, 5 mg $FeSO_4$. Autoclave, cool, then add 0.1 ml 1 M $MgSO_4$, 0.2 ml 50 mg ml^{-1} thiamine, 1.0 ml 20% glucose.

[a]The pH should be checked and, if necessary, adjusted to 7.0–7.2 with NaOH. Media should be sterilized by autoclaving at 121°C, 15 lb in^{-2}, for 20 min. Supplements to be added after autoclaving should be sterilized by filtration.
[b]For agar media add 15 g bacto-agar.

Table 11. Media for overlaying agar plates ('top agar')

Medium	Recipe (for 100 ml)[a,b]
Agarose top	0.8 g agarose
BBL top	1 g BBL trypticase peptone, 0.5 g NaCl, 0.6 g bacto-agar
dYT top	1.6 g bacto-tryptone, 1 g bacto-yeast extract, 0.5 g NaCl, 0.7 g bacto-agar
F top	0.8 g NaCl, 0.7 g bacto-agar
λ top	1 g bacto-tryptone, 0.25 g NaCl, 0.7 g bacto-agar
LB top	1 g bacto-tryptone, 0.5 g bacto-yeast extract, 1 g NaCl, 0.7 g bacto-agar
NZ top[c]	1 g NZ amine A, 0.5 g bacto-yeast extract, 0.2 g $MgSO_4 \cdot 7H_2O$, 0.7 g bacto-agar
Tryptone top	1 g bacto-tryptone, 0.5 g NaCl, 0.7 g bacto-agar
Water top	0.8 g bacto-agar

[a]The pH should be checked and, if necessary, adjusted to 7.0–7.2 with NaOH. Media should be sterilized by autoclaving at 121°C, 15 lb in^{-2}, for 20 min.
[b]The amount of agar added can be varied from 0.3 to 1.2%, according to requirements.
[c]The amount of $MgSO_4 \cdot 7H_2O$ can be reduced to 1 g l^{-1}. NZ amine A is the Type A casein hydrolysate.

Table 12. Media for storage of bacteria and phages

Medium[a]	Recipe (for 100 ml)[b]
Stab agars for storing bacterial cultures	
C-stab agar[c]	1 g bacto-nutrient broth, 0.5 g NaCl, 0.6 g agar. Autoclave, cool, then add 0.1 ml 10 mg ml^{-1} cysteine, 0.1 ml 10 mg ml^{-1} thymine.
N-stab agar	2.5 g bacto-nutrient broth, 0.7 g bacto-agar
Stab agar	1 g bacto-tryptone, 0.7 g bacto-agar
Phage storage media	
λ storage	0.7 g Na_2HPO_4, 0.3 g KH_2PO_4, 0.5 g NaCl, 1 ml 1 M $MgSO_4$, 1 ml 0.01 M $CaCl_2$, 0.1 ml 1% gelatin
M13 storage	1 g bacto-tryptone, 0.5 g bacto-yeast extract, 0.63 g K_2HPO_4, 0.18 g KH_2PO_4, 45 mg Na citrate·$2H_2O$, 90 mg $(NH_4)_2SO_4$, 9 mg $MgSO_4 \cdot 7H_2O$, 3.5 ml glycerol
SM	0.58 g NaCl, 0.2 g $MgSO_4 \cdot 7H_2O$, 5 ml 1 M Tris-HCl (pH 7.5), 0.5 ml 2% gelatin

[a]In addition to the media listed here, all the standard broth media in *Tables 9* and *10* can be used to store *E. coli* cultures. Add 1.5 ml glycerol to 8.5 ml bacterial culture.
[b]The pH should be checked and, if necessary, adjusted to 7.0–7.2 with NaOH. Media should be sterilized by autoclaving at 121°C, 15 lb in^{-2}, for 20 min. Supplements to be added after autoclaving should be sterilized by filtration.
[c]The presence of cysteine increases the storage time for an *E. coli* culture.

Table 13. Indicator media

Medium	Recipe[a]
MacConkey agar	For 1 litre: 40 g bacto-MacConkey agar base. Autoclave, cool, then add 50 ml 20% glucose.
Tetrazolium agar	For 1 litre: 25.5 g bacto-antibiotic medium 2, 50 mg 2,3,5-triphenyl-2H-tetrazolium chloride. Autoclave, cool, then add 50 ml 20% glucose.
X-gal agar[b]	For 3 ml: autoclave top agar (*Table 11*), cool, then add 40 µl freshly prepared 20 mg ml^{-1} X-gal in dimethyl formamide and 4 µl 200 mg ml^{-1} IPTG (aqueous solution, stored at –20°C)

[a]The pH should be checked and, if necessary, adjusted to 7.0–7.2 with NaOH. Media should be sterilized by autoclaving at 121°C, 15 lb in^{-2}, for 20 min. Supplements to be added after autoclaving should be sterilized by filtration.
[b]X-gal (5-bromo-4-chloro-3-indolyl-β-D-galactopyranoside) is a chromogenic substrate of β-galactosidase, cleavage by *lac*$^+$ bacteria gives a blue colour. IPTG (isopropyl-β-D-thiogalacto-pyranoside) is also added to the medium to induce the *lac* genes.

Table 14. Buffers for working with phage λ

Buffer	Recipe (for 100 ml)[a]
λ storage	0.7 g Na_2HPO_4, 0.3 g KH_2PO_4, 0.5 g NaCl, 1 ml 1 M $MgSO_4$, 1 ml 0.01 M $CaCl_2$, 0.1 ml 1% gelatin
SM	0.58 g NaCl, 0.2 g $MgSO_4 \cdot 7H_2O$, 5 ml 1 M Tris-HCl (pH 7.5), 0.5 ml 2% gelatin
TM	5 ml 1 M Tris-HCl (pH 7.5), 0.2 g $MgSO_4 \cdot 7H_2O$
TMG	121 mg Tris-base, 120 mg $MgSO_4 \cdot 7H_2O$, 10 mg gelatin. Adjust the pH to 7.4 with HCl.

[a]Buffers should be sterilized by autoclaving at 121°C, 15 lb in^{-2}, for 20 min.

Table 15. Antibiotics used for the selection of resistant *E. coli* cells

Antibiotic	Stock solution[a]	Working concentration[b]
Ampicillin	50 mg ml^{-1} in water	20–125 μg ml^{-1}
Chloramphenicol	35 mg ml^{-1} in ethanol	25–170 μg ml^{-1} [c]
Kanamycin	10 mg ml^{-1} in water	10–50 μg ml^{-1}
Nalidixic acid	20 mg ml^{-1} in water	20 μg ml^{-1}
Rifampicin	50 mg ml^{-1} in water	10–100 μg ml^{-1}
Streptomycin	10 mg ml^{-1} in water	10–125 μg ml^{-1}
Tetracycline	5 mg ml^{-1} in ethanol	10–50 μg ml^{-1}

[a]Sterilize by filtration and store stock solutions at −20°C. Add to agar media after autoclaving and cooling to 50°C.
[b]The lower concentrations are suitable for the selection of cells in which the resistance gene is chromosomal or carried by a low copy number plasmid. The higher concentrations are used if the gene is present on a high copy number plasmid.
[c]Use 170 μg ml^{-1} chloramphenicol for plasmid amplification.

5. REFERENCES

1. Silhavy, T.J., Berman, M.L. and Enquist, L.W. (1984) *Experiments with Gene Fusions*. Cold Spring Harbor Laboratory Press, New York.

2. Perbal, B. (1988) *A Practical Guide to Molecular Cloning (2nd edn)*. Wiley, New York.

3. Yanisch-Perron, C., Vieira, J. and Messing, J. (1985) *Gene*, **33**, 103.

4. Messing, J., Gronenborn, B., Muller-Hill, B. and Hofschneider, P.H. (1977) *Proc. Natl. Acad. Sci. USA*, **74**, 3642.

5. Dente, L., Cesareni, G. and Cortese, R. (1983) *Nucl. Acids Res.*, **11**, 1645.

6. Ruther, U. and Muller-Hill, B. (1983) *EMBO J.*, **2**, 1791.

7. Clark-Curtiss, J.E. and Curtiss, R. (1983) *Meth. Enzymol.*, **101**, 347.

8. Winnacker, E.-L. (1987) *From Genes to Clones: An Introduction to Gene Technology*. VCH, Weinheim.

9. Greener, A. (1993) *Strategies*, **6**, 7.

10. Bullock, W.O., Fernandez, J.M. and Short, J.M. (1987) *BioTechniques*, **5**, 376.

11. Raleigh, E.A., Lech, K. and Brent, R. (1989) in *Current Protocols in Molecular Biology* (F.M. Ausubel, R. Brent, R.E. Kingston, D.D. Moore, J.G. Seidman, J.A. Smith and K. Struhl, eds). John Wiley, New York, p. 1.4.8.

12. Shatzman, A.R., Gross, M.S. and Rosenberg, M. (1990) in *Current Protocols in Molecular Biology* (F.M. Ausubel, R. Brent, R.E. Kingston, D.D. Moore, J.G. Seidman, J.A. Smith and K. Struhl, eds). John Wiley, New York, Unit 16.3.

13. Geider, K., Hohmeyer, C., Haas, R. and Meyer, T.F. (1985) *Gene*, **33**, 341.

14. Hohn, B. (1979) *Meth. Enzymol.*, **68**, 299.

15. Hohn, B. and Murray, K. (1977) *Proc. Natl. Acad. Sci. USA*, **74**, 3259.

16. Hanahan, D. (1983) *J. Mol. Biol.*, **166**, 557.

17. Weiner, M.P., Anderson, C., Jerpseth, B., Wells, S., Johnson-Browne, B. and Vaillancourt, P. (1994) *Strategies*, **7**, 41.

18. Studier, F.W. and Moffat, B.A. (1986) *J. Mol. Biol.*, **189**, 113.

19. Grodberg, J. and Dunn, J.J. (1988) *J. Bacteriol.*, **170**, 1245.

20. Studier. F.W., Rosenberg, A.H., Dunn, J.J. and Dubendorff, J.W. (1990) *Meth. Enzymol.*, **185**, 60.

21. Young, R.A. and Davis, R.W. (1983) *Proc. Natl. Acad. Sci. USA*, **80**, 1194.

22. Huynh, T.V., Young, R.A. and Davis, R.W. (1985) in *DNA Cloning: A Practical Approach* (D.M. Glover, ed.). IRL Press, Oxford, Vol. 1, p.49.

23. Kunkel, T.A., Roberts, J.D. and Zakour, R.A. (1987) *Meth. Enzymol.*, **154**, 367.

24. Appleyard, R.K. (1954) *Genetics*, **39**, 440.

25. Raleigh, E. and Wilson, G. (1986) *Proc. Natl. Acad. Sci. USA*, **83**, 9070.

26. Bachmann, B.J. (1972) *Bacteriol. Rev.*, **36**, 525.

27. Jendrisak, J., Young, R.A. and Engel, J.D. (1987) *Meth. Enzymol.*, **152**, 359.

28. Baker, T.A., Grossman, A.D. and Gross, C.A. (1984) *Proc. Natl. Acad. Sci. USA*, **81**, 6779.

29. Grossman, A.D., Burgess, R.R., Walter, W. and Gross, C.A. (1983) *Cell*, **32**, 151.

30. Chung, C.H. and Goldberg, A.L. (1981) *Proc. Natl. Acad. Sci. USA*, **78**, 4931.

31. Straus, D.B., Walter, W.A. and Gross, C.A. (1988) *Genes Dev.*, **2**, 1851.

32. Nader, W.F., Edlind, T.D., Huettermann, A. and Sauer, H.W. (1985) *Proc. Natl. Acad. Sci. USA*, **82**, 2698.

33. Wertman, K.F., Wyman, A.R. and Botstein, D. (1986) *Gene*, **49**, 253.

34. Wyman, A.R. and Wertman, K.F. (1987) *Meth. Enzymol.*, **152**, 173.

35. Kunkel, T.A., Roberts, J.D. and Zakour, R.A. (1987) *Meth. Enzymol.*, **154**, 367.

36. Miller, J.H. (1972) *Experiments in Molecular Genetics*. Cold Spring Harbor Laboratory Press, New York.

37. Williams, B.G. and Blattner, F.R. (1979) *J. Virol.*, **29**, 555.

38. Sadler, J.R., Tecklenburg, M. and Betz, J.L. (1980) *Gene*, **8**, 279.

39. Hasan, N. and Szybalski, W. (1987) *Gene*, **56**, 145.

40. Woodcock, D.M., Crowther, P.J., Doherty, J., Jefferson, S., DeCruz, E., Noyer-Weidner, M., Smith, S.S., Michael, M.Z. and Graham, M.W. (1989) *Nucl. Acids Res.*, **17**, 3469.

41. Low, B. (1968) *Proc. Natl. Acad. Sci. USA*, **60**, 160.

42. Whittaker, P.A., Campbell, A.J.B., Southern, E.M. and Murray, N.E. (1988) *Nucl. Acids Res.*, **16**, 6725.

43. Leder, P., Tiemeyer, D. and Enquist, L. (1977) *Science*, **196**, 175.

44. Borck, K., Beggs, J.D., Brammar, W.J., Hopkins, A.S. and Murray, N.E. (1976) *Mol. Gen. Genet.*, **146**, 199.

45. Murray, N.E., Brammar, W.J. and Murray, K. (1977) *Mol. Gen. Genet.*, **150**, 53.

46. Raleigh, E.A. (1987) *Meth. Enzymol.*, **152**, 130.

47. Maurizi, M.R., Trsiter, P. and Gottesman, S. (1985) *J. Bacteriol.*, **164**, 1124.

48. Raleigh, E.A., Trimarchi, R. and Revel, H. (1989) *Genetics*, **122**, 279.

49. Kelleher, J.E. and Raleigh, E.A. (1991) *J. Bacteriol.*, **173**, 5220.

50. Marinus, M., Carraway, M., Frey, A.Z. and Arraj, J.A. (1983) *Mol. Gen. Genet.*, **192**, 288.

51. Boyer, H.W. and Roulland-Dussoix, D. (1969) *J. Mol. Biol.*, **41**, 459.

52. Bolivar, F. and Backman, K. (1979) *Meth. Enzymol.*, **68**, 245.

53. Campbell, J.L., Richardson, C.C. and Studier, F.W. (1978) *Proc. Natl. Acad. Sci. USA*, **75**, 2276.

54. Kushner, S.R., Nagaishi, H., Templin, A. and Clark, A.J. (1971) *Proc. Natl. Acad. Sci. USA*, **68**, 824.

55. Maniatis, T., Fritsch, E.F. and Sambrook, J. (1982) *Molecular Cloning: A Laboratory Manual*. Cold Spring Harbor Laboratory Press, New York.

56. Vieira, J. and Messing, J. (1982) *Gene*, **19**, 259.

57. Messing, J. (1979) *Recomb. DNA Tech. Bull.*, **2(2)**, 43.

58. Messing, J., Crea, R. and Seeburg, P.H. (1981) *Nucl. Acids Res.*, **9**, 309.

59. Anon (1989) *Promega Notes*, **20**, 2.

60. Wood, W.B. (1966) *J. Mol. Biol.*, **16**, 118.

61. Zagursky, R.J. and Berman, M.L. (1984) *Gene*, **27**, 183.

62. Silber, K.R., Keiter, K.C. and Sauer, R.T. (1992) *Proc. Natl. Acad. Sci. USA*, **89**, 295.

63. Guarente, L. and Ptashne, M. (1981) *Proc. Natl. Acad. Sci. USA*, **78**, 2199.

64. Remaut, E., Stanssens, P. and Fiers, W. (1981) *Gene*, **15**, 81.

65. Shimatake, H. and Rosenberg, M. (1981) *Nature*, **292**, 128.

66. Schatz, D.G., Oettinger, M.A. and Baltimore, D. (1989) *Cell*, **59**, 1035.

67. Casadaban, M.J. and Cohen, S.N. (1980) *J. Mol. Biol.*, **138**, 179.

68. Meissner, P.S., Sisk, W.P. and Berman, M.L. (1987) *Proc. Natl. Acad. Sci. USA*, **84**, 4171.

69. Seed, B. (1987) *Nature*, **329**, 840.

70. Zoller, M.J. and Smith, M. (1987) *Meth. Enzymol.*, **154**, 329.

71. Nagai, K. and Thogerson, H.C. (1984) *Nature*, **309**, 810.

72. Gottesman, M.E., Adhya, S. and Das, A. (1980) *J. Mol. Biol.*, **140**, 57.

73. Gross, J. and Gross, M. (1969) *Nature*, **224**, 1166.

74. Arber, W., Enquist, L., Hohn, B., Murray, N.E. and Murray, K. (1983) in *Lambda II* (R.W.Hendrix, J.W.Roberts, F.W.Stahl and R.A.Weisberg, eds). Cold Spring Harbor Laboratory Press, New York, p. 433.

75. Gough, J.A. and Murray, N.E. (1983) *J. Mol. Biol.*, **166**, 1.

76. Frischauf, A.-M., Lehrach, H., Poustka, A. and Murray, N. (1983) *J. Mol. Biol.*, **170**, 827.

77. Raleigh, E.A., Murray, N.E., Revel, H., Blumenthal, R.M., Westaway, D., Reith, A.D., Rigby, P.W.J., Elhai, J. and Hanahan, D. (1988) *Nucl. Acids Res.*, **16**, 1563.

78. Raleigh, E.A., Lech, K. and Brent, R. (1989) in *Current Protocols in Molecular Biology* (F.M. Ausubel, R. Brent, R.E. Kingston, D.D. Moore, J.G. Seidman, J.A. Smith and K. Struhl, eds). John Wiley, New York, p. 1.4.9.

79. Kretz, P.L. and Short, J.M. (1989) *Strategies*, **2**, 25.

80. Kowit, J.D. and Goldberg, A.L. (1977) *J. Biol. Chem.*, **252**, 8350.

81. Karn, J., Brenner, S., Barnett, L. and Cesareni, G. (1980) *Proc. Natl. Acad. Sci. USA*, **77**, 5172.

82. Campbell, A. (1965) *Virology*, **27**, 329.

83. Brent, R. and Ptashne, M. (1981) *Proc. Natl. Acad. Sci. USA*, **78**, 4204.

84. Bolivar, F., Rodriguez, R.L., Greene, P.J., Betlach, M.C., Heyneker, H.L., Boyer, H.W., Crosa, J.H. and Falkow, S. (1977) *Gene*, **2**, 95.

85. Peacock, S.L., McIver, C.M. and Monahan, J.J. (1981) *Biochim. Biophys. Acta*, **655**, 243.

86. Jerpseth, B. and Kretz, P.L. (1993) *Strategies*, **6**, 22.

87. Kushner, S.R. (1978) in *Genetic Engineering* (H.W.Boyer and S. Nicosia, eds). Elsevier, Amsterdam, p. 17.

88. Rosenberg, S.M. (1985) *Gene*, **39**, 313.

89. Greener, A. (1990) *Strategies*, **3**, 5.

90. Hay, B. and Short, J.M. (1992) *Strategies*, **5**, 16.

91. Elgin, E., Mackman, C., Pabst, M., Kretz, P.L. and Greener, A. (1991) *Strategies*, **4**, 8.

92. Elgin, E. (1991) *Strategies*, **4**, 37.

93. Patterson, T.A. and Dean, M. (1987) *Nucleic Acids Res.*, **15**, 6298.

94. Gibson, T.J. (1984) *PhD Thesis*. Cambridge University, UK.

95. Sambrook, J., Fritsch, E.F. and Maniatis, T. (1989) *Molecular Cloning: A Laboratory Manual (2nd edn)*. Cold Spring Harbor Laboratory Press, New York.

96. Simcox, M.E., Huvar, A. and Simcox, T.G. (1994) *Strategies*, **7**, 68.

97. Jerpseth, B., Greener, A., Short, J.M., Viola, J. and Kretz, P.L. (1992) *Strategies*, **5**, 81.

98. Grant, S.G.N., Jessee, J.H., Bloom, F.R. and Hanahan, D. (1990) *Proc. Natl. Acad. Sci. USA*, **87**, 4645.

99. Hatt, J., Callahan, M. and Greener, A. (1992) *Strategies*, **5**, 2.

100. Elish, M.E., Pierce, J.R. and Earhart, C.F. (1988) *J. Gen. Microbiol.*, **134**, 1355.

101. Bachmann, B.J. (1987) in *Escherichia coli and Salmonella typhimurium* (F.C. Neidhardt, J.L. Ingraham, K.B. Low, B. Magasanik, M. Schaechter and H.E. Umbarger, eds). American Society for Microbiology, Washington, D.C., p. 1190.

102. Jerpseth, B., Greener, A., Short, J.M., Viola, J. and Kretz, P.L. (1993) *Strategies*, **6**, 24.

103. Greener, A. and Callahan, M. (1994) *Strategies*, **7**, 32.

104. Anon (1995) *Strategies*, **8**, 57.

105. Levinson, A., Silver, D. and Seed, B. (1984) *J. Mol. Appl. Genet.*, **2**, 507.

106. Young, R.A. and Davis, R.W. (1983) *Science*, **222**, 778.

107. Oka, T., Sakamoto, S., Miyoshi, K., Fuwa, T., Yoda, K., Yamasaki, M., Tamura, G. and Miyake, T. (1985) *Proc. Natl. Acad. Sci. USA*, **82**, 7212.

108. Felton, J. (1983) *Biotechniques*, **1**, 42.

109. Bachmann, B.J. (1983) *Microbiol. Rev.*, **47**, 180.

110. Bachmann, B.J. (1993) *Genetic Maps*, **6**, 2.1–2.33.

111. Bachmann, B.J. and Low, K.B. (1980) *Microbiol. Rev.*, **44**, 1.

112. Bachmann, B.J., Low, K.B. and Taylor, A.L. (1976) *Bacteriol. Rev.*, **40**, 116.

113. Muller-Hill, B., Crapo, L. and Gilbert, W. (1968) *Proc. Natl. Acad. Sci. USA*, **59**, 1259.

114. Ullman, A. and Perrin, D. (1970) in *The Lactose Operon* (J.R. Beckwith and D. Zipser, eds). Cold Spring Harbor Laboratory Press, New York, p. 143.

115. Bedbrook, J.R. and Ausubel, F.M. (1976) *Cell*, **9**, 707.

116. Leach, D.R.F. and Stahl, F.W. (1983) *Nature*, **305**, 448.

117. Kushner, S.R., Nagaishi, H. and Clark, A.J. (1974) *Proc. Natl. Acad. Sci. USA*, **71**, 3593.

118. Raleigh, E.A., Lech, K. and Brent, R. (1989) in *Current Protocols in Molecular Biology* (F.M. Ausubel, R. Brent, R.E. Kingston, D.D. Moore, J.G. Seidman, J.A. Smith and K. Struhl, eds). John Wiley, New York, p. 1.4.6–1.4.10.

119. Bickle, T. (1993) in *Nucleases* (S.M. Linn, R.S. Lloyd and R.J. Roberts, eds). Cold Spring Harbor Laboratory Press, New York, p. 89–109.

120. Raleigh, E.A. (1992) *Mol. Microbiol.*, **6**, 1079.

121. Kelleher, J. and Raleigh, E.A. (1991) *J. Bacteriol.*, **173**, 5220.

122. Sutherland, E., Coe, L. and Raleigh, E.A. (1992) *J. Mol. Biol.*, **225**, 327.

123. Waite-Rees, P.A., Keating, C.J., Moran, L.S., Stalko, B.E., Hornstra, L.J. and Benner, J.S. (1991) *J. Bacteriol.*, **173**, 5207.

124. Hattman, S., Brooks, J.E. and Masurekar, M. (1978) *J. Mol. Biol.*, **126**, 367.

125. Marinus, M.G. and Morris, N.R. (1973) *J. Bacteriol.*, **114**, 1143.

126. May, M.S. and Hattman, S. (1975) *J. Bacteriol.*, **123**, 768.

127. Anon (1996) *Product Catalog*. New England Biolabs, Beverly, p. 243.

128. Sanger, F., Coulson, A.R., Hong, G.-F., Hill, D.F. and Petersen, G.B. (1982) *J. Mol. Biol.*, **162**, 729.

129. Daniels, D.L., Schroeder, J.L., Szybalski, W., Sanger, F. and Blattner, F.R. (1993) *Genetic Maps*, **6**, 1.29–1.49.

130. Barondess, J.J. and Beckwith, J. (1990) *Nature*, **346**, 871.

131. Hendrix, R.W., Roberts, J.W., Stahl, F.W. and Weisberg, R.A. (eds) (1983) *Lambda II*. Cold Spring Harbor Laboratory Press, New York.

132. Lech, K. and Brent, R. (1987) in *Current Protocols in Molecular Biology* (F.M. Ausubel, R. Brent, R.E. Kingston, D.D. Moore, J.G. Seidman, J.A. Smith and K. Struhl, eds). John Wiley, New York, Unit 1.9.

133. Zinder, N.D. and Boeke, J.D. (1982) *Gene*, **19**, 1.

134. Rasched, I. and Oberer, E. (1986) *Microbiol. Rev.*, **50**, 401.

135. Greenstein, D.. and Brent, R. (1993) in *Current Protocols in Molecular Biology* (F.M. Ausubel, R. Brent, R.E. Kingston, D.D. Moore, J.G. Seidman, J.A. Smith and K. Struhl, eds). John Wiley, New York, Unit 1.14.

CHAPTER 2
RESTRICTION AND METHYLATION

Comprehensive information on restriction endonucleases is available at the REBASE internet site (1, 2) maintained by Dr Richard J. Roberts of New England Biolabs. The site is regularly updated and contains complete enzyme listings in a variety of formats and an excellent search facility that enables details of individual enzymes to be obtained. The information held at REBASE includes recognition sequences, isoschizomers, methylation sensitivities, commercial sources and primary references. No attempt is made in this chapter to reproduce all this information: if what you want to know is not described here then consult REBASE.

1. COMPLETE LISTINGS OF RESTRICTION ENDONUCLEASES

(Data kindly provided by Richard J. Roberts, New England Biolabs, 32 Tozer Road, Beverly, MA 01915, USA.)

There are three classes of restriction endonuclease: Types I, II and III. Type II enzymes are the ones that are regularly used in DNA manipulation, these enzymes making specific double-strand cuts, leaving blunt or cohesive ends, cleavage always occurring at the same position relative to the recognition sequence. *Table 1* gives a complete listing of Type II restriction enzymes grouped according to recognition sequence. Each group consists of a 'prototype' enzyme (e.g *Aat*II, *Acc*I), which is usually the first to have been discovered for a particular recognition sequence, and, in most cases, a set of isoschizomers that recognize the same sequence. *Tables 2* and *3* provide equivalent information for Type I and Type III enzymes, respectively.

Table 1. Complete listing of Type II restriction enzymes and their recognition sequences[a]

Enzyme	Isoschizomers	Recognition sequence[b,c]	Enzyme	Isoschizomers	Recognition sequence[b,c]
*Aat*II		GACGT↑C	*Acy*I		GR↑CGYC
	*Asp*JI	GACGTC		*Aha*II	GR↑CGYC
	*Ppu*1253I	GACGTC		*Aos*II	GR↑CGYC
	*Ssp*5230I	GACGT↑C		*Ast*WI	GR↑CGYC
*Acc*I		GT↑MKAC		*Asu*III	GR↑CGYC
	*Dsa*VI	GTMKAC		*Bbi*II	GR↑CGYC
	*Fbl*I	GT↑MKAC		*Bsa*HI	GR↑CGYC
	*Omi*BII	GTMKAC		*Bst*ACI	GR↑CGYC
*Ace*III		CAGCTC(7/11)		*Hgi*I	GR↑CGYC
*Aci*I		CCGC(−3/−1)		*Hgi*DI	GR↑CGYC
*Acl*I		AA↑CGTT		*Hgi*GI	GR↑CGYC
	*Psp*1406I	AA↑CGTT		*Hgi*HII	GR↑CGYC
				*Hin*1I	GR↑CGYC
				*Hin*8I	GRCGYC

Table 1. Type II enzymes (continued)

Enzyme	Isoschizomers	Recognition sequence[b,c]
	*Hsp*92I	GR↑CGYC
	*Msp*17I	GR↑CGYC
	*Nla*SII	GRCGYC
	*Pam*II	GR↑CGYC
	*Ssp*JII	GRCGYC
	*Ssp*M1II	GRCGYC
	*Ssp*M2II	GRCGYC
	*Uba*1381I	GRCGYC
*Afl*II		C↑TTAAG
	*Bfr*I	C↑TTAAG
	*Bsa*FI	CTTAAG
	*Bsp*TI	C↑TTAAG
	*Bst*98I	C↑TTAAG
	*Cfr*92I	CTTAAG
	*Esp*4I	C↑TTAAG
	*Msp*CI	C↑TTAAG
	*Uba*1266I	CTTAAG
	*Uba*1299I	CTTAAG
	*Uba*1312I	CTTAAG
	*Uba*1313I	CTTAAG
	*Uba*1331I	CTTAAG
	*Uba*1374I	CTTAAG
	*Uba*1420I	CTTAAG
	*Uba*1426I	CTTAAG
	*Uba*1443I	CTTAAG
	*Vfi*I	CTTAAG
	*Vha*464I	C↑TTAAG
*Afl*III		A↑CRYGT
*Age*I		A↑CCGGT
	*Asi*AI	A↑CCGGT
	*Csp*AI	A↑CCGGT
	*Pin*AI	A↑CCGGT
*Aha*III		TTT↑AAA
	*Dra*I	TTT↑AAA
	*Pau*AII	TTT↑AAA
	*Sru*I	TTT↑AAA
*Alu*I		AG↑CT
	*Bsa*LI	AGCT
	*Mar*I	AGCT
	*Mlt*I	AG↑CT
	*Otu*I	AGCT
	*Otu*NI	AGCT
	*Oxa*I	AGCT
	*Uba*1433I	AGCT
	*Uba*1441I	AGCT
*Alw*NI		CAGNNN↑CTG
*Apa*I		GGGCC↑C
	*Bsp*120I	G↑GGCCC
	*Eci*EI	GGGCCC
	*Ppe*I	GGGCC↑C
	*Psp*30I	GGGCCC
	*Psp*OMI	G↑GGCCC
	*Uba*1156I	GGGCCC
	*Uba*1157I	GGGCCC
	*Uba*1165I	GGGCCC
	*Uba*1202I	GGGCCC
	*Uba*1241I	GGGCCC
	*Uba*1368I	GGGCCC
*Apa*BI		GCANNNN-N↑TGC
*Apa*LI		G↑TGCAC
	*Aaq*I	GTGCAC
	*Alw*44I	G↑TGCAC
	*Ame*I	GTGCAC
	*Bsp*146I	GTGCAC
	*Pfl*23I	GTGCAC
	*Pfr*12I	GTGCAC
	*Pli*I	GTGCAC
	*Sno*I	G↑TGCAC
	*Uba*1203I	GTGCAC
	*Uba*1387I	GTGCAC
	*Vne*I	G↑TGCAC
*Apo*I		R↑AATTY
	*Acs*I	R↑AATTY
	*Fsi*I	R↑AATTY
*Asc*I		GG↑CGCGCC
*Asu*I		G↑GNCC
	*Aha*B1I	GGNCC
	*Apu*I	GGNCC
	*Asp*S9I	G↑GNCC
	*Avc*I	G↑GNCC
	*Bac*36I	G↑GNCC
	*Bal*228I	G↑GNCC
	*Bav*AII	G↑GNCC
	*Bav*BII	G↑GNCC
	*Bce*22I	G↑GNCC
	*Bsa*SI	GGNCC
	*Bsh*KI	G↑GNCC
	*Bsi*ZI	G↑GNCC

Table 1. Type II enzymes (continued)

Enzyme	Isoschizomers	Recognition sequence[b,c]	Enzyme	Isoschizomers	Recognition sequence[b,c]
	*Bsp*1894I	G↑GNCC		*Uba*1099I	GGNCC
	*Bsp*BII	G↑GNCC		*Uba*1134I	GGNCC
	*Bsu*54I	G↑GNCC		*Uba*1160I	GGNCC
	*Ccu*I	G↑GNCC		*Uba*1164I	GGNCC
	*Cdi*AI	GGNCC		*Vch*O66I	GGNCC
	*Cfr*4I	GGNCC		*Vch*O85I	GGNCC
	*Cfr*8I	GGNCC		*Vch*O90I	GGNCC
	*Cfr*13I	G↑GNCC		*Vpa*K15I	GGNCC
	*Cfr*23I	GGNCC		*Vpa*K25I	GGNCC
	*Cfr*33I	GGNCC		*Vpa*K9AI	GGNCC
	*Cfr*45I	GGNCC		*Vpa*K19AI	GGNCC
	*Cfr*46I	GGNCC		*Vpa*K19BI	GGNCC
	*Cfr*47I	GGNCC		*Vpa*KutAI	GGNCC
	*Cfr*52I	GGNCC		*Vpa*KutBI	GGNCC
	*Cfr*54I	GGNCC		*Vpa*KutJI	GGNCC
	*Cfr*NI	GGNCC			
	*Eco*39I	GGNCC	*Asu*II		TT↑CGAA
	*Eco*47II	GGNCC		*Aca*I	TTCGAA
	*Eco*196II	GGNCC		*Acp*I	TT↑CGAA
	*Eco*201I	GGNCC		*Asp*10HI	TT↑CGAA
	*Fmu*I	GGNC↑C		*Avi*I	TTCGAA
	*Gse*I	GGNCC		*Bim*I	TT↑CGAA
	*Hin*5II	GGNCC		*Bim*19I	TT↑CGAA
	*Mae*K81II	G↑GNCC		*Bpu*14I	TT↑CGAA
	*Mja*II	GGNCC		*Bsi*CI	TT↑CGAA
	*Msp*24I	GGNCC		*Bsp*82I	TTCGAA
	*Mth*BI	GGNCC		*Bsp*90I	TTCGAA
	*Nla*DII	GGNCC		*Bsp*101I	TTCGAA
	*Nmu*EII	GGNCC		*Bsp*102I	TTCGAA
	*Nmu*SI	GGNCC		*Bsp*104I	TTCGAA
	*Nsp*IV	G↑GNCC		*Bsp*119I	TT↑CGAA
	*Nsp*7121I	G↑GNCC		*Bsp*148I	TTCGAA
	*Nsp*LII	GGNCC		*Bsp*151I	TTCGAA
	*Pde*12I	G↑GNCC		*Bsp*241I	TTCGAA
	*Pph*1579I	GGNCC		*Bsp*H22I	TTCGAA
	*Pph*1773I	GGNCC		*Bsp*H103I	TTCGAA
	*Pse*I	GGNCC		*Bsp*H106I	TTCGAA
	*Psp*I	GGNCC		*Bst*BI	TT↑CGAA
	*Psp*PI	G↑GNCC		*Cbi*I	TT↑CGAA
	*Sau*2I	GGNCC		*Csp*45I	TT↑CGAA
	*Sau*5I	GGNCC		*Csp*68KII	TT↑CGAA
	*Sau*13I	GGNCC		*Fsp*II	TT↑CGAA
	*Sau*14I	GGNCC		*Lsp*I	TT↑CGAA
	*Sau*17I	GGNCC		*Mla*I	TT↑CGAA
	*Sau*96I	G↑GNCC		*Mva*16I	TTCGAA
	*Sau*BI	GGNCC		*Nsp*29132I	TTCGAA
	*Sdy*I	GGNCC		*Nsp*BI	TTCGAA

Table 1. Type II enzymes (continued)

Enzyme	Isoschizomers	Recognition sequence[b,c]
	*Nsp*FI	TTCGAA
	*Nsp*JI	TTCGAA
	*Nsp*V	TT↑CGAA
	*Pla*II	TT↑CGAA
	*Ppa*AI	TT↑CGAA
	*Rma*376I	TTCGAA
	*Rma*523I	TTCGAA
	*Rsp*534I	TTCGAA
	*Rsp*556I	TTCGAA
	*Sfu*I	TT↑CGAA
	*Sgr*1839I	TTCGAA
	*Ssp*1I	TT↑CGAA
	*Ssp*152I	TTCGAA
	*Ssp*RFI	TT↑CGAA
	*Svi*I	TT↑CGAA
	*Uba*1385I	TTCGAA
	*Uba*1452I	TTCGAA
*Ava*I		C↑YCGRG
	*Acr*I	CYCGRG
	*Ama*87I	C↑YCGRG
	*Aqu*I	C↑YCGRG
	*Asp*BI	CYCGRG
	*Asp*CI	CYCGRG
	*Asp*DI	CYCGRG
	*Avr*I	CYCGRG
	*Bco*I	C↑YCGRG
	*Bse*15I	C↑YCGRG
	*Bso*BI	C↑YCGRG
	*Bst*7QI	CYCGRG
	*Bst*BAII	CYCGRG
	*Bst*NSII	CYCGRG
	*Bst*SI	C↑YCGRG
	*Bst*Z4I	CYCGRG
	*Eco*88I	C↑YCGRG
	*Eco*27kI	C↑YCGRG
	*Esp*HK29I	CYCGRG
	*Nli*I	CYCGRG
	*Nli*387/7I	CYCGR↑G
	*Nmu*AI	CYCGRG
	*Nsp*III	C↑YCGRG
	*Nsp*DI	CYCGRG
	*Nsp*EI	CYCGRG
	*Nsp*SAI	C↑YCGRG
	*Pla*AI	C↑YCGRG
	*Pun*AI	C↑YCGRG
	*Uba*1205II	CYCGRG
	*Uba*1436I	CYCGRG
	*Uba*1440I	CYCGRG
	*Umi*5I	CYCGRG
*Ava*II		G↑GWCC
	*Afl*I	G↑GWCC
	*Asp*697I	GGWCC
	*Asp*745I	G↑GWCC
	*Asp*BII	GGWCC
	*Asp*CII	GGWCC
	*Asp*DII	GGWCC
	*Bam*NxI	G↑GWCC
	*Bcu*AI	G↑GWCC
	*Bme*18I	G↑GWCC
	*Bme*216I	G↑GWCC
	*Bsp*71I	GGWCC
	*Bsp*100I	GGWCC
	*Bsp*128I	GGWCC
	*Bsp*132I	GGWCC
	*Bsp*133I	GGWCC
	*Bsp*1260I	GGWCC
	*Bsp*F53I	GGWCC
	*Bsp*J105I	GGWCC
	*Bsr*AI	G↑GWCC
	*Bst*4QI	GGWCC
	*Bth*AI	G↑GWCC
	*Bti*I	GGWCC
	*Cau*I	G↑GWCC
	*Cli*I	GGWCC
	*Clm*II	GGWCC
	*Csp*68KI	G↑GWCC
	*Dsa*IV	G↑GWCC
	*Eag*MI	G↑GWCC
	*Eco*47I	G↑GWCC
	*Erp*I	G↑GWCC
	*Fdi*I	G↑GWCC
	*Fsp*MSI	G↑GWCC
	*Fss*I	G↑GWCC
	*Gsp*AI	GGWCC
	*Hgi*BI	G↑GWCC
	*Hgi*CII	G↑GWCC
	*Hgi*EI	G↑GWCC
	*Hgi*HIII	G↑GWCC
	*Hgi*JI	G↑GWCC
	*Hsp*2I	GGWCC
	*Kzo*49I	G↑GWCC
	*Mfo*I	GGWCC
	*Mli*I	GGWCC
	*Msp*AI	GGWCC

Table 1. Type II enzymes (continued)

Enzyme	Isoschizomers	Recognition sequence[b,c]
	*Nli*II	GGWCC
	*Nli*387/7II	GGWCC
	*Nmu*AII	GGWCC
	*Nsp*DII	GGWCC
	*Nsp*GI	GGWCC
	*Nsp*HII	GGWCC
	*Nsp*KI	GGWCC
	*Pfl*19I	GGWCC
	*Pol*I	GGWCC
	*Sfn*I	GGWCC
	*Sgh*1835I	GGWCC
	*Sin*I	G↑GWCC
	*Sin*AI	GGWCC
	*Sin*BI	GGWCC
	*Sin*CI	GGWCC
	*Sin*DI	GGWCC
	*Sin*EI	GGWCC
	*Sin*FI	GGWCC
	*Sin*GI	GGWCC
	*Sin*HI	GGWCC
	*Sin*JI	GGWCC
	*Smu*EI	G↑GWCC
	*Syn*I	GGWCC
	*Tru*I	GGWCC
	*Uba*48I	GGWCC
	*Uba*62I	GGWCC
	*Uba*1131I	GGWCC
	*Uba*1249I	GGWCC
	*Uba*1272I	GGWCC
	*Uba*1278I	GGWCC
	*Uba*1304I	GGWCC
	*Uba*1314I	GGWCC
	*Uba*1373I	GGWCC
	*Uba*1413I	GGWCC
	*Uba*1438I	GGWCC
	*Uth*554I	GGWCC
	*Vpa*K11I	GGWCC
	*Vpa*K65I	GGWCC
	*Vpa*K7AI	GGWCC
	*Vpa*K11AI	↑GGWCC
	*Vpa*K13AI	GGWCC
	*Vpa*K11BI	↑GGWCC
	*Vpa*K11CI	GGWCC
	*Vpa*K11DI	GGWCC
*Ava*III		ATGCAT
	*Bfr*BI	ATG↑CAT
	*Bfr*CI	ATGCAT
	*Csp*68KIII	ATGCA↑T
	*Eco*T22I	ATGCA↑T
	*Mph*1103I	ATGCA↑T
	*Nsi*I	ATGCA↑T
	*Pin*BI	ATGCA↑T
	*Ppu*10I	A↑TGCAT
	*Sep*I	ATGCA↑T
	*Smu*CI	ATGCAT
	*Uba*1353I	ATGCAT
	*Uba*1367I	ATGCAT
	*Uba*1384I	ATGCAT
	*Zsp*2I	ATGCA↑T
*Avr*II		C↑CTAGG
	*Avr*BII	C↑CTAGG
	*Bln*I	C↑CTAGG
	*Bsp*A2I	C↑CTAGG
*Bae*I		(10/15)ACNNNNGTAYC(12/7)
*Bal*I		TGG↑CCA
	*Mls*I	TGG↑CCA
	*Mlu*31I	TGG↑CCA
	*Mlu*NI	TGG↑CCA
	*Msc*I	TGG↑CCA
	*Msp*20I	TGGCCA
*Bam*HI		G↑GATCC
	*Aac*I	GGATCC
	*Aae*I	GGATCC
	*Aca*II	GGATCC
	*Acc*EBI	G↑GATCC
	*Ain*II	GGATCC
	*Ali*I	G↑GATCC
	*Ali*12257I	GGATCC
	*Ali*12258I	GGATCC
	*Apa*CI	G↑GATCC
	*Asi*I	GGATCC
	*Asp*TII	GGATCC
	*Atu*1II	GGATCC
	*Bam*FI	GGATCC
	*Bam*KI	GGATCC
	*Bam*NI	GGATCC
	*Bca*1259I	GGATCC
	*Bce*751I	G↑GATCC
	*Bco*10278I	GGATCC
	*Bna*I	G↑GATCC
	*Bsa*DI	GGATCC

Table 1. Type II enzymes (continued)

Enzyme	Isoschizomers	Recognition sequence[b,c]
	*Bsp*30I	GGATCC
	*Bsp*46I	GGATCC
	*Bsp*98I	GGATCC
	*Bsp*130I	GGATCC
	*Bsp*131I	GGATCC
	*Bsp*144I	GGATCC
	*Bsp*4009I	G↑GATCC
	*Bst*I	G↑GATCC
	*Bst*1126I	GGATCC
	*Bst*2464I	GGATCC
	*Bst*2902I	GGATCC
	*Bst*QI	GGATCC
	*Bsu*90I	GGATCC
	*Bsu*8565I	GGATCC
	*Bsu*8646I	GGATCC
	*Cel*I	GGATCC
	*Dds*I	GGATCC
	*Gdo*I	GGATCC
	*Gin*I	GGATCC
	*Gox*I	GGATCC
	*Gse*III	GGATCC
	*Mle*I	GGATCC
	*Mlu*23I	G↑GATCC
	*Nas*BI	GGATCC
	*Nsp*29132II	G↑GATCC
	*Nsp*SAIV	G↑GATCC
	*Okr*AI	GGATCC
	*Pac*1110I	GGATCC
	*Pae*177I	GGATCC
	*Psp*56I	GGATCC
	*Rhs*I	GGATCC
	*Rlu*4I	GGATCC
	*Sol*I	G↑GATCC
	*Spv*I	GGATCC
	*Sur*I	G↑GATCC
	*Uba*19I	GGATCC
	*Uba*31I	GGATCC
	*Uba*38I	GGATCC
	*Uba*51I	GGATCC
	*Uba*88I	GGATCC
	*Uba*1098I	GGATCC
	*Uba*1163I	GGATCC
	*Uba*1167I	GGATCC
	*Uba*1172I	GGATCC
	*Uba*1173I	GGATCC
	*Uba*1205I	GGATCC
	*Uba*1224I	GGATCC
	*Uba*1242I	GGATCC
	*Uba*1250I	GGATCC
	*Uba*1258I	GGATCC
	*Uba*1297I	GGATCC
	*Uba*1302I	GGATCC
	*Uba*1324I	GGATCC
	*Uba*1325I	GGATCC
	*Uba*1334I	GGATCC
	*Uba*1339I	GGATCC
	*Uba*1346I	GGATCC
	*Uba*1383I	GGATCC
	*Uba*1398I	GGATCC
	*Uba*1402I	GGATCC
	*Uba*1414I	GGATCC
*Bbv*I		GCAGC(8/12)
	*Alw*XI	GCAGC(8/12)
	*Bch*I	GCAGC
	*Bsa*UI	GCAGC
	*Bsp*423I	GCAGC(8/12)
	*Bsr*VI	GCAGC
	*Bst*12I	GCAGC
	*Bst*71I	GCAGC(8/12)
	*Lfe*I	GCAGC
	*Lsp*1109I	GCAGC
*Bbv*II		GAAGAC(2/6)
	*Bbs*I	GAAGAC(2/6)
	*Bbv*16II	GAAGAC(2/6)
	*Bco*102II	GAAGAC
	*Bpi*I	GAAGAC(2/6)
	*Bpu*AI	GAAGAC(2/6)
	*Bsa*VI	GAAGAC
	*Bsc*91I	GAAGAC(2/6)
	*Bsp*BS31I	GAAGAC(2/6)
	*Bsp*IS4I	GAAGAC(2/6)
	*Bsp*TS514I	GAAGAC(2/6)
	*Bsp*VI	GAAGAC
	*Bst*BS32I	GAAGAC(2/6)
	*Bst*TS5I	GAAGAC(2/6)
*Bcc*I		CCATC
*Bce*83I		CTTGAG(16/14)
*Bce*fI		ACGGC(12/13)
	*Bct*I	ACGGC
*Bcg*I		(10/12)GCAN-NNNNNT-CG(12/10)

Table 1. Type II enzymes (continued)

Enzyme	Isoschizomers	Recognition sequence[b,c]	Enzyme	Isoschizomers	Recognition sequence[b,c]
*Bci*VI		GGATAC(6/5)	*Bin*I		GGATC(4/5)
*Bcl*I		T↑GATCA		*Acl*WI	GGATC(4/5)
	*Atu*CI	TGATCA		*Alw*I	GGATC(4/5)
	*Bco*102I	TGATCA		*Bsp*PI	GGATC(4/5)
	*Bme*TI	TGATCA		*Bsr*WI	GGATC
	*Bsi*QI	T↑GATCA		*Bth*II	GGATC
	*Bsp*XII	T↑GATCA		*Ral*8I	GGATC
	*Bst*77I	TGATCA	*Bmg*I		GKGCCC
	*Bst*GI	TGATCA	*Bpl*I		GAGNNNNN-CTC
	*Bst*KI	TGATCA	*Bpu*10I		CCTNAG-C(−5/−2)
	*Bst*Z10II	TGATCA	*Bsa*AI		YAC↑GTR
	*Cpe*I	TGATCA		*Bst*BAI	YAC↑GTR
	*Cth*I	TGATCA		*Msp*YI	YAC↑GTR
	*Fba*I	T↑GATCA		*Ppu*6I	YACGTR
	*Ksp*22I	T↑GATCA		*Ppu*11I	YACGTR
	*Pov*I	TGATCA		*Ppu*21I	YACGTR
	*Sse*I	TGATCA		*Psu*AI	YAC↑GTR
	*Sst*IV	TGATCA	*Bsa*BI		GATNN↑NN-ATC
	*Uba*1282I	TGATCA		*Bco*63I	GATNNNNATC
	*Uba*1283I	TGATCA		*Bco*631I	GATNNNNATC
	*Uba*1431I	TGATCA		*Bse*8I	GATNN↑NN-ATC
	*Uba*1447I	TGATCA		*Bse*631I	GATNNNNATC
	*Umi*7I	TGATCA		*Bsh*1365I	GATNN↑NN-ATC
*Bet*I		W↑CCGGW		*Bsi*BI	GATNN↑NN-ATC
	*Bba*179I	WCCGGW		*Bsr*BRI	GATNN↑NN-ATC
	*Bca*77I	W↑CCGGW		*Mam*I	GATNN↑NN-ATC
	*Bsa*WI	W↑CCGGW	*Bsa*XI		ACNNNNNC-TCC
	*Bst*1473I	WCCGGW		*Bsm*XI	ACNNNNNC-TCC
*Bfi*I		ACTGGG	*Bsb*I		CAACAC
	*Bmr*I	ACTGGG	*Bsc*GI		CCCGT
*Bgl*I		GCCNNNN↑N-GGC	*Bse*PI		G↑CGCGC
	*Bso*JI	GCCNNNNN-GGC		*Bsc*EI	GCGCGC
	*Tsp*8EI	GCCNNNN↑N-GGC		*Bso*PI	GCGCGC
	*Tsp*219I	GCCNNNNN-GGC			
	*Van*I	GCCNNNNN-GGC			
*Bgl*II		A↑GATCT			
	*Ncr*I	A↑GATCT			
	*Nsp*MACI	A↑GATCT			
	*Pae*2kI	A↑GATCT			
	*Pae*18kI	A↑GATCT			

Table 1. Type II enzymes (continued)

Enzyme	Isoschizomers	Recognition sequence[b,c]
	*Bsr*HI	GCGCGC
	*Bss*BI	GCGCGC
	*Bss*HII	G↑CGCGC
	*Cfr*J5I	GCGCGC
	*Eco*143I	GCGCGC
	*Eco*152I	GCGCGC
	*Esp*7I	GCGCGC
	*Esp*8I	GCGCGC
	*Kpn*30I	GCGCGC
	*Pau*I	G↑CGCGC
	*Ttm*II	GCGCGC
	*Uba*69I	GCGCGC
*Bse*RI		GAGGAG(10/8)
*Bsg*I		GTGCAG(16/14)
*Bsi*I		CACGAG(–5/–1)
	*Bss*SI	CACGAG(–5/–1)
	*Bst*2BI	CACGAG(–5/–1)
*Bsi*YI		CCNNNNN↑NNGG
	*Bsc*4I	CCNNNNN↑NNGG
	*Bsc*107I	CCNNNNNNNGG
	*Bse*23I	CCNNNNNNNGG
	*Bse*LI	CCNNNNN↑NNGG
	*Bsl*I	CCNNNNN↑NNGG
	*Bsm*YI	CCNNNNNNNGG
*Bsm*I		GAATGC(1/–1)
	*Asp*26HI	GAATGC(1/–1)
	*Asp*27HI	GAATGC(1/–1)
	*Asp*35HI	GAATGC(1/–1)
	*Asp*36HI	GAATGC(1/–1)
	*Asp*40HI	GAATGC(1/–1)
	*Asp*50HI	GAATGC(1/–1)
	*Bsa*MI	GAATGC(1/–1)
	*Bsc*CI	GAATGC(1/–1)
	*Mva*1269I	GAATGC(1/–1)
	*Uba*1382I	GAATGC
	*Uba*1415I	GAATGC
*Bsm*AI		GTCTC(1/5)
	*Alw*26I	GTCTC(1/5)
*Bsp*24I		(8/13)GACNNNNNNTGG(12/7)
*Bsp*1407I		T↑GTACA
	*Aau*I	T↑GTACA
	*Bsm*GI	TGTACA
	*Bsr*GI	T↑GTACA
	*Bst*170I	TGTACA
	*Ssp*4800I	T↑GTACA
	*Ssp*BI	T↑GTACA
*Bsp*GI		CTGGAC
*Bsp*HI		T↑CATGA
	*Pag*I	TCATGA
	*Rca*I	T↑CATGA
	*Rhc*I	TCATGA
	*Rsp*XI	T↑CATGA
*Bsp*LU11I		A↑CATGT
*Bsp*MI		ACCTGC(4/8)
*Bsp*MII		T↑CCGGA
	*Acc*III	T↑CCGGA
	*Aor*13HI	T↑CCGGA
	*Bbf*7411I	TCCGGA
	*Bbv*AIII	T↑CCGGA
	*Bla*7920I	TCCGGA
	*Bse*AI	T↑CCGGA
	*Bsi*GI	TCCGGA
	*Bsi*MI	T↑CCGGA
	*Bsi*OI	TCCGGA
	*Bsp*13I	T↑CCGGA
	*Bsp*228I	TCCGGA
	*Bsp*233I	TCCGGA
	*Bsp*508I	TCCGGA
	*Bsp*EI	T↑CCGGA
	*Bsp*H226I	TCCGGA
	*Bst*Z3I	TCCGGA
	*Bsu*22I	TCCGGA
	*Bsu*23I	T↑CCGGA
	*Cau*B3I	T↑CCGGA
	*Cfr*57I	TCCGGA
	*Esp*HK26I	TCCGGA

Table 1. Type II enzymes (continued)

Enzyme	Isoschizomers	Recognition sequence[b,c]
	*Kpn*2I	T↑CCGGA
	*Mro*I	T↑CCGGA
	*Pin*BII	T↑CCGGA
	*Pta*I	T↑CCGGA
	*Tsp*507I	TCCGGA
	*Tsp*514I	TCCGGA
	*Uba*1136I	TCCGGA
	*Uba*1279I	TCCGGA
	*Uba*1375I	TCCGGA
	*Uba*1425I	TCCGGA
*Bsr*I		ACTGG(1/–1)
	*Bsc*HI	ACTGG
	*Bse*1I	ACTGG(1/–1)
	*Bse*NI	ACTGG(1/–1)
	*Bso*HI	ACTGG
	*Bsr*SI	ACTGG(1/–1)
	*Bst*11I	ACTGG(1/–1)
	*Tsp*1I	ACTGG(1/–1)
*Bsr*BI		CCGCTC(–3/–3)
	*Acc*BSI	CCGCTC(–3/–3)
	*Bst*D102I	CCGCTC(–3/–3)
	*Mbi*I	CCGCTC
*Bsr*DI		GCAATG(2/0)
	*Bse*3DI	GCAATG(2/0)
	*Bse*MI	GCAATG
*Bst*EII		G↑GTNACC
	*Acr*II	G↑GTNACC
	*Asp*AI	G↑GTNACC
	*Bse*59I	GGTNACC
	*Bse*64I	GGTNACC
	*Bse*T9I	G↑GTNACC
	*Bse*T10I	G↑GTNACC
	*Bsi*KI	G↑GTNACC
	*Bst*31I	GGTNACC
	*Bst*DI	GGTNACC
	*Bst*PI	G↑GTNACC
	*Cfr*7I	GGTNACC
	*Cfr*19I	GGTNACC
	*Eca*I	G↑GTNACC
	*Eci*125I	G↑GTNACC
	*Eco*91I	G↑GTNACC
	*Eco*O65I	G↑GTNACC
	*Eco*O128I	G↑GTNACC
	*Kox*I	GGTNACC
	*Nsp*SAII	G↑GTNACC
	*Psp*EI	G↑GTNACC
	*Sci*AI	GGTNACC
	*Uba*1291I	GGTNACC
*Bst*XI		CCANNNNN↑NTGG
	*Bsc*JI	CCANNNNNNTGG
	*Bss*GI	CCANNNNNNTGG
	*Bst*TI	CCANNNNNNTGG
*Cac*8I		GCN↑NGC
*Cau*II		CC↑SGG
	*Aha*I	CC↑SGG
	*Ase*II	CC↑SGG
	*Asp*1I	CCSGG
	*Asu*C2I	CC↑SGG
	*Bcn*I	CC↑SGG
	*Bsp*7I	CCSGG
	*Bsp*8I	CCSGG
	*Bsp*55I	CCSGG
	*Bsp*F105I	CCSGG
	*Bsp*J67I	CCSGG
	*Eci*DI	CCSGG
	*Eco*121I	CCSGG
	*Eco*179I	CCSGG
	*Eco*190I	CCSGG
	*Eco*1831I	↑CCSGG
	*Eco*HI	↑CCSGG
	*Hgi*S21I	CCSGG
	*Hgi*S22I	CC↑SGG
	*Hin*3I	CCSGG
	*Kpn*49kII	↑CCSGG
	*Mgl*14481I	CC↑SGG
	*Nci*I	CC↑SGG
	*Pae*181I	CCSGG
	*Rsh*II	CCSGG
	*Ssp*2I	CCSGG
	*Tmu*1I	CCSGG
	*Uba*41I	CCSGG
	*Uba*42I	CCSGG
	*Uba*1280I	CCSGG
	*Uba*1318I	CCSGG
	*Uba*1347I	CCSGG
	*Uba*1370I	CCSGG
	*Uba*1372I	CCSGG

Table 1. Type II enzymes (continued)

Enzyme	Isoschizomers	Recognition sequence[b,c]
	*Uba*1376I	CCSGG
	*Uba*1378I	CCSGG
	*Uba*1389I	CCSGG
	*Uba*1401I	CCSGG
	*Uba*1423I	CCSGG
	*Uba*1424I	CCSGG
*Cfr*I		Y↑GGCCR
	*Bfi*89I	Y↑GGCCR
	*Cfr*14I	YGGCCR
	*Cfr*38I	YGGCCR
	*Cfr*39I	YGGCCR
	*Cfr*40I	YGGCCR
	*Cfr*55I	YGGCCR
	*Cfr*59I	YGGCCR
	*Eae*I	Y↑GGCCR
	*Eci*BI	YGGCCR
	*Eco*90I	YGGCCR
	*Eco*164I	YGGCCR
	*Eco*HAI	YGGCCR
	*Eco*HK31I	Y↑GGCCR
	*Esp*HK16I	YGGCCR
	*Esp*HK24I	YGGCCR
	*Ksp*HK15I	YGGCCR
	*Uba*36I	YGGCCR
	*Uba*1188I	YGGCCR
	*Uba*1327I	YGGCCR
*Cfr*10I		R↑CCGGY
	*Bco*118I	R↑CCGGY
	*Bse*118I	R↑CCGGY
	*Bse*634I	R↑CCGGY
	*Bsp*21I	RCCGGY
	*Bsr*FI	R↑CCGGY
	*Bss*AI	R↑CCGGY
	*Tse*DI	RCCGGY
*Cje*I		(8/14)CCANNN-NNNGT(15/9)
	*Cje*PI	(7/13)CCANNN-NNNNTC(14/8)
*Cla*I		AT↑CGAT
	*Aag*I	AT↑CGAT
	*Apu*16I	ATCGAT
	*Asp*707I	ATCGAT
	*Ban*III	AT↑CGAT
	*Bav*CI	ATCGAT
	*Bbv*AII	AT↑CGAT
	*Bci*29I	AT↑CGAT
	*Bci*BI	AT↑CGAT
	*Bcm*I	AT↑CGAT
	*Bdi*I	AT↑CGAT
	*Bfr*AI	ATCGAT
	*Bli*41I	AT↑CGAT
	*Bli*86I	AT↑CGAT
	*Bli*576I	ATCGAT
	*Bli*585I	ATCGAT
	*Bli*RI	AT↑CGAT
	*Bsa*29I	AT↑CGAT
	*Bsc*I	AT↑CGAT
	*Bse*CI	AT↑CGAT
	*Bsi*XI	AT↑CGAT
	*Bsp*2I	ATCGAT
	*Bsp*4I	ATCGAT
	*Bsp*84I	ATCGAT
	*Bsp*106I	AT↑CGAT
	*Bsp*125I	ATCGAT
	*Bsp*126I	ATCGAT
	*Bsp*127I	ATCGAT
	*Bsp*145I	ATCGAT
	*Bsp*DI	AT↑CGAT
	*Bsp*JII	AT↑CGAT
	*Bsp*XI	AT↑CGAT
	*Bsr*CI	ATCGAT
	*Bst*28I	AT↑CGAT
	*Bst*LVI	ATCGAT
	*Bsu*15I	AT↑CGAT
	*Btu*I	ATCGAT
	*Csp*4I	ATCGAT
	*Lca*I	AT↑CGAT
	*Lpl*I	AT↑CGAT
	*Pga*I	AT↑CGAT
	*Rme*21I	ATCGAT
	*Ssp*27144I	AT↑CGAT
	*Uba*22I	ATCGAT
	*Uba*24I	ATCGAT
	*Uba*30I	ATCGAT
	*Uba*34I	ATCGAT
	*Uba*43I	ATCGAT
	*Uba*1096I	ATCGAT
	*Uba*1100I	ATCGAT
	*Uba*1133I	ATCGAT
	*Uba*1137I	ATCGAT
	*Uba*1138I	ATCGAT
	*Uba*1144I	ATCGAT

Table 1. Type II enzymes (continued)

Enzyme	Isoschizomers	Recognition sequence[b,c]
	*Uba*1145I	ATCGAT
	*Uba*1161I	ATCGAT
	*Uba*1168I	ATCGAT
	*Uba*1195I	ATCGAT
	*Uba*1196I	ATCGAT
	*Uba*1197I	ATCGAT
	*Uba*1198I	ATCGAT
	*Uba*1199I	ATCGAT
	*Uba*1200I	ATCGAT
	*Uba*1233I	ATCGAT
	*Uba*1238I	ATCGAT
	*Uba*1246I	ATCGAT
	*Uba*1257I	ATCGAT
	*Uba*1275I	ATCGAT
	*Uba*1286I	ATCGAT
	*Uba*1295I	ATCGAT
	*Uba*1315I	ATCGAT
	*Uba*1342I	ATCGAT
	*Uba*1366II	ATCGAT
	*Uba*1379I	ATCGAT
	*Uba*1380I	ATCGAT
	*Uba*1394I	ATCGAT
	*Uba*1412I	ATCGAT
	*Uba*1416I	ATCGAT
	*Uba*1427I	ATCGAT
	*Uba*1430I	ATCGAT
	*Uba*1451I	ATCGAT
	*Uba*1453I	ATCGAT
*Cvi*JI		RG↑CY
	*Cvi*KI	RGCY
	*Cvi*LI	RGCY
	*Cvi*MI	RGCY
	*Cvi*NI	RGCY
	*Cvi*OI	RGCY
*Cvi*RI		TG↑CA
*Dde*I		C↑TNAG
	*Bst*295I	CTNAG
	*Bst*DEI	C↑TNAG
*Dpn*I		GA↑TC
	*Cfu*I	GA↑TC
	*Mph*1103II	GATC
	*Nan*II	GATC
	*Ngo*DIII	GATC
	*Nmu*DI	GATC
	*Nmu*EI	GATC
	*Nsu*DI	GATC
*Dra*II		RG↑GNCCY
	*Eco*O109I	RG↑GNCCY
	*Kaz*48kI	RGGNC↑CY
	*Pss*I	RGGNC↑CY
	*Uba*1326I	RGGNCCY
	*Vne*AI	RGGNCCY
*Dra*III		CACNNN↑GTG
	*Ade*I	CACNNNCTG
*Drd*I		GACNNNN↑N-NGTC
	*Dse*DI	GACNNNN↑N-NGTC
*Drd*II		GAACCA
*Dsa*I		C↑CRYGG
	*Bst*DSI	C↑CRYGG
*Eam*1105I		GACNNN↑NN-GTC
	*Ahd*I	GACNNN↑NN-GTC
	*Asp*EI	GACNNN↑NN-GTC
	*Bsp*OVI	GACNNN↑NN-GTC
	*Bst*Z2I	GACNNNNN-GTC
	*Ecl*HKI	GACNNN↑NN-GTC
	*Nru*GI	GACNNN↑NN-GTC
	*Uba*1190I	GACNNNNN-GTC
	*Uba*1191I	GACNNNNN-GTC
*Eci*I		TCCGCC
*Eco*31I		GGTCTC(1/5)
	*Bli*49I	GGTCTC
	*Bli*161I	GGTCTC
	*Bli*576II	GGTCTC
	*Bli*736I	GGTCTC(1/5)
	*Bsa*I	GGTCTC(1/5)
	*Cfr*56I	GGTCTC
	*Eco*42I	GGTCTC

Table 1. Type II enzymes (continued)

Enzyme	Isoschizomers	Recognition sequence[b,c]
	*Eco*51I	GGTCTC
	*Eco*95I	GGTCTC
	*Eco*97I	GGTCTC
	*Eco*101I	GGTCTC
	*Eco*120I	GGTCTC
	*Eco*127I	GGTCTC
	*Eco*129I	GGTCTC
	*Eco*155I	GGTCTC
	*Eco*156I	GGTCTC
	*Eco*157I	GGTCTC
	*Eco*162I	GGTCTC
	*Eco*185I	GGTCTC
	*Eco*191I	GGTCTC
	*Eco*203I	GGTCTC
	*Eco*204I	GGTCTC
	*Eco*205I	GGTCTC
	*Eco*217I	GGTCTC
	*Eco*225I	GGTCTC
	*Eco*233I	GGTCTC
	*Eco*239I	GGTCTC
	*Eco*240I	GGTCTC
	*Eco*241I	GGTCTC
	*Eco*246I	GGTCTC
	*Eco*247I	GGTCTC
	*Eco*263I	GGTCTC
	*Eco*A4I	GGTCTC(1/5)
	*Eco*O44I	GGTCTC(1/5)
	*Ppa*I	GGTCTC
	*Rle*69I	GGTCTC
	*Sau*12I	GGTCTC
	*Slb*I	GGTCTC
	*Uba*65I	GGTCTC
	*Uba*84I	GGTCTC
	*Uba*1316I	GGTCTC
	*Uba*1343I	GGTCTC
	*Vpa*K57I	GGTCTC
	*Vpa*K57AI	GGTCTC
	*Vpa*KutHI	GGTCTC
*Eco*47III		AGC↑GCT
	*Afe*I	AGC↑GCT
	*Ait*I	AGC↑GCT
	*Aor*51HI	AGC↑GCT
	*Fun*I	AGC↑GCT
*Eco*57I		CTGAAG(16/14)
	*Bsp*6II	CTGAAG
	*Bsp*KT5I	CTGAAG(16/14)
	*Eco*112I	CTGAAG
	*Eco*125I	CTGAAG
	*Fsf*I	CTGAAG
*Eco*NI		CCTNN↑NNN-AGG
	*Bpu*1268I	CCTNNNNN-AGG
	*Bso*EI	CCTNNNNN-AGG
	*Bst*WI	CCTNNNNN-AGG
	*Uba*1289I	CCTNNNNN-AGG
	*Uba*1290I	CCTNNNNN-AGG
	*Uba*1308I	CCTNNNNN-AGG
	*Uba*1309I	CCTNNNNN-AGG
	*Uba*1310I	CCTNNNNN-AGG
*Eco*RI		G↑AATTC
	*Eco*82I	GAATTC
	*Eco*159I	GAATTC
	*Eco*228I	GAATTC
	*Eco*237I	GAATTC
	*Eco*252I	GAATTC
	*Hal*I	G↑AATTC
	*Kpn*49kI	G↑AATTC
	*Ppu*111I	G↑AATTC
	*Rsr*I	G↑AATTC
	*Sso*I	G↑AATTC
	*Uba*58I	GAATTC
	*Van*91II	GAATTC
	*Vch*N100I	GAATTC
	*Vch*O2I	GAATTC
*Eco*RII		↑CCWGG
	*Acc*38I	CCWGG
	*Aeu*I	CC↑WGG
	*Aor*I	CC↑WGG
	*Apa*ORI	CC↑WGG
	*Apy*I	CC↑WGG
	*Asp*2HI	CCWGG
	*Atu*II	CCWGG
	*Atu*1I	CCWGG
	*Atu*BI	CCWGG

Table 1. Type II enzymes (continued)

Enzyme	Isoschizomers	Recognition sequence[b,c]	Enzyme	Isoschizomers	Recognition sequence[b,c]
	*Bci*BII	CC↑WGG		*Eca*II	CCWGG
	*Bin*SI	CCWGG		*Ecl*II	CCWGG
	*Bsa*NI	CCWGG		*Ecl*66I	CCWGG
	*Bse*16I	CC↑WGG		*Ecl*136I	CCWGG
	*Bse*17I	CC↑WGG		*Ecl*137II	CCWGG
	*Bse*24I	CC↑WGG		*Ecl*37kII	CCWGG
	*Bse*BI	CC↑WGG		*Ecl*54kI	CCWGG
	*Bsh*GI	CC↑WGG		*Ecl*57kI	CCWGG
	*Bsi*LI	CC↑WGG		*Ecl*1zII	CCWGG
	*Bsi*UI	CCWGG		*Ecl*S39I	CCWGG
	*Bsi*VI	CCWGG		*Eco*38I	CCWGG
	*Bso*GI	CCWGG		*Eco*40I	CCWGG
	*Bsp*44I	CCWGG		*Eco*41I	CCWGG
	*Bsp*56I	CCWGG		*Eco*60I	CCWGG
	*Bsp*103I	CCWGG		*Eco*61I	CCWGG
	*Bsp*317I	CCWGG		*Eco*67I	CCWGG
	*Bsp*H43I	CCWGG		*Eco*70I	CCWGG
	*Bsp*NI	CC↑WGG		*Eco*71I	CCWGG
	*Bst*1I	CC↑WGG		*Eco*128I	CCWGG
	*Bst*2I	CC↑WGG		*Eco*170I	CCWGG
	*Bst*38I	CC↑WGG		*Eco*193I	CCWGG
	*Bst*100I	CC↑WGG		*Eco*206I	CCWGG
	*Bst*2UI	CC↑WGG		*Eco*207I	CCWGG
	*Bst*7QII	CCWGG		*Eco*254I	CCWGG
	*Bst*GII	CCWGG		*Eco*256I	CCWGG
	*Bst*M6I	CC↑WGG		*Ese*6II	CCWGG
	*Bst*NI	CC↑WGG		*Esp*2I	CCWGG
	*Bst*OI	CC↑WGG		*Esp*24I	CCWGG
	*Bth*DI	CC↑WGG		*Esp*HK7I	CCWGG
	*Bth*EI	CC↑WGG		*Esp*HK22I	CCWGG
	*Cdi*27I	CCWGG		*Esp*HK30I	CCWGG
	*Cfr*5I	CCWGG		*Fsp*1604I	CC↑WGG
	*Cfr*11I	CCWGG		*Hhd*I	CCWGG
	*Cfr*20I	CCWGG		*Kox*165I	CCWGG
	*Cfr*22I	CCWGG		*Kpn*10I	CCWGG
	*Cfr*24I	CCWGG		*Kpn*13I	CCWGG
	*Cfr*25I	CCWGG		*Kpn*14I	CCWGG
	*Cfr*27I	CCWGG		*Kpn*16I	CCWGG
	*Cfr*28I	CCWGG		*Ksp*HK12I	CCWGG
	*Cfr*29I	CCWGG		*Ksp*HK14I	CCWGG
	*Cfr*30I	CCWGG		*Lla*497I	CCWGG
	*Cfr*31I	CCWGG		*Mlu*2300I	CCWGG
	*Cfr*35I	CCWGG		*Mph*I	CCWGG
	*Cfr*58I	CCWGG		*Mva*I	CC↑WGG
	*Cfr*S37I	CCWGG		*Psp*GI	CCWGG
	*Cth*II	CC↑WGG		*Sau*16I	CCWGG
	*Eag*KI	CCWGG		*Scg*2I	CCWGG

Table 1. Type II enzymes (continued)

Enzyme	Isoschizomers	Recognition sequence[b,c]
	*Sfl*2aI	CCWGG
	*Sfl*2bI	CCWGG
	*Sgr*20I	CCWGG
	*Sle*I	↑CCWGG
	*Sni*I	CC↑WGG
	*Ssl*I	CC↑WGG
	*Ssp*AI	↑CCWGG
	*Sth*117I	CC↑WGG
	*Sth*455I	CCWGG
	*Taq*XI	CC↑WGG
	*Tsp*AI	CCWGG
	*Uba*11I	CCWGG
	*Uba*13I	CCWGG
	*Uba*20I	CCWGG
	*Uba*81I	CCWGG
	*Uba*82I	CCWGG
	*Uba*1114I	CCWGG
	*Uba*1118I	CCWGG
	*Uba*1120I	CCWGG
	*Uba*1121I	CCWGG
	*Uba*1125I	CCWGG
	*Uba*1171I	CCWGG
	*Uba*1181I	CCWGG
	*Uba*1185I	CCWGG
	*Uba*1189I	CCWGG
	*Uba*1193I	CCWGG
	*Uba*1218I	CCWGG
	*Uba*1243I	CCWGG
	*Uba*1410I	CCWGG
	*Uba*1428I	CCWGG
	*Zan*I	CC↑WGG
*Eco*RV		GAT↑ATC
	*Bsh*LI	GATATC
	*Bso*AI	GATATC
	*Bsp*16I	GATATC
	*Bst*RI	GATATC
	*Ceq*I	GAT↑ATC
	*Eco*32I	GAT↑ATC
	*Eco*178I	GATATC
	*Hja*I	GAT↑ATC
	*Nan*I	GATATC
	*Nfl*AI	GATATC
	*Nsi*CI	GAT↑ATC
	*Pac*1110II	GATATC
	*Pfl*16I	GATATC
	*Rma*495II	GATATC
	*Rma*496II	GATATC
	*Rma*497II	GATATC
	*Tsp*273I	GATATC
	*Uba*1400I	GATATC
*Esp*I		GC↑TNAGC
	*Blp*I	GC↑TNAGC
	*Bpu*1102I	GC↑TNAGC
	*Bpu*GCI	GCTNAGC
	*Bsp*1720I	GC↑TNAGC
	*Cel*II	GC↑TNAGC
	*Uba*1221I	GCTNAGC
	*Uba*1222I	GCTNAGC
	*Uba*1284I	GCTNAGC
	*Uba*1320I	GCTNAGC
*Esp*3I		CGTCTC(1/5)
	*Bsm*BI	CGTCTC(1/5)
	*Esp*16I	CGTCTC
	*Esp*23I	CGTCTC
*Fau*I		CCCGC(4/6)
	*Bst*Z11I	CCCGC
*Fin*I		GGGAC
	*Bsm*FI	GGGAC(10/14)
	*Bsp*LU11III	GGGAC(10/14)
*Fnu*4HI		GC↑NGC
	*Bso*FI	GC↑NGC
	*Bsp*6I	GC↑NGC
	*Bss*FI	GCNGC
	*Bss*XI	GCNGC
	*Cac*824I	GCNGC
	*Cco*P215I	GCNGC
	*Cco*P216I	GCNGC
	*Fbr*I	GC↑NGC
	*Fsp*4HI	GC↑NGC
	*Ita*I	GC↑NGC
	*Uur*960I	GC↑NGC
*Fnu*DII		CG↑CG
	*Acc*II	CG↑CG
	*Bce*FI	CGCG
	*Bce*RI	CGCG
	*Bep*I	CG↑CG
	*Bpu*95I	CG↑CG
	*Bsh*1236I	CG↑CG
	*Bsp*50I	CG↑CG
	*Bsp*70I	CGCG
	*Bsp*J76I	CGCG

Table 1. Type II enzymes (continued)

Enzyme	Isoschizomers	Recognition sequence[b,c]
	*Bst*UI	CG↑CG
	*Bsu*1192II	CGCG
	*Bsu*1193I	CGCG
	*Bsu*1532I	CG↑CG
	*Bsu*6633I	CGCG
	*Bsu*EII	CGCG
	*Btk*I	CG↑CG
	*Cpa*1150I	CGCG
	*Cpa*AI	CGCG
	*Csp*KVI	CG↑CG
	*Fau*BII	CG↑CG
	*Fsp*MI	CGCG
	*Hin*1056I	CGCG
	*Mva*AI	CGCG
	*Mvn*I	CG↑CG
	*Pfl*AI	CGCG
	*Sce*I	CGCG
	*Sel*I	↑CGCG
	*Tha*I	CG↑CG
	*Tma*I	CGCG
	*Uba*1321I	CGCG
	*Uba*1404I	CGCG
	*Uba*1405I	CGCG
	*Uba*1446I	CGCG
*Fok*I		GGATG(9/13)
	*Bse*GI	GGATG(2/0)
	*Bst*F5I	GGATG(2/0)
	*Hin*GUII	GGATG
	*Sts*I	GGATG(10/14)
*Fse*I		GGCCGG↑CC
*Gdi*II		CGGCCR(–5/–1)
*Gsu*I		CTGGAG(16/14)
	*Bco*35I	CTGGAG
	*Bpm*I	CTGGAG(16/14)
	*Bsp*22I	CTGGAG
	*Bsp*28I	CTGGAG
	*Bsp*J74I	CTGGAG
	*Uba*1437I	CTGGAG
	*Uba*1444I	CTGGAG
*Hae*I		WGG↑CCW
*Hae*II		RGCGC↑Y
	*Acc*B2I	RGCGC↑Y
	*Bme*142I	RGC↑GCY
	*Bsm*HI	RGCGCY
	*Bsp*143II	RGCGC↑Y
	*Bst*16I	RGCGCY
	*Bst*1473II	RGCGCY
	*Bst*H2I	RGCGC↑Y
	*Btu*34II	RGCGCY
	*Hin*HI	RGCGCY
	*Lpn*I	RGC↑GCY
	*Ngo*I	RGCGCY
	*Ngo*AI	RGCGCY
*Hae*III		GG↑CC
	*Aca*IV	GGCC
	*Asp*742I	GGCC
	*Asp*TIII	GGCC
	*Avr*BI	GGCC
	*Bal*475I	GGCC
	*Bal*3006I	GGCC
	*Bce*71I	GGCC
	*Bco*33I	GGCC
	*Bec*AII	GG↑CC
	*Bfi*458I	GGCC
	*Bim*19II	GG↑CC
	*Bli*I	GGCC
	*Blu*II	GGCC
	*Bsa*RI	GGCC
	*Bse*I	GGCC
	*Bse*QI	GG↑CC
	*Bsh*I	GG↑CC
	*Bsh*AI	GGCC
	*Bsh*BI	GGCC
	*Bsh*CI	GGCC
	*Bsh*DI	GGCC
	*Bsh*EI	GGCC
	*Bsh*FI	GG↑CC
	*Bsi*AI	GGCC
	*Bsi*DI	GGCC
	*Bsi*HI	GGCC
	*Bsp*23I	GGCC
	*Bsp*44II	GGCC
	*Bsp*137I	GGCC
	*Bsp*211I	GG↑CC
	*Bsp*226I	GGCC
	*Bsp*1261I	GGCC
	*Bsp*BRI	GG↑CC
	*Bsp*H106II	GGCC
	*Bsp*KI	GG↑CC
	*Bsp*RI	GG↑CC
	*Bss*CI	GGCC

Restriction/Methylation

Table 1. Type II enzymes (continued)

Enzyme	Isoschizomers	Recognition sequence[b,c]	Enzyme	Isoschizomers	Recognition sequence[b,c]
	*Bst*CI	GGCC		*Uba*1097I	GGCC
	*Bst*JI	GGCC		*Uba*1140I	GGCC
	*Bsu*1076I	GGCC		*Uba*1146I	GGCC
	*Bsu*1114I	GGCC		*Uba*1147I	GGCC
	*Bsu*RI	GG↑CC		*Uba*1150I	GGCC
	*Bte*I	GG↑CC		*Uba*1152I	GGCC
	*Clm*I	GGCC		*Uba*1153I	GGCC
	*Clt*I	GG↑CC		*Uba*1155I	GGCC
	*Csp*2I	GGCC		*Uba*1169I	GGCC
	*Dsa*II	GG↑CC		*Uba*1174I	GGCC
	*Fin*SI	GGCC		*Uba*1175I	GGCC
	*Fnu*DI	GG↑CC		*Uba*1176I	GGCC
	*Hhg*I	GGCC		*Uba*1178I	GGCC
	*Mch*AII	GG↑CC		*Uba*1179I	GGCC
	*Mfo*AI	GG↑CC		*Uba*1207I	GGCC
	*Mni*I	GGCC		*Uba*1208I	GGCC
	*Mnn*II	GGCC		*Uba*1209I	GGCC
	*Mth*TI	GGCC		*Uba*1210I	GGCC
	*Ngo*II	GGCC		*Uba*1214I	GGCC
	*Ngo*PII	GG↑CC		*Uba*1223I	GGCC
	*Ngo*SI	GGCC		*Uba*1228I	GGCC
	*Nla*I	GGCC		*Uba*1230I	GGCC
	*Pai*I	GGCC		*Uba*1231I	GGCC
	*Pal*I	GG↑CC		*Uba*1235I	GGCC
	*Pde*133I	GG↑CC		*Uba*1288I	GGCC
	*Pfl*KI	GG↑CC		*Uba*1292I	GGCC
	*Pla*I	GG↑CC		*Uba*1293I	GGCC
	*Ple*214I	GGCC		*Uba*1319I	GGCC
	*Ppu*I	GGCC		*Uba*1322I	GGCC
	*Psp*29I	GGCC		*Uba*1336I	GGCC
	*Sag*I	GGCC		*Uba*1377I	GGCC
	*Sbv*I	GG↑CC		*Uba*1388I	GGCC
	*Sfa*I	GG↑CC		*Uba*1392I	GGCC
	*Spl*III	GGCC		*Uba*1395I	GGCC
	*Sua*I	GG↑CC		*Uba*1408I	GGCC
	*Sul*I	GGCC		*Uba*1418I	GGCC
	*Tsp*132I	GGCC		*Uba*1422I	GGCC
	*Tsp*266I	GGCC		*Uba*1429I	GGCC
	*Tsp*273II	GGCC		*Uba*1449I	GGCC
	*Tsp*281I	GGCC		*Uba*1450I	GGCC
	*Tsp*560I	GGCC		*Uth*549I	GGCC
	*Tsp*ZNI	GGCC		*Uth*555I	GGCC
	*Tte*AI	GGCC		*Uth*557I	GGCC
	*Ttn*I	GGCC		*Van*91III	GGCC
	*Uba*9I	GGCC		*Vha*I	GGCC
	*Uba*54I	GGCC		*Vha*1168I	GGCC
	*Uba*61I	GGCC		*Vni*I	GGCC

Table 1. Type II enzymes (continued)

Enzyme	Isoschizomers	Recognition sequence[b,c]
*Hga*I		GACGC(5/10)
*Hgi*AI		GWGCW↑C
	*Alw*21I	GWGCW↑C
	*Asp*HI	GWGCW↑C
	*Bbv*12I	GWGCW↑C
	*Bsa*GI	GWGCWC
	*Bsh*45I	GWGCW↑C
	*Bsi*HKAI	GWGCW↑C
	*Bsm*PI	GWGCWC
	*Msp*V281I	GWGCW↑C
	*Pph*3215I	GWGCWC
*Hgi*CI		G↑GYRCC
	*Acc*B1I	G↑GYRCC
	*Ban*I	G↑GYRCC
	*Bbv*BI	G↑GYRCC
	*Bsh*NI	G↑GYRCC
	*Eco*50I	GGYRCC
	*Eco*64I	G↑GYRCC
	*Eco*168I	GGYRCC
	*Eco*169I	GGYRCC
	*Eco*171I	GGYRCC
	*Eco*173I	GGYRCC
	*Eco*195I	GGYRCC
	*Esp*1I	GGYRCC
	*Esp*6I	GGYRCC
	*Esp*9I	GGYRCC
	*Esp*10I	GGYRCC
	*Esp*11I	GGYRCC
	*Esp*12I	GGYRCC
	*Esp*13I	GGYRCC
	*Esp*14I	GGYRCC
	*Esp*15I	GGYRCC
	*Esp*21I	GGYRCC
	*Esp*22I	GGYRCC
	*Esp*25I	GGYRCC
	*Hgi*HI	G↑GYRCC
	*Msp*B4I	G↑GYRCC
	*Pfa*AI	G↑GYRCC
	*Pph*14I	GGYRCC
	*Ssp*M1III	GGYRCC
	*Uba*1127I	GGYRCC
	*Vsp*2246I	GGYRCC
*Hgi*EII		ACCNNNNNN-GGT
*Hgi*JII		GRGCY↑C
	*Ban*II	GRGCY↑C
	*Bpu*I	GRGCY↑C
	*Bsp*117I	GRGCYC
	*Bsp*519I	GRGCY↑C
	*Bst*Z7I	GRGCYC
	*Bsu*1854I	GRGCY↑C
	*Bvu*I	GRGCY↑C
	*Cfr*48I	GRGCYC
	*Eco*24I	GRGCY↑C
	*Eco*25I	GRGCYC
	*Eco*26I	GRGCYC
	*Eco*35I	GRGCYC
	*Eco*68I	GRGCYC
	*Eco*113I	GRGCYC
	*Eco*180I	GRGCYC
	*Eco*211I	GRGCYC
	*Eco*215I	GRGCYC
	*Eco*216I	GRGCYC
	*Eco*232I	GRGCYC
	*Eco*249I	GRGCYC
	*Eco*262I	GRGCYC
	*Eco*75KI	GRGCY↑C
	*Eco*T38I	GRGCY↑C
	*Eco*T88I	GRGCYC
	*Eco*T93I	GRGCYC
	*Eco*T95I	GRGCYC
	*Ese*4I	GRGCYC
	*Fri*OI	GRGCY↑C
	*Kox*II	GRGCY↑C
	*Omi*AI	GRGCYC
	*Omi*BI	GRGCYC
	*Pae*HI	GRGCY↑C
	*Ppu*20I	GRGCYC
	*Psp*31I	GRGCYC
	*Sac*NI	GRGCY↑C
	*Uba*39I	GRGCYC
	*Uba*57I	GRGCYC
	*Uba*1124I	GRGCYC
	*Uba*1142I	GRGCYC
	*Uba*1159I	GRGCYC
	*Uba*1206I	GRGCYC
	*Uba*1263I	GRGCYC
	*Uba*1264I	GRGCYC
	*Uba*1307I	GRGCYC
	*Uba*1329I	GRGCYC
	*Uba*1330I	GRGCYC
	*Uba*1357I	GRGCYC

Table 1. Type II enzymes (continued)

Enzyme	Isoschizomers	Recognition sequence[b,c]
	*Uba*1363I	GRGCYC
	*Uba*1409I	GRGCYC
	*Uba*1421I	GRGCYC
*Hha*I		GCG↑C
	*Asp*LEI	GCG↑C
	*Bca*I	GCGC
	*Cco*P95I	GCGC
	*Cfo*I	GCG↑C
	*Csp*1470I	GCGC
	*Fnu*DIII	GCG↑C
	*Hin*6I	G↑CGC
	*Hin*7I	GCGC
	*Hin*GUI	GCGC
	*Hin*P1I	G↑CGC
	*Hin*S1I	GCGC
	*Hin*S2I	GCGC
	*Hsp*AI	G↑CGC
	*Mnn*IV	GCGC
	*Sci*NI	G↑CGC
*Hin*4I		GABNNNNN-VTC
*Hind*II		GTY↑RAC
	*Chu*II	GTYRAC
	*Hin*1160II	GTYRAC
	*Hin*1161II	GTYRAC
	*Hin*JCI	GTY↑RAC
	*Hinc*II	GTY↑RAC
	*Mnn*I	GTYRAC
*Hind*III		A↑AGCTT
	*Asp*52I	AAGCTT
	*Asp*3065I	AAGCTT
	*Bbr*AI	AAGCTT
	*Bpe*I	AAGCTT
	*Bse*HI	AAGCTT
	*Bsm*GII	AAGCTT
	*Bsp*KT8I	A↑AGCTT
	*Bst*170II	AAGCTT
	*Bst*FI	A↑AGCTT
	*Cfr*32I	AAGCTT
	*Chu*I	AAGCTT
	*Eco*65I	AAGCTT
	*Eco*98I	AAGCTT
	*Eco*188I	AAGCTT
	*Eco*231I	AAGCTT
	*Eco*VIII	A↑AGCTT
	*Hin*5III	AAGCTT
	*Hin*173I	AAGCTT
	*Hin*1076III	AAGCTT
	*Hin*bIII	AAGCTT
	*Hin*fII	AAGCTT
	*Hin*JCII	AAGCTT
	*Hin*SAFI	AAGCTT
	*Hsu*I	A↑AGCTT
	*Mki*I	AAGCTT
	*Spa*PIV	AAGCTT
	*Ssb*I	A↑AGCTT
	*Uba*83I	AAGCTT
	*Uba*1164II	AAGCTT
	*Uba*1219I	AAGCTT
	*Uba*1435I	AAGCTT
*Hin*fI		G↑ANTC
	*Cvi*BI	G↑ANTC
	*Cvi*CI	GANTC
	*Cvi*DI	GANTC
	*Cvi*EI	GANTC
	*Cvi*FI	GANTC
	*Cvi*GI	GANTC
	*Fnu*AI	G↑ANTC
	*Hha*II	G↑ANTC
	*Nca*I	GANTC
	*Nov*II	GANTC
	*Nsi*HI	GANTC
*Hpa*I		GTT↑AAC
	*Bsa*KI	GTTAAC
	*Bse*II	GTTAAC
	*Bst*HPI	GTT↑AAC
	*Ksp*AI	GTT↑AAC
	*Mwh*I	GTTAAC
	*Ssr*I	GTT↑AAC
	*Uba*1408II	GTTAAC
*Hpa*II		C↑CGG
	*Asp*748I	CCGG
	*Bco*27I	C↑CGG
	*Bsa*ZI	CCGG
	*Bsh*MI	CCGG
	*Bsi*SI	C↑CGG
	*Bsp*5I	CCGG
	*Bsp*47I	CCGG
	*Bsp*48I	CCGG
	*Bsp*116I	CCGG
	*Bst*40I	C↑CGG

Table 1. Type II enzymes (continued)

Enzyme	Isoschizomers	Recognition sequence[b,c]
	*Bsu*1192I	CCGG
	*Bsu*FI	C↑CGG
	*Fin*II	CCGG
	*Hap*II	C↑CGG
	*Hin*2I	C↑CGG
	*Hin*5I	CCGG
	*Mni*II	CCGG
	*Mno*I	C↑CGG
	*Msp*I	C↑CGG
	*Pde*137I	C↑CGG
	*Pme*35I	CCGG
	*Sec*II	CCGG
	*Sfa*GUI	CCGG
	*Sth*134I	C↑CGG
	*Uba*1128I	CCGG
	*Uba*1141I	CCGG
	*Uba*1338I	CCGG
	*Uba*1355I	CCGG
	*Uba*1439I	CCGG
*Hph*I		GGTGA(8/7)
	*Asu*HPI	GGTGA(8/7)
	*Ngo*BI	GGTGA
*Kpn*I		GGTAC↑C
	*Acc*65I	G↑GTACC
	*Aha*B8I	G↑GTACC
	*Asp*718I	G↑GTACC
	*Bsp*J106I	GGTACC
	*Eco*149I	GGTACC
	*Esp*19I	GGTACC
	*Kpn*K14I	GGTACC
	*Mvs*I	GGTACC
	*Mvs*AI	GGTACC
	*Mvs*BI	GGTACC
	*Mvs*CI	GGTACC
	*Mvs*DI	GGTACC
	*Mvs*EI	GGTACC
	*Nmi*I	GGTACC
	*Sau*10I	GGTACC
	*Sth*I	G↑GTACC
	*Sth*AI	GGTACC
	*Sth*BI	GGTACC
	*Sth*CI	GGTACC
	*Sth*DI	GGTACC
	*Sth*EI	GGTACC
	*Sth*FI	GGTACC
	*Sth*GI	GGTACC
	*Sth*HI	GGTACC
	*Sth*JI	GGTACC
	*Sth*KI	GGTACC
	*Sth*LI	GGTACC
	*Sth*MI	GGTACC
	*Sth*NI	GGTACC
	*Uba*76I	GGTACC
	*Uba*85I	GGTACC
	*Uba*86I	GGTACC
	*Uba*87I	GGTACC
	*Uba*1201I	GGTACC
*Ksp*632I		CTCTTC(1/4)
	*Bco*5I	CTCTTC(1/4)
	*Bco*116I	CTCTTC(1/4)
	*Bco*KI	CTCTTC(1/4)
	*Bcr*AI	CTCTTC
	*Bse*ZI	CTCTTC(1/4)
	*Bsr*EI	CTCTTC
	*Bst*158I	CTCTTC
	*Bsu*6I	CTCTTC(1/4)
	*Eam*1104I	CTCTTC(1/4)
	*Ear*I	CTCTTC(1/4)
	*Uba*1192I	CTCTTC
	*Uba*1276I	CTCTTC
	*Vpa*KutEI	CTCTTC
	*Vpa*KutFI	CTCTTC
	*Vpa*O5I	CTCTTC
*Mae*I		C↑TAG
	*Bfa*I	C↑TAG
	*Mja*I	CTAG
	*Mth*FI	CTAG
	*Mth*ZI	C↑TAG
	*Rma*I	C↑TAG
	*Rma*485I	CTAG
	*Rma*486I	CTAG
	*Rma*490I	CTAG
	*Rma*495I	CTAG
	*Rma*496I	CTAG
	*Rma*497I	CTAG
	*Rma*500I	CTAG
	*Rma*501I	CTAG
	*Rma*503I	CTAG
	*Rma*506I	CTAG
	*Rma*509I	CTAG
	*Rma*510I	CTAG
	*Rma*515I	CTAG

Table 1. Type II enzymes (continued)

Enzyme	Isoschizomers	Recognition sequence[b,c]
	*Rma*516I	CTAG
	*Rma*517I	CTAG
	*Rma*518I	CTAG
	*Rma*519I	CTAG
	*Rma*522I	CTAG
*Mae*II		A↑CGT
	*Tai*I	ACGT↑
	*Tsc*I	ACGT↑
	*Tsp*49I	ACGT↑
	*Tsp*IDSI	ACGT
	*Tsp*WAM8AI	ACGT
	*Ttm*I	ACGT
*Mae*III		↑GTNAC
*Mbo*I		↑GATC
	*Asp*MDI	↑GATC
	*Bce*243I	↑GATC
	*Bfi*57I	↑GATC
	*Bme*12I	↑GATC
	*Bsa*PI	GATC
	*Bsc*FI	GATC
	*Bsm*XII	GATC
	*Bsp*I	GATC
	*Bsp*9I	GATC
	*Bsp*18I	GATC
	*Bsp*49I	GATC
	*Bsp*51I	GATC
	*Bsp*52I	GATC
	*Bsp*54I	GATC
	*Bsp*57I	GATC
	*Bsp*58I	GATC
	*Bsp*59I	GATC
	*Bsp*60I	GATC
	*Bsp*61I	GATC
	*Bsp*64I	GATC
	*Bsp*65I	GATC
	*Bsp*66I	GATC
	*Bsp*67I	↑GATC
	*Bsp*72I	GATC
	*Bsp*74I	GATC
	*Bsp*76I	GATC
	*Bsp*91I	GATC
	*Bsp*105I	↑GATC
	*Bsp*122I	GATC
	*Bsp*135I	GATC
	*Bsp*136I	GATC
	*Bsp*138I	GATC
	*Bsp*143I	↑GATC
	*Bsp*147I	GATC
	*Bsp*2095I	↑GATC
	*Bsp*AI	↑GATC
	*Bsp*FI	↑GATC
	*Bsp*JI	↑GATC
	*Bsp*J64I	GATC
	*Bsp*KT6I	GAT↑C
	*Bsr*MI	GATC
	*Bsr*PII	GATC
	*Bss*GII	GATC
	*Bst*1274I	GATC
	*Bst*EIII	GATC
	*Bst*XII	GATC
	*Btc*I	GATC
	*Btk*II	↑GATC
	*Btu*33I	GATC
	*Btu*34I	GATC
	*Btu*36I	GATC
	*Btu*37I	GATC
	*Btu*39I	GATC
	*Btu*41I	GATC
	*Cac*I	↑GATC
	*Cco*P31I	GATC
	*Cco*P76I	GATC
	*Cco*P84I	GATC
	*Cco*P95II	GATC
	*Cco*P219I	GATC
	*Ccy*I	↑GATC
	*Cha*I	GATC↑
	*Cin*1467I	GATC
	*Cpa*I	GATC
	*Cpf*I	↑GATC
	*Cpf*AI	GATC
	*Csp*5I	GATC
	*Cte*1179I	GATC
	*Cte*1180I	GATC
	*Cty*I	GATC
	*Cvi*AI	↑GATC
	*Cvi*HI	GATC
	*Dpn*II	↑GATC
	*Fnu*AII	GATC
	*Fnu*CI	↑GATC
	*Fnu*EI	↑GATC
	*Hac*I	↑GATC
	*Kzo*9I	↑GATC
	*Lla*AI	↑GATC

Table 1. Type II enzymes (continued)

Enzyme	Isoschizomers	Recognition sequence[b,c]
	*Lla*DCHI	GATC
	*Lsp*1109II	GATC
	*Meu*I	GATC
	*Mgo*I	↑GATC
	*Mkr*AI	↑GATC
	*Mme*II	GATC
	*Mno*III	GATC
	*Mos*I	GATC
	*Msp*67II	GATC
	*Msp*BI	GATC
	*Mth*I	GATC
	*Mth*1047I	GATC
	*Mth*AI	GATC
	*Nci*AI	GATC
	*Nde*II	↑GATC
	*Nfl*I	GATC
	*Nfl*AII	GATC
	*Nfl*BI	GATC
	*Nla*II	↑GATC
	*Nla*DI	GATC
	*Nme*CI	↑GATC
	*Nph*I	↑GATC
	*Nsi*AI	GATC
	*Nsp*AI	GATC
	*Nsu*I	GATC
	*Pei*9403I	GATC
	*Pfa*I	GATC
	*Pph*288I	GATC
	*Rlu*1I	GATC
	*Sal*AI	GATC
	*Sal*HI	GATC
	*Sau*15I	GATC
	*Sau*3AI	↑GATC
	*Sau*6782I	GATC
	*Sau*CI	GATC
	*Sau*DI	GATC
	*Sau*EI	GATC
	*Sau*FI	GATC
	*Sau*GI	GATC
	*Sau*MI	↑GATC
	*Sin*MI	GATC
	*Tru*II	GATC
	*Tsp*133I	GATC
	*Uba*4I	GATC
	*Uba*59I	GATC
	*Uba*1101I	GATC
	*Uba*1177I	GATC
	*Uba*1182I	GATC
	*Uba*1183I	GATC
	*Uba*1204I	GATC
	*Uba*1259I	GATC
	*Uba*1317I	GATC
	*Uba*1323I	GATC
	*Uba*1366I	GATC
	*Vha*44I	GATC
*Mbo*II		GAAGA(8/7)
	*Ncu*I	GAAGA
	*Tce*I	GAAGA
*Mcr*I		CGRY↑CG
	*Bsa*OI	CGRY↑CG
	*Bsh*1285I	CGRY↑CG
	*Bsi*EI	CGRY↑CG
	*Bst*MCI	CGRY↑CG
	*Bst*Z5I	CGRYCG
	*Uba*1303I	CGRYCG
*Mfe*I		C↑AATTG
	*Mun*I	C↑AATTG
*Mlu*I		A↑CGCGT
	*Ape*I	ACGCGT
	*Bbi*24I	A↑CGCGT
	*Bst*Z9I	ACGCGT
	*Uba*6I	ACGCGT
*Mme*I		TCCRAC(20/18)
*Mnl*I		CCTC(7/6)
*Mse*I		T↑TAA
	*Tru*1I	T↑TAA
	*Tru*9I	T↑TAA
*Msl*I		CAYNN↑NN-RTG
*Mst*I		TGC↑GCA
	*Aca*III	TGCGCA
	*Acc*16I	TGC↑GCA
	*Aos*I	TGC↑GCA
	*Avi*II	TGC↑GCA
	*Bco*6I	TGCGCA
	*Bsa*TI	TGCGCA
	*Clc*II	TGCGCA
	*Cli*II	TGCGCA
	*Fdi*II	TGC↑GCA
	*Fsp*I	TGC↑GCA

Table 1. Type II enzymes (continued)

Enzyme	Isoschizomers	Recognition sequence[b,c]
	*Gsp*AII	TGCGCA
	*Nsb*I	TGC↑GCA
	*Nsp*HIII	TGCGCA
	*Nsp*LI	TGCGCA
	*Nsp*MI	TGCGCA
	*Pam*I	TGC↑GCA
*Mwo*I		GCNNNNN↑N-NGC
	*Bce*1247I	GCNNNNNNN-GC
	*Bsp*WI	GCNNNNN↑N-NGC
*Nae*I		GCC↑GGC
	*Afa*24RI	GCCGGC
	*Ame*II	GCCGGC
	*Ani*MI	GCCGGC
	*Ape*AI	GCCGGC
	*Apr*I	GCCGGC
	*Bhe*I	GCCGGC
	*Cco*I	GCC↑GGC
	*Eco*56I	G↑CCGGC
	*Esp*5I	GCCGGC
	*Mau*AI	GCCGGC
	*Mis*I	GCCGGC
	*Mlu*9273II	GCCGGC
	*Mro*NI	G↑CCGGC
	*Nas*WI	GCCGGC
	*Nba*I	GCCGGC
	*Nbr*I	GCCGGC
	*Ngo*AIV	G↑CCGGC
	*Ngo*MI	G↑CCGGC
	*Nmu*I	GCCGGC
	*Nmu*FI	GCCGGC
	*Nsp*WI	GCCGGC
	*Nta*SII	GCCGGC
	*Pgl*I	GCCGGC
	*Psp*61I	GCCGGC
	*Rlu*I	GCCGGC
	*Sac*AI	GCCGGC
	*Sal*CI	GCCGGC
	*Sao*I	GCCGGC
	*Sau*AI	GCCGGC
	*Sau*BMKI	GCC↑GGC
	*Sau*HPI	GCC↑GGC
	*Sau*LPI	GCC↑GGC
	*Sau*NI	GCC↑GGC
	*Sau*SI	GCC↑GGC
	*Sce*III	G↑CCGGC
	*Ska*I	GCCGGC
	*Slu*1777I	GCC↑GGC
	*Smo*40529I	GCCGGC
	*Uba*1122I	GCCGGC
*Nar*I		GG↑CGCC
	*Bbe*I	GGCGC↑C
	*Bbe*AI	GGCGCC
	*Bin*SII	GGCGCC
	*Eco*78I	GGC↑GCC
	*Ege*I	GGC↑GCC
	*Ehe*I	GGC↑GCC
	*Kas*I	G↑GCGCC
	*Mca*AI	GGCGCC
	*Mch*I	GG↑CGCC
	*Mly*113I	GG↑CGCC
	*Msa*I	GGCGCC
	*Nam*I	GGCGCC
	*Nda*I	GG↑CGCC
	*Nun*II	GG↑CGCC
	*Pat*AI	GGCGCC
	*Pmn*I	GGCGCC
	*Sfo*I	GGCGCC
	*Sse*AI	GG↑CGCC
*Nco*I		C↑CATGG
	*Ate*I	CCATGG
	*Bsp*19I	C↑CATGG
	*Nsp*SAIII	CCATGG
*Nde*I		CA↑TATG
	*Fau*NDI	CA↑TATG
	*Pfa*AII	CA↑TATG
*Nhe*I		G↑CTAGC
	*Ace*II	GCTAG↑C
	*Asu*NHI	G↑CTAGC
	*Pst*NHI	G↑CTAGC
*Nla*III		CATG↑
	*Cvi*AII	C↑ATG
	*Hin*1II	CATG↑
	*Hin*8II	CATG
	*Hsp*92II	CATG↑
*Nla*IV		GGN↑NCC
	*Asp*NI	GGN↑NCC
	*Bcr*I	GGNNCC

Table 1. Type II enzymes (continued)

Enzyme	Isoschizomers	Recognition sequence[b,c]
	*Bsa*EI	GGNNCC
	*Bsc*BI	GGN↑NCC
	*Bsp*29I	GGNNCC
	*Bsp*LI	GGN↑NCC
	*Bss*I	GGNNCC
	*Psp*N4I	GGN↑NCC
	*Rlu*3I	GGNNCC
	*Uba*1305I	GGNNCC
	*Uba*1445I	GGNNCC
*Not*I		GC↑GGCCGC
	*Cci*NI	GC↑GGCCGC
	*Csp*BI	GC↑GGCCGC
	*Mch*AI	GC↑GGCCGC
*Nru*I		TCG↑CGA
	*Ama*I	TCGCGA
	*Bsp*68I	TCG↑CGA
	*Mlu*9273I	TCGCGA
	*Mlu*B2I	TCG↑CGA
	*Sal*DI	TCGCGA
	*Sbo*13I	TCG↑CGA
	*Sna*3286I	TCGCGA
	*Spo*I	TCG↑CGA
	*Uba*1117I	TCGCGA
	*Uba*1386I	TCGCGA
	*Vch*O70I	TCGCGA
*Nsp*I		RCATG↑Y
	*Bst*NSI	RCATG↑Y
	*Lsp*1270I	RCATGY
	*Nsp*HI	RCATG↑Y
	*Pau*AI	RCATG↑Y
	*Pun*AII	RCATG↑Y
*Nsp*BII		CMG↑CKG
	*Msp*A1I	CMG↑CKG
*Pac*I		TTAAT↑TAA
*Pfl*1108I		TCGTAG
*Pfl*MI		CCANNNN↑N-TGG
	*Acc*B7I	CCANNNN↑N-TGG
	*Acp*II	CCANNNN↑N-TGG
	*Asp*10HII	CCANNNN↑N-TGG
	*Esp*1396I	CCANNNN↑N-TGG
	*Van*91I	CCANNNN↑N-TGG
*Ple*I		GAGTC(4/5)
	*Bsm*EI	GAGTC
	*Mly*I	GACTC(5/5)
*Pma*CI		CAC↑GTG
	*Bbr*PI	CAC↑GTG
	*Bco*AI	CAC↑GTG
	*Bsp*87I	CACGTG
	*Eco*72I	CAC↑GTG
	*Pgl*34I	CACGTG
	*Pml*I	CAC↑GTG
	*Psh*CI	CACGTG
	*Psh*DI	CACGTG
	*Psp*38I	CACGTG
	*Psp*BI	CACGTG
	*Vpa*K3AI	CACGTG
	*Vpa*K3BI	CACGTG
*Pme*I		GTTT↑AAAC
*Ppu*MI		RG↑GWCCY
	*Mlu*1106I	RGGWCCY
	*Pfl*27I	RG↑GWCCY
	*Ppu*XI	RG↑GWCCY
	*Psp*5II	RG↑GWCCY
	*Psp*PPI	RG↑GWCCY
*Psh*AI		GACNN↑NN-GTC
*Pst*I		CTGCA↑G
	*Ain*I	CTGCAG
	*Ajo*I	CTGCA↑G
	*Ali*2882I	CTGCAG
	*Ali*AJI	CTGCA↑G
	*Api*I	CTGCA↑G
	*Asp*36I	CTGCAG
	*Asp*708I	CTGCAG
	*Asp*713I	CTGCA↑G
	*Asp*TI	CTGCAG
	*Bbi*I	CTGCAG
	*Bce*170I	CTGCAG
	*Blo*HIII	CTGCAG
	*Bme*BI	CTGCAG
	*Bsa*NII	CTGCAG
	*Bsa*QI	CTGCAG

Table 1. Type II enzymes (continued)

Enzyme	Isoschizomers	Recognition sequence[b,c]
	*Bsc*DI	CTGCAG
	*Bsp*17I	CTGCAG
	*Bsp*43I	CTGCAG
	*Bsp*63I	CTGCA↑G
	*Bsp*78I	CTGCAG
	*Bsp*81I	CTGCAG
	*Bsp*93I	CTGCAG
	*Bsp*107I	CTGCAG
	*Bsp*108I	CTGCAG
	*Bsp*268I	CTGCAG
	*Bsp*BI	CTGCA↑G
	*Bsu*BI	CTGCA↑G
	*Cau*III	CTGCAG
	*Cfl*I	CTGCA↑G
	*Cfr*A4I	CTGCA↑G
	*Cfu*II	CTGCA↑G
	*Clc*I	CTGCAG
	*Cst*I	CTGCA↑G
	*Eae*PI	CTGCAG
	*Ecl*77I	CTGCAG
	*Ecl*133I	CTGCAG
	*Ecl*593I	CTGCAG
	*Ecl*37kI	CTGCA↑G
	*Ecl*699kI	CTGCAG
	*Ecl*1zI	CTGCAG
	*Ecl*2zI	CTGCA↑G
	*Eco*48I	CTGCAG
	*Eco*49I	CTGCAG
	*Eco*83I	CTGCAG
	*Eco*161I	CTGCAG
	*Eco*167I	CTGCAG
	*Eco*260I	CTGCAG
	*Eco*261I	CTGCAG
	*Esp*5II	CTGCAG
	*Esp*141I	CTGCAG
	*Gse*II	CTGCAG
	*Hal*II	CTGCA↑G
	*Kpn*12I	CTGCAG
	*Mau*I	CTGCAG
	*Mha*AI	CTGCA↑G
	*Miz*I	CTGCAG
	*Mkr*I	CTGCAG
	*Mma*I	CTGCAG
	*Nas*I	CTGCAG
	*Ngb*I	CTGCAG
	*Noc*I	CTGCAG
	*Pae*8I	CTGCAG
	*Pae*9I	CTGCAG
	*Pae*14I	CTGCAG
	*Pae*15I	CTGCAG
	*Pae*22I	CTGCAG
	*Pae*24I	CTGCAG
	*Pae*25I	CTGCAG
	*Pae*26I	CTGCAG
	*Pae*39I	CTGCAG
	*Pae*40I	CTGCAG
	*Pae*41I	CTGCAG
	*Pae*PI	CTGCA↑G
	*Pfl*37I	CTGCAG
	*Pma*I	CTGCAG
	*Pma*44I	CTGCAG
	*Pmy*I	CTGCAG
	*Pph*2059I	CTGCAG
	*Pph*2066I	CTGCAG
	*Psh*EI	CTGCAG
	*Psp*28I	CTGCAG
	*Psp*46I	CTGCAG
	*Psp*SI	CTGCAG
	*Sal*13I	CTGCAG
	*Sal*PI	CTGCA↑G
	*Sco*AI	CTGCAG
	*Sfl*I	CTGCA↑G
	*Sgi*I	CTGCAG
	*Ska*II	CTGCAG
	*Spr*I	CTGCAG
	*Sri*I	CTGCAG
	*Uba*46I	CTGCAG
	*Uba*71I	CTGCAG
	*Uba*72I	CTGCAG
	*Uba*1112I	CTGCAG
	*Uba*1115I	CTGCAG
	*Uba*1116I	CTGCAG
	*Uba*1119I	CTGCAG
	*Uba*1123I	CTGCAG
	*Uba*1149I	CTGCAG
	*Uba*1184I	CTGCAG
	*Uba*1186I	CTGCAG
	*Uba*1211I	CTGCAG
	*Uba*1212I	CTGCAG
	*Uba*1213I	CTGCAG
	*Uba*1215I	CTGCAG
	*Uba*1216I	CTGCAG
	*Uba*1225I	CTGCAG
	*Uba*1232I	CTGCAG

Table 1. Type II enzymes (continued)

Enzyme	Isoschizomers	Recognition sequence[b,c]
	*Uba*1256I	CTGCAG
	*Uba*1262I	CTGCAG
	*Uba*1287I	CTGCAG
	*Uba*1294II	CTGCAG
	*Uba*1296I	CTGCAG
	*Uba*1328I	CTGCAG
	*Uba*1337I	CTGCAG
	*Uba*1399I	CTGCAG
	*Uba*1411I	CTGCAG
	*Uba*1417I	CTGCAG
	*Uba*HKBI	CTGCAG
	*Vch*O87I	CTGCAG
	*Vpa*K4AI	CTGCAG
	*Vpa*K29AI	CTGCAG
	*Vpa*K4BI	CTGCAG
	*Vpa*KutGI	CTGCAG
	*Xma*II	CTGCAG
	*Xor*I	CTGCAG
	*Xph*I	CTGCAG
	*Yen*I	CTGCA↑G
	*Yen*AI	CTGCAG
	*Yen*BI	CTGCAG
	*Yen*CI	CTGCAG
	*Yen*DI	CTGCAG
	*Yen*EI	CTGCAG
*Pvu*I		CGAT↑CG
	*Afa*22MI	CGAT↑CG
	*Afa*16RI	CGAT↑CG
	*Bma*I	CGATCG
	*Bma*AI	CGATCG
	*Bma*BI	CGATCG
	*Bma*CI	CGATCG
	*Bma*DI	CGATCG
	*Bsp*CI	CGAT↑CG
	*Bst*Z8I	CGATCG
	*Cas*2I	CGATCG
	*Cfr*51I	CGATCG
	*Drd*III	CGATCG
	*Eag*BI	CGAT↑CG
	*Ecl*JI	CGATCG
	*Erh*B9I	CGAT↑CG
	*Kpl*79I	CGATCG
	*Nbl*I	CGAT↑CG
	*Ple*19I	CGAT↑CG
	*Pnt*I	CGATCG
	*Psu*161I	CGAT↑CG
	*Pvu*84I	CGATCG
	*Rsh*I	CGAT↑CG
	*Rsp*I	CGATCG
	*Sma*AIII	CGATCG
	*Spa*PII	CGATCG
	*Spl*AIII	CGATCG
	*Uba*1129I	CGATCG
	*Uba*1139I	CGATCG
	*Xgl*3216I	CGATCG
	*Xgl*3217I	CGATCG
	*Xgl*3218I	CGATCG
	*Xgl*3219I	CGATCG
	*Xgl*3220I	CGATCG
	*Xml*I	CGATCG
	*Xml*AI	CGATCG
	*Xni*I	CGATCG
	*Xor*II	CGAT↑CG
*Pvu*II		CAG↑CTG
	*Bav*I	CAG↑CTG
	*Bav*AI	CAG↑CTG
	*Bav*BI	CAG↑CTG
	*Bsp*153AI	CAG↑CTG
	*Bsp*M39I	CAG↑CTG
	*Bsp*O4I	CAG↑CTG
	*Cfr*6I	CAG↑CTG
	*Dma*I	CAG↑CTG
	*Ecl*I	CAG↑CTG
	*Gsp*I	CAGCTG
	*Mzi*I	CAGCTG
	*Nme*RI	CAG↑CTG
	*Psp*3I	CAGCTG
	*Psp*5I	CAGCTG
	*Pvu*84II	CAG↑CTG
	*Pvu*HKUI	CAGCTG
	*Sba*I	CAGCTG
	*Sci*AII	CAGCTG
	*Sma*AIV	CAGCTG
	*Sol*3335I	CAGCTG
	*Spa*PIII	CAGCTG
	*Spl*AIV	CAGCTG
	*Uba*1227I	CAGCTG
	*Uba*1245I	CAGCTG
*Rle*AI		CCCACA(12/9)
*Rsa*I		GT↑AC
	*Afa*I	GT↑AC
	*Asp*16HI	GTAC
	*Asp*17HI	GTAC

Table 1. Type II enzymes (continued)

Enzyme	Isoschizomers	Recognition sequence[b,c]
	*Asp*18HI	GTAC
	*Asp*29HI	GTAC
	*Cco*P73I	GTAC
	*Csp*6I	G↑TAC
	*Cvi*QI	G↑TAC
	*Cvi*RII	G↑TAC
	*Pla*AII	GT↑AC
*Rsr*II		CG↑GWCCG
	*Cpo*I	CG↑GWCCG
	*Csp*I	CG↑GWCCG
*Sac*I		GAGCT↑C
	*Ecl*136II	GAG↑CTC
	*Ecl*137I	GAGCTC
	*Eco*ICRI	GAG↑CTC
	*Mxa*I	GAG↑CTC
	*Nas*SI	GAGCTC
	*Pfl*18I	GAGCTC
	*Psp*124BI	GAGCT↑C
	*Sco*I	GAGCTC
	*Sst*I	GAGCT↑C
*Sac*II		CCGC↑GG
	*Aos*III	CCGCGG
	*Asp*32HI	CCGCGG
	*Bac*I	CCGCGG
	*Bac*465I	CCGCGG
	*Bsp*12I	CCGCGG
	*Cfr*37I	CCGCGG
	*Cfr*41I	CCGCGG
	*Cfr*42I	CCGC↑GG
	*Cfr*43I	CCGCGG
	*Cfr*45II	CCGCGG
	*Csc*I	CCGC↑GG
	*Cte*1I	CCGCGG
	*Drd*AI	CCGCGG
	*Drd*BI	CCGCGG
	*Drd*CI	CCGCGG
	*Drd*EI	CCGCGG
	*Drd*FI	CCGCGG
	*Dsp*1I	CCGCGG
	*Eae*46I	CCGC↑GG
	*Ecc*I	CCGCGG
	*Ecl*1I	CCGCGG
	*Ecl*28I	CCGCGG
	*Ecl*37I	CCGCGG
	*Eco*55I	CCGCGG
	*Eco*92I	CCGCGG
	*Eco*96I	CCGCGG
	*Eco*99I	CCGCGG
	*Eco*100I	CCGCGG
	*Eco*104I	CCGCGG
	*Eco*134I	CCGCGG
	*Eco*135I	CCGCGG
	*Eco*151I	CCGCGG
	*Eco*158I	CCGCGG
	*Eco*182I	CCGCGG
	*Eco*196I	CCGCGG
	*Eco*208I	CCGCGG
	*Eco*29kI	CCGC↑GG
	*Ese*3I	CCGCGG
	*Ese*6I	CCGCGG
	*Fsc*I	CCGCGG
	*Gal*I	CCGC↑GG
	*Gce*I	CCGC↑GG
	*Gce*GLI	CCGC↑GG
	*Kpn*19I	CCGCGG
	*Kpn*378I	CCGC↑GG
	*Ksp*I	CCGC↑GG
	*Mlu*113I	CC↑GCGG
	*Mra*I	CCGCGG
	*Ngo*III	CCGCGG
	*Ngo*AIII	CCGC↑GG
	*Ngo*DI	CCGCGG
	*Ngo*PIII	CCGC↑GG
	*Nla*DIII	CCGCGG
	*Nla*SI	CCGCGG
	*Pae*7I	CCGCGG
	*Pae*17I	CCGCGG
	*Pae*36I	CCGCGG
	*Pae*42I	CCGCGG
	*Pae*43I	CCGCGG
	*Pae*44I	CCGCGG
	*Pae*AI	CCGC↑GG
	*Pae*QI	CCGC↑GG
	*Pfl*1108II	CCGCGG
	*Saa*I	CCGCGG
	*Sab*I	CCGCGG
	*Sak*I	CCGCGG
	*Sbo*I	CCGCGG
	*Sen*PT14bI	CCGC↑GG
	*Sex*BI	CCGC↑GG
	*Sex*CI	CCGC↑GG
	*Sfr*I	CCGCGG
	*Sfr*303I	CCGC↑GG

Table 1. Type II enzymes (continued)

Enzyme	Isoschizomers	Recognition sequence[b,c]
	*Sfr*382I	CCGCGG
	*Sgr*BI	CCGC↑GG
	*Shy*I	CCGCGG
	*Spu*I	CCGC↑GG
	*Sse*II	CCGCGG
	*Ssp*1725I	CCGCGG
	*Sst*II	CCGC↑GG
	*Sta*I	CCGCGG
	*Tgl*I	CCGCGG
	*Tto*I	CCGCGG
	*Uba*66I	CCGCGG
	*Uba*77I	CCGCGG
	*Uba*90I	CCGCGG
	*Uba*1093I	CCGCGG
	*Uba*1095I	CCGCGG
	*Uba*1111I	CCGCGG
	*Uba*1113I	CCGCGG
	*Uba*1126I	CCGCGG
	*Uba*1187I	CCGCGG
	*Uba*1229I	CCGCGG
	*Uba*1234I	CCGCGG
	*Uba*1244I	CCGCGG
	*Uba*1306I	CCGCGG
	*Uba*1364I	CCGCGG
	*Uba*1369I	CCGCGG
	*Uba*HKAI	CCGCGG
*Sal*I		G↑TCGAC
	*Acs*1371I	GTCGAC
	*Acs*1372I	GTCGAC
	*Acs*1373I	GTCGAC
	*Acs*1421I	GTCGAC
	*Acs*1422I	GTCGAC
	*Bst*Z16I	GTCGAC
	*Btg*AI	GTCGAC
	*Cgl*AII	GTCGAC
	*Hgi*CIII	G↑TCGAC
	*Hgi*DII	G↑TCGAC
	*Koy*I	GTCGAC
	*Nop*I	G↑TCGAC
	*Psp*32I	GTCGAC
	*Psp*33I	GTCGAC
	*Psp*89I	GTCGAC
	*Rfl*FI	G↑TCGAC
	*Rhe*I	GTCGAC
	*Rhp*I	GTCGAC
	*Rrh*I	GTCGAC
	*Rrh*4273I	GTCGAC
	*Rro*I	GTCGAC
	*Rtr*I	G↑TCGAC
	*Xam*I	GTCGAC
	*Xci*I	G↑TCGAC
*San*DI		GG↑GWCCC
	*Sse*1825I	GG↑GWCCC
*Sap*I		GCTCTTC(1/4)
	*Vpa*K32I	GCTCTTC(1/4)
*Sau*I		CC↑TNAGG
	*Aoc*I	CC↑TNAGG
	*Axy*I	CC↑TNAGG
	*Bli*643I	CCTNAGG
	*Bli*HKI	CC↑TNAGG
	*Bse*21I	CC↑TNAGG
	*Bsp*R7I	CC↑TNAGG
	*Bst*29I	CCTNAGG
	*Bst*30I	CCTNAGG
	*Bst*Z6I	CCTNAGG
	*Bsu*36I	CC↑TNAGG
	*Cvn*I	CC↑TNAGG
	*Eci*CI	CCTNAGG
	*Eco*76I	CCTNAGG
	*Eco*81I	CC↑TNAGG
	*Eco*115I	CCTNAGG
	*Eco*118I	CCTNAGG
	*Eco*110kI	CCTNAGG
	*Lmu*60I	CC↑TNAGG
	*Mst*II	CC↑TNAGG
	*Oxa*NI	CC↑TNAGG
	*Sau*HI	CCTNAGG
	*Sec*III	CCTNAGG
	*Ssh*AI	CC↑TNAGG
	*Uba*1184II	CCTNAGG
	*Uba*1294I	CCTNAGG
	*Uba*1332I	CCTNAGG
	*Uba*1333I	CCTNAGG
*Sca*I		AGT↑ACT
	*Acc*113I	AGT↑ACT
	*Afl*IV	AGTACT
	*Asp*763I	AGTACT
	*Bsh*HI	AGTACT
	*Bso*SI	AGTACT
	*Bst*MI	AGTACT
	*Dpa*I	AGT↑ACT
	*Eco*255I	AGT↑ACT

Table 1. Type II enzymes (continued)

Enzyme	Isoschizomers	Recognition sequence[b,c]
	*Pin*I	AGTACT
	*Rfl*FII	AGT↑ACT
	*Uba*1094I	AGTACT
	*Uba*1158I	AGTACT
	*Vch*O49I	AGTACT
*Scr*FI		CC↑NGG
	*Bme*1390I	CC↑NGG
	*Bsa*CI	CCNGG
	*Bso*I	CCNGG
	*Bsp*53I	CCNGG
	*Bsp*73I	CCNGG
	*Bsp*548I	CCNGG
	*Bss*KI	↑CCNGG
	*Dsa*V	↑CCNGG
	*Eco*43I	CCNGG
	*Eco*51II	CCNGG
	*Eco*80I	CCNGG
	*Eco*85I	CCNGG
	*Eco*93I	CCNGG
	*Eco*153I	CCNGG
	*Eco*200I	CCNGG
	*Msp*67I	CC↑NGG
	*Msp*R9I	CC↑NGG
	*Sen*PI	CCNGG
	*Sso*II	↑CCNGG
	*Sty*D4I	↑CCNGG
	*Uba*17I	CCNGG
	*Uba*1391I	CCNGG
*Sdu*I		GDGCH↑C
	*Aoc*II	GDGCH↑C
	*Bka*1125I	GDGCHC
	*Bmy*I	GDGCH↑C
	*Bso*CI	GDGCH↑C
	*Bsp*1286I	GDGCH↑C
	*Bsp*LS2I	GDGCH↑C
	*Bst*Z15I	GDGCHC
	*Nsp*II	GDGCH↑C
	*Tse*AI	GDGCHC
	*Uba*1362I	GDGCHC
*Sec*I		C↑CNNGG
	*Bsa*JI	C↑CNNGG
	*Bse*DI	C↑CNNGG
	*Bst*Z10I	CCNNGG
	*Uba*1442I	CCNNGG
*Sex*AI		A↑CCWGGT
*Sfa*NI		GCATC(5/9)
	*Bsc*AI	GCATC(4/6)
	*Bsm*NI	GCATC
	*Bsp*ST5I	GCATC(5/9)
	*Pha*I	GCATC(5/9)
*Sfe*I		C↑TRYAG
	*Bco*163I	CTRYAG
	*Bdi*SI	C↑TRYAG
	*Bfm*I	C↑TRYAG
	*Bst*SFI	C↑TRYAG
	*Lla*BI	C↑TRYAG
	*Sfc*I	C↑TRYAG
*Sfi*I		GGCCNNNN↑-NGGCC
	*Sdi*I	GGCCNNNN↑-NGGCC
*Sgf*I		GCGAT↑CGC
*Sgr*AI		CR↑CCGGYG
*Sim*I		GGGTC(–3/0)
*Sma*I		CCC↑GGG
	*Ahy*I	C↑CCGGG
	*Cfr*9I	C↑CCGGG
	*Cfr*J4I	CCC↑GGG
	*Eae*AI	C↑CCGGG
	*Ecl*RI	C↑CCGGG
	*Kte*AI	CCCGGG
	*Pac*25I	C↑CCGGG
	*Pae*BI	CCC↑GGG
	*Psp*AI	C↑CCGGG
	*Psp*ALI	CCC↑GGG
	*Uba*1220I	CCCGGG
	*Uba*1393I	CCCGGG
	*Xcy*I	C↑CCGGG
	*Xma*I	C↑CCGGG
	*Xma*CI	C↑CCGGG
*Sml*I		CTYRAG
*Sna*I		GTATAC
	*Bsp*M90I	GTA↑TAC
	*Bss*NAI	GTA↑TAC
	*Bst*1107I	GTA↑TAC
	*Bst*BSI	GTA↑TAC
	*Bst*Z17I	GTATAC
	*Vch*O25I	GTATAC

Table 1. Type II enzymes (continued)

Enzyme	Isoschizomers	Recognition sequence[b,c]
	*Xca*I	GTA↑TAC
*Sna*BI		TAC↑GTA
	*Bst*SNI	TAC↑GTA
	*Eci*AI	TACGTA
	*Eco*105I	TAC↑GTA
	*Eco*158II	TACGTA
	*Ssp*JI	TACGTA
	*Ssp*M1I	TACGTA
	*Ssp*M2I	TACGTA
	*Uba*1240I	TACGTA
*Spe*I		A↑CTAGT
	*Acl*NI	A↑CTAGT
	*Bcu*I	ACTAGT
*Sph*I		GCATG↑C
	*Asp*5HI	GCATGC
	*Bbu*I	GCATG↑C
	*Bsp*121I	GCATGC
	*Btg*AII	GCATGC
	*Cgl*AI	GCATGC
	*Pae*I	GCATG↑C
	*Pae*CI	GCATGC
	*Pfa*AIII	GCATG↑C
	*Spa*HI	GCATG↑C
	*Spa*XI	GCATGC
	*Uba*1162I	GCATGC
	*Uba*1226I	GCATGC
	*Vch*O68I	GCATGC
*Spl*I		C↑GTACG
	*Bpu*B5I	C↑GTACG
	*Bsi*WI	C↑GTACG
	*Bsm*WI	CGTACG
	*Bvu*BI	C↑GTACG
	*Mae*K81I	C↑GTACG
	*Pfl*23II	C↑GTACG
	*Pfu*I	CGTACG
	*Ppu*AI	C↑GTACG
	*Psp*LI	C↑GTACG
	*Sma*AI	CGTACG
	*Spl*AI	CGTACG
	*Ssp*KI	CGTACG
	*Sun*I	C↑GTACG
*Srf*I		GCCC↑GGGC
*Sse*8387I		CCTGCA↑GG
	*Sbf*I	CCTGCA↑GG
	*Sda*I	CCTGCA↑GG
*Sse*8647I		AG↑GWCCT
*Ssp*I		AAT↑ATT
*Stu*I		AGG↑CCT
	*Aat*I	AGG↑CCT
	*Asp*78I	AGGCCT
	*Asp*MI	AGG↑CCT
	*Chy*I	AGGCCT
	*Eco*147I	AGG↑CCT
	*Gdi*I	AGG↑CCT
	*Gob*AI	AGGCCT
	*Nta*SI	AGGCCT
	*Plu*I	AGGCCT
	*Pme*55I	AGG↑CCT
	*Ppu*13I	AGGCCT
	*Sar*I	AGG↑CCT
	*Sru*30DI	AGG↑CCT
	*Sse*BI	AGG↑CCT
	*Ssv*I	AGGCCT
	*Ste*I	AGG↑CCT
	*Uba*40I	AGGCCT
	*Uba*1170I	AGGCCT
	*Uba*1180I	AGGCCT
	*Uba*1217I	AGGCCT
	*Uba*1239I	AGGCCT
	*Uba*1371I	AGGCCT
	*Uba*1403I	AGGCCT
	*Uba*1419I	AGGCCT
	*Vch*O44I	AGGCCT
*Sty*I		C↑CWWGG
	*Bsm*SI	C↑CWWGG
	*Bss*T1I	C↑CWWGG
	*Bst*224I	CCWWGG
	*Cfr*BI	C↑CWWGG
	*Eco*130I	C↑CWWGG
	*Eco*208II	CCWWGG
	*Eco*T14I	C↑CWWGG
	*Eco*T104I	CCWWGG
	*Erh*I	C↑CWWGG
	*Erh*B9II	C↑CWWGG
	*Sbl*AI	CCWWGG
	*Sbl*BI	CCWWGG
	*Sbl*CI	CCWWGG
	*Uba*1311I	CCWWGG
*Swa*I		ATTT↑AAAT
	*Smi*I	ATTT↑AAAT

Table 1. Type II enzymes (continued)

Enzyme	Isoschizomers	Recognition sequence[b,c]
*Taq*I		T↑CGA
	*Ppa*AII	T↑CGA
	*Tfl*I	TCGA
	*Tsp*32I	T↑CGA
	*Tsp*32II	T↑CGA
	*Tsp*358I	TCGA
	*Tsp*505I	TCGA
	*Tsp*510I	TCGA
	*Tsp*NI	TCGA
	*Tth*HB8I	T↑CGA
*Taq*II[d]		GACCGA(11/9)
*Taq*II[d]		CACCCA(11/9)
*Tat*I		WGTACW
*Tau*I		GCSGC
*Tfi*I		G↑AWTC
*Tse*I		G↑CWGC
	*Ace*I	G↑CWGC
	*Taq*52I	G↑CWGC
	*Tse*BI	GCWGC
*Tsp*45I		↑GTSAC
*Tsp*4CI		ACN↑GT
	*Bst*4CI	ACN↑GT
*Tsp*EI		↑AATT
	*Sse*9I	↑AATT
	*Tse*CI	AATT
	*Tsp*509I	↑AATT
*Tsp*RI		CAGTG(2/–7)
*Tth*111I		GACN↑NNGTC
	*Asp*I	GACN↑NNGTC
	*Ats*I	GACN↑NNGTC
	*Fsu*I	GACNNNGTC
	*Nta*I	GACNNNGTC
	*Pfl*FI	GACNNNGTC
	*Sma*AII	GACNNNGTC
	*Spa*PI	GACNNNGTC
	*Spl*II	GACNNNGTC
	*Spl*AII	GACNNNGTC
	*Tel*I	GACN↑NNGTC
	*Tsp*I	GACNNNGTC
	*Tte*I	GACNNNGTC
	*Ttr*I	GACNNNGTC
*Tth*111II		CAARCA(11/9)

Enzyme	Isoschizomers	Recognition sequence[b,c]
*Uba*DI		GAACNNNNN-NTCC
*Uba*EI		CACCTGC
*Vsp*I		AT↑TAAT
	*Ase*I	AT↑TAAT
	*Asn*I	AT↑TAAT
	*Bpo*AI	AT↑TAAT
	*Psh*BI	AT↑TAAT
	*Sru*4DI	AT↑TAAT
*Xba*I		T↑CTAGA
	*Bsp*LU11II	TCTAGA
	*Bsr*XI	TCTAGA
	*Msp*23I	TCTAGA
*Xcm*I		CCANNNNN↑-NNNNTGG
*Xho*I		C↑TCGAG
	*Abr*I	C↑TCGAG
	*Aer*AI	CTCGAG
	*Ahy*AI	CTCGAG
	*Asp*15I	CTCGAG
	*Asp*47I	CTCGAG
	*Asp*703I	CTCGAG
	*Bad*I	CTCGAG
	*Bbf*I	CTCGAG
	*Bbi*III	CTCGAG
	*Blu*I	C↑TCGAG
	*Bsp*92I	CTCGAG
	*Bsp*129I	CTCGAG
	*Bsp*139I	CTCGAG
	*Bsp*140I	CTCGAG
	*Bsp*141I	CTCGAG
	*Bsp*142I	CTCGAG
	*Bss*HI	CTCGAG
	*Bst*HI	CTCGAG
	*Bst*LI	CTCGAG
	*Bst*VI	C↑TCGAG
	*Bsu*MI	CTCGAG
	*Bth*I	CTCGAG
	*Ccr*I	C↑TCGAG
	*Cja*I	CTCGAG
	*Dde*II	CTCGAG
	*Drd*DI	CTCGAG
	*Eae*2I	CTCGAG
	*Mav*I	C↑TCGAG
	*Mca*I	CTCGAG

Table 1. Type II enzymes (continued)

Enzyme	Isoschizomers	Recognition sequence[b,c]
	*Mec*I	CTCGAG
	*Mha*I	CTCGAG
	*Mla*AI	C↑TCGAG
	*Mpu*I	CTCGAG
	*Mrh*I	CTCGAG
	*Msc*AI	CTCGAG
	*Msi*I	CTCGAG
	*Msp*23II	CTCGAG
	*Oco*I	CTCGAG
	*Pae*R7I	C↑TCGAG
	*Pan*I	C↑TCGAG
	*Pfl*67I	CTCGAG
	*Pfl*NI	CTCGAG
	*Pfl*WI	CTCGAG
	*Psp*4I	CTCGAG
	*Psp*NI	CTCGAG
	*Sal*1974I	CTCGAG
	*Sau*3239I	C↑TCGAG
	*Sau*LPII	C↑TCGAG
	*Sbi*68I	C↑TCGAG
	*Sca*1827I	CTCGAG
	*Sci*I	CTC↑GAG
	*Sci*1831I	CTCGAG
	*Sci*BI	CTCGAG
	*Scu*I	CTCGAG
	*Sdi*AI	CTCGAG
	*Sex*I	CTCGAG
	*Sfr*274I	C↑TCGAG
	*Sfu*1762I	CTCGAG
	*Sga*I	CTCGAG
	*Sgo*I	CTCGAG
	*Sgr*1841I	CTCGAG
	*Shy*1766I	CTCGAG
	*Sla*I	C↑TCGAG
	*Slu*I	CTCGAG
	*Sol*10179I	C↑TCGAG
	*Spa*I	CTCGAG
	*Sph*1719I	CTCGAG
	*Sri*fpI	CTCGAG
	*Ssp*4I	CTCGAG
	*Sta*AI	CTCGAG
	*Sve*194I	CTCGAG
	*Tli*I	CTCGAG
	*Uba*1130I	CTCGAG
	*Uba*1148I	CTCGAG
	*Uba*1154I	CTCGAG
	*Uba*1166I	CTCGAG
	*Uba*1237I	CTCGAG
	*Uba*1248I	CTCGAG
	*Uba*1271I	CTCGAG
	*Uba*1298I	CTCGAG
	*Uba*1335I	CTCGAG
	*Uba*1397I	CTCGAG
	*Uba*1448I	CTCGAG
	*Xpa*I	C↑TCGAG
*Xho*II		R↑GATCY
	*Ait*II	RGATCY
	*Ait*AI	RGATCY
	*Asp*17I	RGATCY
	*Asp*22I	RGATCY
	*Asp*1HI	RGATCY
	*Asp*6HI	RGATCY
	*Asp*8HI	RGATCY
	*Asp*14HI	RGATCY
	*Asp*21HI	RGATCY
	*Blo*HI	RGATCY
	*Bst*X2I	R↑GATCY
	*Bst*YI	R↑GATCY
	*Dsa*III	R↑GATCY
	*Mfl*I	R↑GATCY
	*Tru*201I	R↑GATCY
	*Uba*1432I	RGATCY
*Xma*III		C↑GGCCG
	*Aaa*I	C↑GGCCG
	*Bse*X3I	C↑GGCCG
	*Bso*DI	CGGCCG
	*Bst*ZI	C↑GGCCG
	*Eag*I	C↑GGCCG
	*Ecl*XI	C↑GGCCG
	*Eco*52I	C↑GGCCG
	*Sen*PT16I	C↑GGCCG
	*Tsp*504I	CGGCCG
*Xmn*I		GAANN↑NN-TTC
	*Asp*700I	GAANN↑NN-TTC
	*Bbv*AI	GAANN↑NN-TTC
	*Mro*XI	GAANN↑NN-TTC
	*Syn*II	GAANNNN-TTC

[a]For the primary sources of data, see REBASE (1).
[b]Convention for describing recognition sequences. An arrow (↑) indicates the position of the cut site in the 5′ to 3′ strand of a palindromic recognition sequence:

*Aat*II	GACGT↑C	5′-G ACGT↑C-3′ 3′-C↑TGCA G-5′	3′ overhang of four nucleotides
*Acl*I	AA↑CGTT	5′-AA↑CG TT-3′ 3′-TT GC↑AA-5′	5′ overhang of two nucleotides
*Aha*III	TTT↑AAA	5′-TTT↑AAA-3′ 3′-AAA↑TTT-5′	Blunt end

Numbers in brackets after the recognition sequence indicate that the cut site is not palindromic. The first number gives the cut position relative to the recognition sequence in the 5′ to 3′ strand, the second number gives the cut position in the 3′ to 5′ strand. If the numbers are positive then the cut position is downstream of the recognition sequence, if negative then the cut position is upstream:

*Ace*III	CAGCTC(7/11)	5′-CAGCTCNNNNNNN↑NNNN N-3′ 3′-GTCGAGNNNNNNN NNNN↑N-5′
*Aci*I	CCGC(–3/–1)	5′-N↑NN NCCGC-3′ 3′-N NN↑NGGCG-5′

If numbers are listed at both ends of the recognition sequence then the enzyme cuts both upstream and downstream, releasing the recognition site as a small fragment:

*Bcg*I (10/12)GCANNNNNNTCG(12/10)

5′-N NN↑NNNNNNNNNNGCANNNNNNTCGNNNNNNNNNN NN↑N-3′
3′-N↑NN NNNNNNNNNNCGTNNNNNNAGCNNNNNNNNNN↑NN N-5′

The recognition site is released as a 32 bp fragment with 3′ overhangs of two nucleotides at either end.

If the recognition sequence is given with neither an arrow nor bracketed numbers then the cut position is unknown.
[c]Abbreviations: R = G or A; Y = C or T; M = A or C; K = G or T; S = G or C; W = A or T; B = not A (C or G or T); D = not C (A or G or T); H = not G (A or C or T); V = not T (A or C or G); N = A or C or G or T.
[d]*Taq*II has two distinct recognition sequences.

Table 2. Type I restriction enzymes and their recognition sequences[a]

Enzyme	Recognition sequence[b]
*Cfr*AI	GCANNNNNNNNGTGG
*Eco*AI	GAGNNNNNNNGTCA
*Eco*BI	TGANNNNNNNNTGCT
*Eco*DI	TTANNNNNNNGTCY
*Eco*DR2	TCANNNNNNGTCG
*Eco*DR3	TCANNNNNNNATCG
*Eco*DXXI	TCANNNNNNNRTTC
*Eco*EI	GAGNNNNNNNATGC
*Eco*KI	AACNNNNNNGTGC
*Eco*R124I	GAANNNNNNRTCG
*Eco*R124II	GAANNNNNNNRTCG
*Eco*RD2	GAANNNNNNRTTC
*Eco*RD3	GAANNNNNNNRTTC

Table 2. Type I enzymes (continued)

Enzyme	Recognition sequence[b]
*Eco*prrI	CCANNNNNNNRTGC
*Sty*LTIII	GAGNNNNNNRTAYG
*Sty*SJ	GAGNNNNNNGTRC
*Sty*SKI	CGATNNNNNNNGTTA
*Sty*SPI	AACNNNNNNGTRC
*Sty*SQ	AACNNNNNNRTAYG

[a]For primary sources of data see REBASE (1).
[b]For a description of the convention used in describing recognition sequences see *Table 1*, footnote b. For abbreviations see *Table 1*, footnote c.

Table 3. Type III restriction enzymes and their recognition sequences[a]

Enzyme	Isoschizomer	Recognition sequence[b]
*Eco*PI		AGACC
*Eco*P15I		CAGCAG(25/27)
*Hin*fIII		CĠAAT
	*Hin*eI	CGAAT
*Sty*LTI		CAGAG

[a]For primary sources of data see REBASE (1).
[b]For a description of the convention used in describing recognition sequences see *Table 1*, footnote b. For abbreviations see *Table 1*, footnote c.

2. RESTRICTION IN THE LABORATORY

Several hundred Type II enzymes are available commercially. The most commonly used of these are listed in *Table 4* with their recognition sequences and details of the ends produced by each enzyme. The availability from commercial suppliers of ready-made buffers for restriction enzymes means that few laboratories now make their own buffers, but it should be appreciated that most suppliers have only a limited set of buffers and the one provided with the enzyme may not be completely optimal. Often this is not a problem but for some applications (e.g. complete cleavage of a cloning vector, restriction of DNA minipreps that are not completely free of contaminants) the use of optimized buffer conditions can have a significant effect on the amount of restriction that is achieved. *Table 5* gives recommended reaction conditions for the commonly used enzymes. This information should not be considered sacrosanct, especially as the best buffer for one type of DNA may not give the maximum restriction with DNA from another source. *Table 5* also gives the optimal reaction temperature for each enzyme and an inactivation temperature for those enzymes that can be heat-killed.

Often, it is necessary to cut a sample of DNA with two enzymes. This is possible in a single reaction if the two enzymes have similar reaction conditions, in particular if the salt requirements of the two buffers overlap. *Table 6* shows the effects of NaCl and KCl concentrations on enzyme activities. If the salt requirements of the enzymes being used do not overlap then carry out the first restriction with the enzyme with the lower salt optimum, and then add more salt before carrying out the second restriction. Alternatively, if complete restriction is required, purify the DNA by phenol extraction and ethanol precipitation between restrictions.

An increasing number of restriction enzymes are now known to exhibit 'star' activity, recognizing relaxed versions of their recognition sequences under certain circumstances (*Table 7*). This is useful for some applications but for routine restrictions should obviously be avoided. It is not simply a case of avoiding gross errors in setting up a restriction reaction. For some enzymes star activity is promoted by high enzyme concentrations (so addition of extra enzyme to cut a difficult piece of DNA may be counterproductive) or by high glycerol concentrations (easily achieved if restriction is carried out in too small a total volume). For several enzymes carry-over of ethanol from an earlier DNA precipitation can result in star activity.

A few restriction enzymes are known to have limited activity with single-stranded DNA. A list of these is given in *Table 8*.

Table 4. Commonly used restriction enzymes

Enzyme	Recognition sequence[a]	Ends resulting from cleavage	
		Type	Overhang sequence (5′ to 3′)
*Aat*I	AGG↑CCT	Blunt	
*Aat*II	GACGT↑C	3′ overhang	-ACGT
*Acc*I	GT↑MKAC	5′ overhang	MK-
*Acc*II	CG↑CG	Blunt	
*Acc*III	T↑CCGGA	5′ overhang	CCGG-
*Acc*65I	G↑GTACC	5′ overhang	GTAC-
*Acc*B7I	CCANNNN↑NTGG	3′ overhang	-NNN
*Aci*I	CCGC(−3/−1)	5′ overhang	NN-
*Acs*I	R↑AATTY	5′ overhang	AATT-
*Acy*I	GR↑CGYC	5′ overhang	CG-
*Afa*I	GT↑AC	Blunt	
*Afl*I	G↑GWCC	5′ overhang	GWC-
*Afl*II	C↑TTAAG	5′ overhang	TTAA-
*Afl*III	A↑CRYGT	5′ overhang	CRYG-
*Age*I	A↑CCGGT	5′ overhang	CCGG-
*Aha*I	CC↑SGG	5′ overhang	S-
*Aha*II	GR↑CGYC	5′ overhang	CG-
*Aha*III	TTT↑AAA	Blunt	
*Ahd*I	GACNNN↑NNGTC	3′ overhang	-N
*Alu*I	AG↑CT	Blunt	
*Alw*I	GGATC(4/5)	5′ overhang	N-
*Alw*26I	GTCTC(1/5)	5′ overhang	NNNN-
*Alw*44I	G↑TGCAC	5′ overhang	TGCA-
*Alw*NI	CAGNNN↑CTG	3′ overhang	-NNN
*Aoc*I	CC↑TNAGG	5′ overhang	TNA-
*Aoc*II	GDGCH↑C	3′ overhang	-DGCH

Table 4. Commonly used restriction enzymes (continued)

Enzyme	Recognition sequence[a]	Ends resulting from cleavage		
		Type	Overhang sequence (5′ to 3′)	
*Aor*51HI	AGC↑GCT	Blunt		
*Aos*I	TGC↑GCA	Blunt		
*Apa*I	GGGCC↑C	3′ overhang	-GGCC	
*Apa*LI	G↑TGCAC	5′ overhang		TGCA-
*Apo*I	R↑AATTY	5′ overhang		AATT-
*Apy*I	CC↑WGG	5′ overhang		W-
*Aqu*I	C↑YCGRG	5′ overhang		YCGR-
*Asc*I	GG↑CGCGCC	5′ overhang		CGCG-
*Ase*I	AT↑TAAT	5′ overhang		TA-
*Asn*I	AT↑TAAT	5′ overhang		TA-
*Asp*I	GACN↑NNGTC	5′ overhang		N-
*Asp*700I	GAANN↑NNTTC	Blunt		
*Asp*718I	G↑GTACC	5′ overhang		GTAC-
*Asp*EI	GACNNN↑NNGTC	3′ overhang	-N	
*Asp*HI	GWGCW↑C	3′ overhang	-WGCW	
*Asu*I	G↑GNCC	5′ overhang		GNC-
*Asu*II	TT↑CGAA	5′ overhang		CG-
*Ava*I	C↑YCGRG	5′ overhang		YCGR-
*Ava*II	G↑GWCC	5′ overhang		GWC-
*Avi*II	TGC↑GCA	Blunt		
*Avr*II	C↑CTAGG	5′ overhang		CTAG-
*Axy*I	CC↑TNAGG	5′ overhang		TNA-
*Bal*I	TGG↑CCA	Blunt		
*Bam*HI	G↑GATCC	5′ overhang		GATC-
*Ban*I	G↑GYRCC	5′ overhang		GYRC-
*Ban*II	GRGCY↑C	3′ overhang	-RGCY	
*Ban*III	AT↑CGAT	5′ overhang		CG-
*Bbe*I	GGCGC↑C	3′ overhang	-GCGC	
*Bbi*II	GR↑CGYC	5′ overhang		CG-
*Bbr*PI	CAC↑GTG	Blunt		
*Bbs*I	GAAGAC(2/6)	5′ overhang		NNNN-
*Bbu*I	GCATG↑C	3′ overhang	-CATG	
*Bbv*I	GCAGC(8/12)	5′ overhang		NNNN-
*Bcg*I	(10/12)GCANNNNNNTC-G(12/10)	See *Table 1*, footnote b		
*Bci*VI	GGATAC(6/5)	3′ overhang	-N	
*Bcl*I	T↑GATCA	5′ overhang		GATC-
*Bcn*I	CC↑SGG	5′ overhang		S-
*Bfa*I	C↑TAG	5′ overhang		TA-
*Bfr*I	C↑TTAAG	5′ overhang		TTAA-
*Bgl*I	GCCNNNN↑NGGC	3′ overhang	-NNN	
*Bgl*II	A↑GATCT	5′ overhang		GATC-
*Bln*I	C↑CTAGG	5′ overhang		CTAG-
*Blp*I	GC↑TNAGC	5′ overhang		TNA-
*Bmy*I	GDGCH↑C	3′ overhang	-DGCH	

Table 4. Commonly used restriction enzymes (continued)

Enzyme	Recognition sequence[a]	Ends resulting from cleavage	
		Type	Overhang sequence (5′ to 3′)
*Bpm*I	CTGGAG(16/14)	3′ overhang	-NN
*Bpu*1102I	GC↑TNAGC	5′ overhang	TNA-
*Bpu*AI	GAAGAC(2/6)	5′ overhang	NNNN-
*Bsa*I	GGTCTC(1/5)	5′ overhang	NNNN-
*Bsa*AI	YAC↑GTR	Blunt	
*Bsa*BI	GATNN↑NNATC	Blunt	
*Bsa*HI	GR↑CGYC	5′ overhang	CG-
*Bsa*JI	C↑CNNGG	5′ overhang	CNNG-
*Bsa*MI	GAATGC(1/–1)	3′ overhang	-NGAATGCN
*Bsa*OI	CGRY↑CG	3′ overhang	-RY
*Bsa*WI	W↑CCGGW	5′ overhang	CCGG-
*Bse*AI	T↑CCGGA	5′ overhang	CCGG-
*Bse*RI	GAGGAG(10/8)	3′ overhang	-NN
*Bsg*I	GTGCAG(16/14)	3′ overhang	-NNNN
*Bsh*1236I	CG↑CG	Blunt	
*Bsi*CI	TT↑CGAA	5′ overhang	CG-
*Bsi*EI	CGRY↑CG	3′ overhang	-RY
*Bsi*HKAI	GWGCW↑C	3′ overhang	-WGCW
*Bsi*WI	C↑GTACG	5′ overhang	GTAC-
*Bsi*YI	CCNNNNN↑NNGG	3′ overhang	-NNN
*Bsl*I	CCNNNNN↑NNGG	3′ overhang	-NNN
*Bsm*I	GAATGC(1/–1)	3′ overhang	-NGAATGCN
*Bsm*AI	GTCTC(1/5)	5′ overhang	NNNN-
*Bsm*BI	CGTCTC(1/5)	5′ overhang	NNNN-
*Bsm*FI	GGGAC(10/14)	5′ overhang	NNNN-
*Bso*BI	C↑YCGRG	5′ overhang	YCGR-
*Bsp*106I	AT↑CGAT	5′ overhang	CG-
*Bsp*120I	G↑GGCCC	5′ overhang	GGCC-
*Bsp*1286I	GDGCH↑C	3′ overhang	-DGCH
*Bsp*CI	CGAT↑CG	3′ overhang	-T
*Bsp*DI	AT↑CGAT	5′ overhang	CG-
*Bsp*EI	T↑CCGGA	5′ overhang	CCGG-
*Bsp*HI	T↑CATGA	5′ overhang	CATG-
*Bsp*LU11I	A↑CATGT	5′ overhang	CATG-
*Bsp*MI	ACCTGC(4/8)	5′ overhang	NNNN-
*Bsp*MII	T↑CCGGA	5′ overhang	CCGG-
*Bsr*I	ACTGG(1/–1)	3′ overhang	-NACTGGN
*Bsr*BI	CCGCTC(–3/–3)	Blunt	
*Bsr*BRI	GATNN↑NNATC	Blunt	
*Bsr*DI	GCAATG(2/0)	3′ overhang	-NN
*Bsr*FI	R↑CCGGY	5′ overhang	CCGG-
*Bsr*GI	T↑GTACA	5′ overhang	GTAC-
*Bsr*SI	ACTGG(1/–1)	3′ overhang	-NACTGGN
*Bss*HII	G↑CGCGC	5′ overhang	CGCG-
*Bss*KI	↑CCNGG	5′ overhang	CCNGG-

Table 4. Commonly used restriction enzymes (continued)

Enzyme	Recognition sequence[a]	Ends resulting from cleavage	
		Type	Overhang sequence (5′ to 3′)
*Bss*SI	CACGAG(–5/–1)	5′ overhang	NNNN-
*Bst*I	G↑GATCC	5′ overhang	GATC-
*Bst*71I	GCAGC(8/12)	5′ overhang	NNNN-
*Bst*98I	C↑TTAAG	5′ overhang	TTAA-
*Bst*1107I	GTA↑TAC	Blunt	
*Bst*BI	TT↑CGAA	5′ overhang	CG-
*Bst*EII	G↑GTNACC	5′ overhang	GTNAC-
*Bst*NI	CC↑WGG	5′ overhang	W-
*Bst*OI	CC↑WGG	5′ overhang	W-
*Bst*PI	G↑GTNACC	5′ overhang	GTNAC-
*Bst*UI	CG↑CG	Blunt	
*Bst*XI	CCANNNNN↑NTGG	3′ overhang	-NNNN
*Bst*YI	R↑GATCY	5′ overhang	GATC-
*Bst*ZI	C↑GGCCG	5′ overhang	GGCC-
*Bst*Z17I	GTA↑TAC	Blunt	
*Bsu*36I	CC↑TNAGG	5′ overhang	TNA-
*Cac*8I	GCN↑NGC	Blunt	
*Cfo*I	GCG↑C	3′ overhang	-CG
*Cfr*9I	C↑CCGGG	5′ overhang	CCGG-
*Cfr*10I	R↑CCGGY	5′ overhang	CCGG-
*Cfr*13I	G↑GNCC	5′ overhang	GNC-
*Cla*I	AT↑CGAT	5′ overhang	CG-
*Cpo*I	CG↑GWCCG	5′ overhang	GWC-
*Csp*I	CG↑GWCCG	5′ overhang	GWC-
*Csp*45I	TT↑CGAA	5′ overhang	CG-
*Cvn*I	CC↑TNAGG	5′ overhang	TNA-
*Dde*I	C↑TNAG	5′ overhang	TNA-
*Dpn*I	GA↑TC	Blunt	
*Dpn*II	↑GATC	5′ overhang	GATC-
*Dra*I	TTT↑AAA	Blunt	
*Dra*II	RG↑GNCCY	5′ overhang	GNC-
*Dra*III	CACNNN↑GTG	3′ overhang	-NNN
*Drd*I	GACNNNN↑NNGTC	3′ overhang	-NN
*Eae*I	Y↑GGCCR	5′ overhang	GGCC-
*Eag*I	C↑GGCCG	5′ overhang	GGCC-
*Eam*1104I	CTCTTC(1/4)	5′ overhang	NNN-
*Eam*1105I	GACNNN↑NNGTC	3′ overhang	-N
*Ear*I	CTCTTC(1/4)	5′ overhang	NNN-
*Ecl*136II	GAG↑CTC	Blunt	
*Ecl*HKI	GACNNN↑NNGTC	3′ overhang	-N
*Eco*47I	G↑GWCC	5′ overhang	GWC-
*Eco*47III	AGC↑GCT	Blunt	
*Eco*52I	C↑GGCCG	5′ overhang	GGCC-
*Eco*57I	CTGAAG(16/14)	3′ overhang	-NN
*Eco*72I	CAC↑GTG	Blunt	

Restriction/Methylation

Table 4. Commonly used restriction enzymes (continued)

Enzyme	Recognition sequence[a]	Ends resulting from cleavage	
		Type	Overhang sequence (5′ to 3′)
*Eco*81I	CC↑TNAGG	5′ overhang	TNA-
*Eco*105I	TAC↑GTA	Blunt	
*Eco*ICRI	GAG↑CTC	Blunt	
*Eco*NI	CCTNN↑NNNAGG	5′ overhang	N-
*Eco*O109I	RG↑GNCCY	5′ overhang	GNC-
*Eco*RI	G↑AATTC	5′ overhang	AATT-
*Eco*RII	↑CCWGG	5′ overhang	CCWGG-
*Eco*RV	GAT↑ATC	Blunt	
*Eco*T14I	C↑CWWGG	5′ overhang	CWWG-
*Eco*T22I	ATGCA↑T	3′ overhang	-TGCA
*Ehe*I	GGC↑GCC	Blunt	
*Esp*I	GC↑TNAGC	5′ overhang	TNA-
*Fba*I	T↑GATCA	5′ overhang	GATC-
*Fdi*II	TGC↑GCA	Blunt	
*Fnu*4HI	GC↑NGC	5′ overhang	N-
*Fnu*DII	CG↑CG	Blunt	
*Fok*I	GGATG(9/13)	5′ overhang	NNNN-
*Fse*I	GGCCGG↑CC	3′ overhang	-CCGG
*Fsp*I	TGC↑GCA	Blunt	
*Hae*II	RGCGC↑Y	3′ overhang	-GCGC
*Hae*III	GG↑CC	Blunt	
*Hap*II	C↑CGG	5′ overhang	CG-
*Hga*I	GACGC(5/10)	5′ overhang	NNNNN-
*Hgi*AI	GWGCW↑C	3′ overhang	-WGCW
*Hha*I	GCG↑C	3′ overhang	-CG
*Hin*1I	GR↑CGYC	5′ overhang	CG-
*Hinc*II	GTY↑RAC	Blunt	
*Hind*II	GTY↑RAC	Blunt	
*Hind*III	A↑AGCTT	5′ overhang	AGCT-
*Hin*fI	G↑ANTC	5′ overhang	ANT-
*Hin*P1I	G↑CGC	5′ overhang	CG-
*Hpa*I	GTT↑AAC	Blunt	
*Hpa*II	C↑CGG	5′ overhang	CG-
*Hph*I	GGTGA(8/7)	3′ overhang	-N
*Hsp*92I	GR↑CGYC	5′ overhang	CG-
*Hsp*92II	CATG↑	3′ overhang	-CATG
*Ita*I	GC↑NGC	5′ overhang	N-
*Kas*I	G↑GCGCC	5′ overhang	GCGC-
*Kpn*I	GGTAC↑C	3′ overhang	-GTAC
*Kpn*2I	T↑CCGGA	5′ overhang	CCGG-
*Ksp*I	CCGC↑GG	3′ overhang	-GC
*Ksp*632I	CTCTTC(1/4)	5′ overhang	NNN-
*Mae*I	C↑TAG	5′ overhang	TA-
*Mae*II	A↑CGT	5′ overhang	CG-
*Mae*III	↑GTNAC	5′ overhang	GTNAC-

Table 4. Commonly used restriction enzymes (continued)

Enzyme	Recognition sequence[a]	Ends resulting from cleavage	
		Type	Overhang sequence (5′ to 3′)
*Mam*I	GATNN↑NNATC	Blunt	
*Mbo*I	↑GATC	5′ overhang	GATC-
*Mbo*II	GAAGA(8/7)	3′ overhang	-N
*Mfe*I	C↑AATTG	5′ overhang	AATT-
*Mfl*I	R↑GATCY	5′ overhang	GATC-
*Mlu*I	A↑CGCGT	5′ overhang	CGCG-
*Mlu*NI	TGG↑CCA	Blunt	
*Mnl*I	CCTC(7/6)	3′ overhang	-N
*Mro*I	T↑CCGGA	5′ overhang	CCGG-
*Msc*I	TGG↑CCA	Blunt	
*Mse*I	T↑TAA	5′ overhang	TA-
*Msl*I	CAYNN↑NNRTG	Blunt	
*Msp*I	C↑CGG	5′ overhang	CG-
*Msp*A1I	CMG↑CKG	Blunt	
*Mst*I	TGC↑GCA	Blunt	
*Mst*II	CC↑TNAGG	5′ overhang	TNA-
*Mun*I	C↑AATTG	5′ overhang	AATT-
*Mva*I	CC↑WGG	5′ overhang	W-
*Mvn*I	CG↑CG	Blunt	
*Mwo*I	GCNNNNN↑NNGC	3′ overhang	-NNN
*Nae*I	GCC↑GGC	Blunt	
*Nar*I	GG↑CGCC	5′ overhang	CG-
*Nci*I	CC↑SGG	5′ overhang	S-
*Nco*I	C↑CATGG	5′ overhang	CATG-
*Nde*I	CA↑TATG	5′ overhang	TA-
*Nde*II	↑GATC	5′ overhang	GATC-
*Ngo*AIV	G↑CCGGC	5′ overhang	CCGG-
*Ngo*MI	G↑CCGGC	5′ overhang	CCGG-
*Nhe*I	G↑CTAGC	5′ overhang	CTAG-
*Nla*III	CATG↑	3′ overhang	-CATG
*Nla*IV	GGN↑NCC	Blunt	
*Not*I	GC↑GGCCGC	5′ overhang	GGCC-
*Nru*I	TCG↑CGA	Blunt	
*Nsi*I	ATGCA↑T	3′ overhang	-TGCA
*Nsp*I	RCATG↑Y	3′ overhang	-CATG
*Nsp*II	GDGCH↑C	3′ overhang	-DGCH
*Nsp*III	C↑YCGRG	5′ overhang	YCGR-
*Nsp*IV	G↑GNCC	5′ overhang	GNC-
*Nsp*V	TT↑CGAA	5′ overhang	CG-
*Nsp*BII	CMG↑CKG	Blunt	
*Nsp*HI	RCATG↑Y	3′ overhang	-CATG
*Nun*II	GG↑CGCC	5′ overhang	CG-
*Pac*I	TTAAT↑TAA	3′ overhang	-AT
*Pae*R7I	C↑TCGAG	5′ overhang	TCGA-
*Pal*I	GG↑CC	Blunt	

Table 4. Commonly used restriction enzymes (continued)

Enzyme	Recognition sequence[a]	Ends resulting from cleavage	
		Type	Overhang sequence (5′ to 3′)
*Pfl*FI	GACN↑NNGTC	5′ overhang	N-
*Pfl*MI	CCANNNN↑NTGG	3′ overhang	-NNN
*Ple*I	GAGTC(4/5)	5′ overhang	N-
*Pma*CI	CAC↑GTG	Blunt	
*Pme*I	GTTT↑AAAC	Blunt	
*Pml*I	CAC↑GTG	Blunt	
*Ppu*10I	A↑TGCAT	5′ overhang	TGCA-
*Ppu*MI	RG↑GWCCY	5′ overhang	GWC-
*Psh*AI	GACNN↑NNGTC	Blunt	
*Psp*5II	RG↑GWCCY	5′ overhang	GWC-
*Psp*1406I	AA↑CGTT	5′ overhang	CG-
*Psp*AI	C↑CCGGG	5′ overhang	CCGG-
*Pss*I	RGGNC↑CY	3′ overhang	-GNC
*Pst*I	CTGCA↑G	3′ overhang	-TGCA
*Pvu*I	CGAT↑CG	3′ overhang	-AT
*Pvu*II	CAG↑CTG	Blunt	
*Rsa*I	GT↑AC	Blunt	
*Rsp*XI	T↑CATGA	5′ overhang	CATG-
*Rsr*I	G↑AATTC	5′ overhang	AATT-
*Rsr*II	CG↑GWCCG	5′ overhang	GWC-
*Sac*I	GAGCT↑C	3′ overhang	-AGCT
*Sac*II	CCGC↑GG	3′ overhang	-GC
*Sal*I	G↑TCGAC	5′ overhang	TCGA-
*San*DI	GG↑GWCCC	5′ overhang	GWC-
*Sap*I	GCTCTTC(1/4)	5′ overhang	NNN-
*Sau*I	CC↑TNAGG	5′ overhang	TNA-
*Sau*3AI	↑GATC	5′ overhang	GATC-
*Sau*96I	G↑GNCC	5′ overhang	GNC-
*Sca*I	AGT↑ACT	Blunt	
*Scr*FI	CC↑NGG	5′ overhang	N-
*Sdu*I	GDGCH↑C	3′ overhang	-DGCH
*Sex*AI	A↑CCWGGT	5′ overhang	CCWGG-
*Sfa*NI	GCATC(5/9)	5′ overhang	NNNN-
*Sfc*I	C↑TRYAG	5′ overhang	TRYA-
*Sfi*I	GGCCNNNN↑NGGCC	3′ overhang	-NNN
*Sfu*I	TT↑CGAA	5′ overhang	CG-
*Sgf*I	GCGAT↑CGC	3′ overhang	-AT
*Sgr*AI	CR↑CCGGYG	5′ overhang	CCGG-
*Sin*I	G↑GWCC	5′ overhang	GWC-
*Sma*I	CCC↑GGG	Blunt	
*Sml*I	C↑TYRAG	5′ overhang	TY-
*Sna*BI	TAC↑GTA	Blunt	
*Spe*I	A↑CTAGT	5′ overhang	CTAG-
*Sph*I	GCATG↑C	3′ overhang	-CATG
*Spl*I	C↑GTACG	5′ overhang	GTAC-

Table 4. Commonly used restriction enzymes (continued)

Enzyme	Recognition sequence[a]	Ends resulting from cleavage	
		Type	Overhang sequence (5′ to 3′)
*Spo*I	TCG↑CGA	Blunt	
*Srf*I	GCCC↑GGGC	Blunt	
*Sse*8387I	CCTGCA↑GG	3′overhang	-TGCA
*Ssp*I	AAT↑ATT	Blunt	
*Ssp*BI	T↑GTACA	5′ overhang	GTAC-
*Sst*I	GAGCT↑C	3′ overhang	-AGCT
*Sst*II	CCGC↑GG	3′ overhang	-GC
*Stu*I	AGG↑CCT	Blunt	
*Sty*I	C↑CWWGG	5′ overhang	CWWG-
*Swa*I	ATTT↑AAAT	Blunt	
*Tai*I	ACGT↑	3′ overhang	-ACGT
*Taq*I	T↑CGA	5′ overhang	CG-
*Tfi*I	G↑AWTC	5′ overhang	AWT-
*Tha*I	CG↑CG	Blunt	
*Tru*9I	T↑TAA	5′ overhang	TA-
*Tse*I	G↑CWGC	5′ overhang	CWG-
*Tsp*45I	↑GTSAC	5′ overhang	GTSAC-
*Tsp*509I	↑AATT	5′ overhang	AATT-
*Tsp*RI	CAGTG(2/–7)	3′ overhang	-NNNNNNNCAGTGNN
*Tth*111I	GACN↑NNGTC	5′ overhang	N-
*Tth*HB8I	T↑CGA	5′ overhang	CG-
*Van*91I	CCANNNN↑NTGG	3′ overhang	-NNN
*Vsp*I	AT↑TAAT	5′ overhang	TA-
*Xba*I	T↑CTAGA	5′ overhang	CTAG-
*Xcm*I	CCANNNNN↑NNNNTGG	3′ overhang	-N
*Xcy*I	C↑CCGGG	5′ overhang	CCGG-
*Xho*I	C↑TCGAG	5′ overhang	TCGA-
*Xho*II	R↑GATCY	5′ overhang	GATC-
*Xma*I	C↑CCGGG	5′ overhang	CCGG-
*Xma*III	C↑GGCCG	5′ overhang	GGCC-
*Xma*CI	C↑CCGGG	5′ overhang	CCGG-
*Xmn*I	GAANN↑NNTTC	Blunt	
*Xor*II	CGAT↑CG	3′ overhang	-AT

[a]For a description of the convention used in describing recognition sequences see *Table 1*, footnote b. For abbreviations see *Table 1*, footnote c.

Table 5. Reaction conditions for restriction enzymes[a,b]

Enzyme	Tris-HCl		$MgCl_2$ (mM)	NaCl (mM)	KCl (mM)	DTT[c] (mM)	β-ME[c] (mM)	Temperature (°C)	
	mM	pH						Reactn[d]	Inactn[d]
*Aat*I	10	7.5	7	–	60	–	6	37	75
*Aat*II	10	7.5	10	–	50	1	–	37	85
*Acc*I[e]	6	8.0	6	6	–	–	6	37	90
*Acc*II	10	7.5	10	60	–	–	10	37	

Table 5. Reaction conditions (continued)

Enzyme	Tris-HCl		$MgCl_2$ (mM)	NaCl (mM)	KCl (mM)	DTT[c] (mM)	β-ME[c] (mM)	Temperature (°C)	
	mM	pH						Reactn[d]	Inactn[d]
*Acc*III	10	7.7	10	100	–	1	–	60	
*Acc*65I	10	7.5	50	100	–	–	–	37	65
*Acc*B7I	6	7.5	6	100	–	1	–	37	
*Aci*I	50	7.9	10	100	–	1	–	37	65
*Acs*I	10	8.0	5	100	–	–	1	50	
*Acy*I	10	8.5	7	100	–	1	–	37	90
*Afa*I	10	8.0	7	50	–	–	7	37	70
*Afl*I	10	8.0	10	50	–	–	10	37	
*Afl*II	10	8.0	10	50	–	–	10	37	65
*Afl*III	10	8.0	10	150	–	–	6	37	100
*Age*I	10[f]	7.0	10	–	–	1	–	37	65
*Aha*I	10	7.5	10	25	–	–	10	37	
*Aha*II	10	8.0	10	100	–	–	10	37	
*Aha*III	25	7.7	10	–	–	1	–	37	
*Ahd*I	20[g]	7.9	10[g]	–	50[g]	1	–	37	65
*Alu*I	10	7.5	6	50	–	–	6	37	70
*Alw*I	10	7.4	10	–	–	–	10	37	65
*Alw*26I	33[g]	7.9	10[g]	–	66[g]	–	–	37	65
*Alw*44I	10	8.0	5	20	–	1	–	37	80
*Alw*NI	10	7.4	10	50	–	–	10	37	65
*Aoc*I	10	7.7	10	25	–	1	–	37	
*Aoc*II	10	7.5	10	25	–	–	10	37	
*Aor*51HI	10	8.0	7	–	60	–	7	37	
*Aos*I	10	7.5	10	–	50	1	–	37	
*Apa*I	10	7.5	6	6	–	–	6	37	80
*Apa*LI	10	7.5	10	–	–	–	10	37	65
*Apo*I	50	7.9	10	100	–	1	–	50	
*Apy*I	50	7.5	10	100	–	1	–	37	
*Aqu*I	25	7.7	10	50	–	–	10	37	
*Asc*I	20[g]	7.9	10[g]	–	50[g]	1	–	37	65
*Ase*I	10	7.5	10	–	100	–	–	37	65
*Asn*I	50	7.5	10	100	–	1	–	37	
*Asp*I	10	8.0	5	100	–	–	1	37	
*Asp*700I	50	7.5	10	50	–	1	–	37	
*Asp*718I	50	7.5	10	100	–	1	–	37	
*Asp*EI	10	7.5	10	–	–	1	–	37	65
*Asp*HI	10	8.0	5	100	–	–	1	37	
*Asu*I	6	7.6	6	50	–	–	6	37	
*Asu*II	10	7.5	10	–	–	1	–	37	
*Ava*I[e]	10	8.0	10	50	–	–	6	37	100
*Ava*II	10	8.0	10	60	–	–	6	37	65
*Avi*II	50	7.5	10	100	–	1	–	37	
*Avr*II	10	7.4	10	50	–	–	10	37	
*Axy*I	50	7.5	10	–	100	1	–	37	
*Bal*I	50	8.5	5	–	–	–	10	37	65
*Bam*HI	20	7.4	7	100	–	–	6	37	85

Table 5. Reaction conditions (continued)

Enzyme	Tris-HCl		MgCl$_2$ (mM)	NaCl (mM)	KCl (mM)	DTT[c] (mM)	β-ME[c] (mM)	Temperature (°C)	
	mM	pH						Reactn[d]	Inactn[d]
*Ban*I	10	8.0	7	–	–	–	6	50	70
*Ban*II	10	7.5	7	50	–	–	6	37	60
*Ban*III	10	7.5	7	–	80	–	6	37	70
*Bbe*I	10	7.5	10	–	–	1	–	37	
*Bbi*II	10	7.4	5	–	–	–	7	37	
*Bbr*PI	10	8.0	5	100	–	–	1	37	
*Bbs*I	10[f]	7.0	10	–	–	1	–	37	65
*Bbu*I	6	7.5	6	6	–	–	6	37	65
*Bbv*I	10	8.0	10	50	–	–	10	37	65
*Bcg*I	10	8.4	10	100	–	1	–	37	65
*Bcl*I	10	8.0	10	–	75	–	6	50	100
*Bcn*I	10	7.5	10	50	–	1	–	37	
*Bfa*I	20[g]	7.9	10[g]	–	50[g]	1	–	37	
*Bfr*I	10	7.5	10	50	–	1	–	37	
*Bgl*I	100	8.0	10	60	–	–	–	37	65
*Bgl*II	100	8.0	5[h]	60	–	–	6	37	100
*Bln*I	10	7.5	7	–	150	–	7	37	
*Blp*I	20[g]	7.9	10[g]	–	50[g]	1	–	37	
*Bmy*I	33[g]	7.9	10[g]	–	66[g]	0.5	–	37	
*Bpm*I	50	7.9	10	100	–	1	–	37	65
*Bpu*1102I	10	7.5	10	50	–	1	–	37	
*Bpu*AI	10	8.0	5	100	–	–	1	37	
*Bsa*I	20[g]	7.9	10[g]	–	50[g]	1	–	50	65
*Bsa*AI	50	7.9	10	100	–	1	–	37	
*Bsa*BI	10	7.9	10	50	–	1	–	60	
*Bsa*HI	20[g]	7.9	10[g]	–	50[g]	1	–	37	
*Bsa*JI	10	7.9	10	50	–	1	–	60	
*Bsa*MI	6	7.9	6	150	–	1	–	65	
*Bsa*OI	10	7.9	10	50	–	1	–	50	
*Bsa*WI	10	7.9	10	50	–	1	–	60	
*Bse*AI	10	8.0	5	100	–	–	1	55	
*Bse*RI	10	7.9	10	50	–	1	–	37	65
*Bsg*I	20[g]	7.9	10[g]	–	50[g]	1	–	37	65
*Bsh*1236I	10	8.5	10	100	–	1	–	37	65
*Bsi*CI	10	7.5	10	–	10	–	–	65	
*Bsi*EI	10	7.9	10	50	–	1	–	60	
*Bsi*HKAI	50	7.9	10	100	–	1	–	65	
*Bsi*WI	50	7.5	10	100	–	–	10	55	
*Bsi*YI	10	7.5	10	50	–	–	10	55	
*Bsl*I	50	7.9	10	100	–	1	–	55	
*Bsm*I	10	7.4	10	50	–	–	10	65	90
*Bsm*AI	10	8.4	10	100	–	–	10	50	
*Bsm*BI	50	7.9	10	100	–	1	–	55	
*Bsm*FI	20[g]	7.9	10[g]	–	50[g]	1	–	65	
*Bso*BI	10	7.9	10	50	–	1	–	65	
*Bsp*106I	25[g]	7.6	10[g]	–	100[g]	1	0.5	37	65

Table 5. Reaction conditions (continued)

Enzyme	Tris-HCl		MgCl$_2$ (mM)	NaCl (mM)	KCl (mM)	DTT[c] (mM)	β-ME[c] (mM)	Temperature (°C)	
	mM	pH						Reactn[d]	Inactn[d]
*Bsp*120I	10	7.5	10	–	–	–	–	37	
*Bsp*1286I	10	7.5	10	–	–	–	10	37	65
*Bsp*CI	20[g]	7.9	10[g]	–	50[g]	1	–	37	
*Bsp*EI	50	7.9	10	100	–	1	–	37	
*Bsp*HI	10	7.4	10	–	100	–	–	37	65
*Bsp*LU11I	50	7.5	10	100	–	1	–	48	
*Bsp*MI	10	7.5	10	150	–	–	–	37	65
*Bsp*MII	10	7.5	10	150	–	–	–	60	
*Bsr*I	10	7.8	10	–	150	–	–	65	
*Bsr*BI	10	7.9	10	50	–	1	–	37	
*Bsr*BRI	90	7.5	10	50	–	–	–	37	
*Bsr*DI	10	7.9	10	50	–	1	–	60	
*Bsr*FI	10	7.9	10	50	–	1	–	37	
*Bsr*GI	10	7.9	10	50	–	1	–	60	
*Bsr*SI	6	7.9	6	150	–	1	–	65	
*Bss*HII	10	7.5	10	25	–	–	10	50	90
*Bss*KI	50	7.9	10	100	–	1	–	60	
*Bss*SI	50	7.9	10	100	–	1	–	37	
*Bst*I	10	7.5	10	50	–	–	1	55	85[i]
*Bst*71I	6	7.9	6	150	–	1	–	50	
*Bst*98I	6	7.9	6	150	–	1	–	37	
*Bst*1107I	20	8.5	10	–	100	1	–	37	65
*Bst*BI	10	7.5	10	50	–	–	10	65	
*Bst*EII	6	8.0	6	150	–	–	6	60	100
*Bst*NI	10	7.7	10	150	–	1	–	60	
*Bst*OI	10	7.9	10	50	–	1	–	60	
*Bst*PI	50	8.0	7	100	–	–	7	60	
*Bst*UI	10	8.0	10	–	–	–	–	60	
*Bst*XI	10	7.6	7	150	–	–	6	55	100
*Bst*YI	10[f]	7.0	10	–	–	–	–	60	
*Bst*ZI	6	7.9	6	150	–	1	–	50	
*Bsu*36I	10	7.4	10	100	–	–	–	37	
*Cac*8I	50	7.9	10	100	–	1	–	37	65
*Cfo*I	6	7.6	6	50	–	–	6	37	
*Cfr*9I	10	7.5	5	200[j]	–	1	–	37	
*Cfr*10I[e]	20	8.5	3[e]	–	100	–	–	37	100
*Cfr*13I	10	8.5	5	50	–	–	–	37	100
*Cla*I	10	7.9	10	50	–	–	6	37	65
*Cpo*I	10	8.0	7	100	–	–	7	30	60
*Csp*I	10	7.4	10	–	150	–	10	30	65
*Csp*45I	10	7.5	7	60	–	–	–	37	65
*Cvn*I	10	7.5	10	50	–	1	–	37	65
*Dde*I	10	7.5	6	150	–	–	6	37	70
*Dpn*I	10	7.5	7	150	–	–	10	37	65
*Dpn*II	50	6.5	10	100	–	1	–	37	65
*Dra*I	10	7.5	10	50	–	–	6	37	65

Table 5. Reaction conditions (continued)

Enzyme	Tris-HCl		$MgCl_2$ (mM)	NaCl (mM)	KCl (mM)	DTT[c] (mM)	β-ME[c] (mM)	Temperature (°C)	
	mM	pH						Reactn[d]	Inactn[d]
*Dra*II	10	7.5	10	25	–	–	10	37	65
*Dra*III	10	8.5	10	100	–	1	–	37	65
*Drd*I	20[g]	7.9	10[g]	–	50[g]	1	–	37	85
*Eae*I	50	7.4	10	–	50	–	10	37	65
*Eag*I	10	8.2	10	150	–	–	10	37	65
*Eam*1104I	25[g]	7.6	10[g]	–	100[g]	–	0.5	37	65
*Eam*1105I	20	8.5	10	–	100	1	–	37	65
*Ear*I	10[f]	7.0	10	–	–	1	–	37	65
*Ecl*136II	33[g]	7.9	10[g]	–	66[g]	–	–	37	65
*Ecl*HKI	6	7.5	6	100	–	1	–	37	
*Eco*47I	10	7.5	7	100	–	–	7	37	100
*Eco*47III	10	8.5	7	100	–	–	7	37	100
*Eco*52I	10	9.0	3	100	–	–	–	37	80
*Eco*57I[e]	10	7.5	10	–	–	–	–	37	65
*Eco*72I	10	7.4	10	–	150	–	–	37	65
*Eco*81I	10	8.5	7	–	20	–	7	37	90
*Eco*105I	10	7.5	5	20	–	–	10	37	65
*Eco*ICRI	6	7.5	6	50	–	1	–	37	65
*Eco*NI	10	7.5	10	50	–	–	10	37	65
*Eco*O109I	40	8.0	10	–	–	–	10	37	65
*Eco*RI	100	7.5	10	50	–	–	6	37	70
*Eco*RII	25	7.7	10	50	–	–	10	37	60
*Eco*RV	10	8.0	6	100	–	–	6	37	100
*Eco*T14I	50	7.5	10	100	–	1	–	37	
*Eco*T22I	10	7.5	7	125	–	–	7	37	100
*Ehe*I	10	8.0	10	–	–	1	–	37	70
*Esp*I	7	7.5	7	100	–	–	6	37	100
*Fba*I	10	8.0	7	–	150	–	7	37	
*Fdi*II	10	8.0	10	60	–	–	7	50	100
*Fnu*4HI[e]	10	7.4	10	10	–	–	10	37	65
*Fnu*DII	10	7.5	10	–	–	1	–	37	
*Fok*I	10	7.7	10	–	10	–	10	37	65
*Fse*I	20[g]	7.9	10[g]	–	50[g]	1	–	37	65
*Fsp*I	10	7.4	10	50	–	–	10	37	65
*Hae*II	50	7.5	6	50	–	–	6	37	70
*Hae*III	50	7.5	6	50	–	1	–	37	90
*Hap*II	10	7.5	10	–	–	1	–	37	
*Hga*I	10	7.4	10	50	–	1	–	37	65
*Hgi*AI	10	8.0	10	150	–	–	10	37	65
*Hha*I	10	8.0	6	100	–	–	6	37	90
*Hin*1I	10	8.5	5	25	–	–	–	37	80
*Hinc*II	10	7.5	7	100	–	–	6	37	70
*Hind*II	10	7.5	10	50	–	1	–	37	
*Hind*III	50	8.0	10	60	–	–	–	37	90
*Hinf*I	10	7.5	7	60	–	–	6	37	80
*Hin*P1I	10	7.9	10	50	–	1	–	37	65

Restriction/Methylation

Table 5. Reaction conditions (continued)

Enzyme	Tris-HCl		$MgCl_2$ (mM)	NaCl (mM)	KCl (mM)	DTT[c] (mM)	β-ME[c] (mM)	Temperature (°C)	
	mM	pH						Reactn[d]	Inactn[d]
*Hpa*I	10	7.4	10	–	20	1	–	37	90
*Hpa*II	10	7.5	10	–	10	–	6	37	90
*Hph*I	10	7.5	10	–	10	1	–	37	65
*Hsp*92I	10	8.5	10	100	–	1	–	37	
*Hsp*92II	10	7.4	10	–	150	–	–	37	
*Ita*I	50	7.5	10	100	–	1	–	37	
*Kas*I	10	7.9	10	50	–	1	–	37	65
*Kpn*I	10	7.5	10	10	–	–	6	37	85
*Kpn*2I	20	7.4	5	–	50	–	–	55	
*Ksp*I	10	7.5	10	–	–	1	–	37	
*Ksp*632I	33[g]	7.9	10[g]	–	66[g]	0.5	–	37	
*Mae*I	50	7.5	10	100	–	1	–	45	
*Mae*II	50	7.5	10	100	–	1	–	50	
*Mae*III	50	7.5	10	100	–	1	–	45	85
*Mam*I	50	7.5	10	100	–	1	–	37	
*Mbo*I	10	7.5	10	100	–	1	–	37	65
*Mbo*II	10	7.4	10	–	10	1	–	37	65
*Mfe*I	20[g]	7.9	10[g]	–	50[g]	1	–	37	65
*Mfl*I	10	7.5	10	–	–	1	–	37	
*Mlu*I	10	7.5	7	150	–	–	7	37	100
*Mlu*NI	33[g]	7.9	10[g]	–	66[g]	0.5	–	37	
*Mnl*I	12	7.6	12	50	–	–	10	37	65
*Mro*I	10	8.0	12	20	–	–	–	37	100
*Msc*I	20[g]	7.9	10[g]	–	50[g]	1	–	37	65
*Mse*I	10	7.4	10	50	–	–	10	37	65
*Msl*I	10	7.9	10	50	–	1	–	37	65
*Msp*I[e]	10	7.5	10	50	–	–	6	37	90
*Msp*A1I	20[g]	7.9	10[g]	–	50[g]	1	–	37	65
*Mst*I	10	7.7	10	150	–	1	–	37	
*Mst*II	10	7.7	10	150	–	1	–	37	65
*Mun*I	10	7.5	10	50	–	1	–	37	65
*Mva*I	10	8.5	15	–	150	1	–	37	100
*Mvn*I	10	7.5	10	50	–	1	–	37	
*Mwo*I	50	7.9	10	150	–	1	–	60	
*Nae*I	10	8.0	10	20	–	–	6	37	100
*Nar*I	10	7.5	10	–	10	–	6	37	65
*Nci*I	10	7.5	10	25	–	–	6	37	80
*Nco*I	10	7.5	10	150	–	–	–	37	65
*Nde*I	10	7.8	7	150	–	–	6	37	65
*Nde*II	50	7.5	10	100	–	1	–	37	
*Ngo*AIV	200[g]	7.9	10[g]	–	50[g]	–	–	37	
*Ngo*MI	20[g]	7.9	10[g]	–	50[g]	1	–	37	
*Nhe*I	10	8.0	10	50	–	–	10	37	65
*Nla*III[e]	20[g]	7.9	10[g]	–	50[g]	1	–	37	65
*Nla*IV[e]	20[g]	7.9	10[g]	–	50[g]	1	–	37	65
*Not*I[e]	10	7.5	10	150	–	–	6	37	100

Table 5. Reaction conditions (continued)

Enzyme	Tris-HCl		$MgCl_2$ (mM)	NaCl (mM)	KCl (mM)	DTT[c] (mM)	β-ME[c] (mM)	Temperature (°C)	
	mM	pH						Reactn[d]	Inactn[d]
*Nru*I	10	8.0	10	150	–	–	6	37	80
*Nsi*I	10	7.7	10	150	–	–	6	37	65
*Nsp*I	10	8.0	10	20	–	–	7	50	
*Nsp*II	20	8.5	10	–	–	–	10	37	
*Nsp*III	10	8.0	10	–	25	–	10	37	85
*Nsp*IV	10	8.0	10	–	–	–	10	37	
*Nsp*V	10	8.0	10	25	–	–	6	50	100
*Nsp*BII	10	7.5	10	–	50	1	–	37	
*Nsp*HI	10	7.5	10	–	50	1	–	37	
*Nun*II	10	7.5	10	–	50	1	–	37	
*Pac*I	10[f]	7.0	10	–	–	1	–	37	65
*Pae*R7I	10	7.4	10	–	–	–	10	37	
*Pal*I	10	7.5	10	–	–	–	1	37	85[i]
*Pfl*MI	10	7.4	10	50	–	–	10	37	65
*Ple*I	6	7.8	6	–	–	–	–	37	65
*Pma*CI	10	7.5	7	20	–	–	7	37	60
*Pme*I	20[g]	7.9	10[g]	–	50[g]	1	–	37	65
*Pml*I	10[f]	7.0	10	–	–	1	–	37	65
*Ppu*10I	33[g]	7.9	10[g]	–	66[g]	–	–	37	
*Ppu*MI	20[g]	7.9	10[g]	–	50[g]	1	–	37	
*Psh*AI	10	8.5	7	–	60	–	7	37	65
*Psp*5II	50	8.0	10	50	–	–	–	37	
*Psp*1406I	33[g]	7.9	10[g]	–	66[g]	–	–	37	65
*Psp*AI	25[g]	7.6	10[g]	–	100[g]	–	0.5	37	65
*Pss*I	10	7.5	10	50	–	1	–	37	
*Pst*I	10	7.5	10	100	–	–	6	37	85
*Pvu*I	10	7.4	7		150		6	37	100
*Pvu*II	10	7.5	6	60	–	–	6	37	95
*Rsa*I	10	8.0	10	50	–	–	6	37	100
*Rsp*XI	50	7.5	10	–	100	1	–	37	
*Rsr*I	25	7.7	10	–	–	1	–	37	
*Rsr*II	10	8.0	5	10	–	1	–	37	65
*Sac*I	6	7.4	6	20	–	–	6	37	65
*Sac*II	10	7.5	10	–	10	–	10	37	80
*Sal*I	6	7.9	7[e]	150	–	–	6	37	80
*San*DI	50[g]	7.6	20[g]	–	200[g]	–	1	37	65
*Sap*I	20[g]	7.9	10[g]	–	50[g]	1	–	37	65
*Sau*I	25	7.7	10	–	–	1	–	37	
*Sau*3AI	10	7.5	7	100	–	–	–	37	85
*Sau*96I	10	7.7	10	50	–	–	10	37	65
*Sca*I	10	7.4	6	100	–	–	6	37	100
*Scr*FI	10	7.6	10	50	–	–	6	37	65
*Sdu*I	10	7.5	10	–	–	1	–	37	
*Sex*AI	10	8.0	5	100	–	–	1	37	
*Sfa*NI	10	7.5	10	150	–	–	–	37	65
*Sfc*I	20[g]	7.9	10[g]	–	50[g]	1	–	37	65

Table 5. Reaction conditions (continued)

Enzyme	Tris-HCl		$MgCl_2$ (mM)	NaCl (mM)	KCl (mM)	DTT[c] (mM)	β-ME[c] (mM)	Temperature (°C)	
	mM	pH						Reactn[d]	Inactn[d]
*Sfi*I	10	7.8	10	50	–	–	10	50	90
*Sfu*I	50	7.5	10	100	–	1	–	37	
*Sgf*I	10	7.9	10	50	–	1	–	37	
*Sgr*AI	33[g]	7.9	10[g]	–	66[g]	0.5	–	37	
*Sin*I	6	7.4	6	20	–	–	6	37	65
*Sma*I	10	8.0	6	–	20	–	6	30	60
*Sna*BI	10	7.7	10	50	–	–	10	37	
*Spe*I	6	7.5	10	50	–	–	6	37	65
*Sph*I	8	7.4	7	150	–	–	6	37	100
*Spl*I	50	7.5	10	100	–	1	–	55	
*Spo*I	15	7.5	7	–	50	–	6	37	
*Srf*I	25[g]	7.6	10[g]	–	100[g]	–	0.5	37	65
*Sse*8387I	10	7.5	7	–	80	–	7	37	60
*Ssp*I	10	7.4	10	100	–	1	–	37	65
*Ssp*BI	10	8.0	5	100	–	–	1	50	
*Sst*I	10	7.5	10	50	–	1	–	37	
*Sst*II	10	7.5	10	50	–	1	–	37	
*Stu*I	10	8.0	10	50	–	–	6	37	85
*Sty*I	10	8.5	100	10	–	1	–	37	65
*Swa*I	50	7.5	10	100	–	1	–	37	
*Tai*I	10	8.5	10	–	100	–	–	65	
*Taq*I	6	8.4	6	100	–	–	6	65	90
*Tfi*I	50	7.9	10	100	–	1	–	60	
*Tha*I	10	7.5	10	–	–	1	–	60	
*Tru*9I	10	8.5	10	100	–	1	–	65	
*Tse*I	50	7.9	10	100	–	1	–	65	
*Tsp*45I	10[f]	7.0	10	–	–	1	–	65	
*Tsp*509I	10[f]	7.0	10	–	–	1	–	65	
*Tsp*RI	20[g]	7.9	10[g]	–	50[g]	1	–	65	
*Tth*111I	20[g]	7.9	10[g]	–	50[g]	1	–	65	
*Tth*HB8I	50	7.5	10	100	–	1	–	65	
*Van*91I	10	8.0	5	100	–	–	1	37	65
*Vsp*I	20	8.5	10	150	–	–	–	37	
*Xba*I	10	8.0	6	100	–	–	6	37	70
*Xcm*I	10	7.9	10	50	–	1	–	37	65
*Xcy*I	10	7.5	10	50	–	1	–	37	
*Xho*I	10	8.0	6	150	–	–	6	37	80
*Xho*II	25	7.7	10	–	–	1	–	37	
*Xma*I	10	7.5	10	25	–	–	10	37	65
*Xma*III	10	7.5	10	–	–	1	–	25	
*Xma*CI	10	7.5	5	–	–	1	–	37	
*Xmn*I	10	8.0	10	6	–	–	10	37	65
*Xor*II	10	7.5	10	–	–	1	–	37	

[a]Data taken from various sources: personal experience, colleagues' recommendations, commercial catalogues, original descriptions of the enzymes.

Table 5. Reaction conditions (continued)

[b]The addition of 100 µg ml^{-1} BSA is recommended for all enzymes, especially for lengthy digestions (3, 4).
[c]Abbreviations: DTT, dithiothreitol; β-ME, β-mercaptoethanol.
[d]Reactn = recommended reaction temperature. Inactn = temperature needed to achieve inactivation of the enzyme after 20 min. The inactivation temperature can be affected by trace amounts of contaminating compounds able to stabilize the enzyme. Phenol extraction is recommended if 100% inactivation is essential. If there is no entry under 'Inactn' then assume that heat inactivation is ineffective.
[e]Reaction buffers for these enzymes should have additional supplements as follows: *Acc*I, 0.01% Triton X-100; *Ava*I, 0.01% Triton X-100; *Cfr*10I, 0.02% Triton X-100; *Eco*57I, 10 µM *S*-adenosylmethionine; *Fnu*4HI, 5 mM KPO_4; *Msp*I, 0.02% Triton X-100; *Nla*III, 50 mM $(NH_4)_2SO_4$; *Nla*IV, 50 mM $(NH_4)_2SO_4$; *Not*I, 0.01% Triton X-100.
[f]Use Bis-Tris-propane-HCl.
[g]Use Tris-acetate, magnesium acetate and potassium acetate.
[h]Use $MgSO_4$.
[i]Requires 30 min incubation.
[j]Use sodium glutamate.

Table 6. Effects of salt on restriction enzyme activities[a]

Enzyme	Effects of NaCl[b] (mM)				Effects of KCl[b] (mM)			
	0	50	100	150	0	50	100	150
*Aat*I	++	++	++	−	++	++	++	−
*Aat*II	−	+	+	−	+	++	+	−
*Acc*I	++	++	−	−	++	++	−	−
*Acc*II	+	+	++	++				
*Acc*III	−	+	++	+	nd	+	+	++
*Acc*65I	+	++	++	nd	nd	+	nd	nd
*Acc*B7I	+	++	++	++				
*Aci*I	+	++	++	nd	nd	++	nd	nd
*Acs*I	−	++	++	nd				
*Acy*I	++	++	++	−				
*Afl*II	+	++	+	−	++	++	++	+
*Afl*III	+	++	++	++	nd	++	nd	nd
*Age*I	++	++	++	nd	nd	++	nd	nd
*Aha*II	−	+	++	++				
*Aha*III	++	++	++	++				
*Ahd*I	+	+	−	nd	nd	++	nd	nd
*Alu*I	++	++	++	+	+	++	++	+
*Alw*I	++	++	+	−	nd	++	nd	nd
*Alw*26I	++	++	+	nd	nd	++	nd	nd
*Alw*44I	++	++	++	−	+	++	++	−
*Alw*NI	++	++	++	−	nd	++	nd	nd
*Aoc*I	++	+	−	nd				
*Aor*51HI	++	++	−	nd	nd	++	nd	nd
*Apa*I	++	+	+	−	++	++	+	−
*Apa*LI	++	++	−	−	nd	++	+	−
*Apo*I	−	++	++	nd	nd	++	nd	nd
*Asc*I	−	−	−	nd	nd	++	nd	nd
*Ase*I	+	++	++	+	+	++	++	+
*Asn*I	+	++	++	nd				

Restriction/Methylation

Table 6. Effects of salt (continued)

Enzyme	Effects of NaCl[b] (mM)				Effects of KCl[b] (mM)			
	0	50	100	150	0	50	100	150
*Asp*700I	+	++	–	nd				
*Asp*718I	–	+	++	nd				
*Asp*EI	++	+	–	nd				
*Asp*HI	+	++	++	nd				
*Asu*I	–	++	++	–				
*Ava*I	+	++	++	–	+	++	++	–
*Ava*II	++	++	+	–	nd	++	++	++
*Avi*II	+	++	++	nd				
*Avr*II	++	++	++	+	nd	++	nd	nd
*Bal*I	++	+	+	–				
*Bam*HI	–	+	++	++	–	+	++	++
*Ban*I	++	++	+	–	++	++	+	–
*Ban*II	++	++	++	+	++	++	++	+
*Ban*III	+	++	++	++	+	++	++	++
*Bbi*II	++	–	–	nd				
*Bbr*PI	++	++	++	nd				
*Bbs*I	++	++	+	nd	nd	++	nd	nd
*Bbu*I	++	++	–	–				
*Bbv*I	++	++	+	+	nd	++	nd	nd
*Bcg*I	+	++	++	nd	nd	+	nd	nd
*Bcl*I	+	+	+	+	–	++	++	–
*Bcn*I	+	++	++	nd				
*Bfa*I	++	+	–	nd	nd	++	nd	nd
*Bfr*I	++	++	+	nd				
*Bgl*I	–	+	++	++	–	+	++	++
*Bgl*II	+	+	++	++	–	+	++	++
*Bln*I	–	+	++	nd	nd	nd	nd	++
*Bmy*I	++	+	–	nd	nd	++	nd	nd
*Bpm*I	++	++	++	nd	nd	++	nd	nd
*Bpu*1102I	nd	++	nd	nd	nd	+	++	++
*Bpu*AI	+	+	++	nd				
*Bsa*I	++	++	++	nd	nd	++	nd	nd
*Bsa*AI	++	++	++	nd	nd	++	nd	nd
*Bsa*BI	++	++	++	nd	nd	++	nd	nd
*Bsa*HI	+	++	++	nd	nd	++	nd	nd
*Bsa*JI	++	++	++	nd	nd	++	nd	nd
*Bsa*MI	+	+	nd	++				
*Bsa*OI	+	++	nd	+				
*Bsa*WI	+	++	+	nd	nd	+	nd	nd
*Bse*AI	–	++	++	nd				
*Bse*RI	++	++	++	nd	nd	++	nd	nd
*Bsg*I	+	++	+	nd	nd	++	nd	nd
*Bsh*1236I					nd	–	–	++
*Bsi*CI	++	++	++	nd				
*Bsi*EI	+	++	–	nd	nd	++	nd	nd
*Bsi*HKAI	+	++	++	nd	nd	++	nd	nd

Table 6. Effects of salt (continued)

Enzyme	Effects of NaCl[b] (mM)				Effects of KCl[b] (mM)			
	0	50	100	150	0	50	100	150
*Bsi*WI	–	+	++	nd	nd	+	nd	nd
*Bsi*YI	++	++	++	nd				
*Bsl*I	–	+	++	nd	nd	++	nd	nd
*Bsm*I	+	++	++	+	+	++	++	+
*Bsm*AI	++	++	++	nd	nd	++	nd	nd
*Bsm*BI	++	++	++	nd	nd	++	nd	nd
*Bsm*FI	–	++	++	nd	nd	++	nd	nd
*Bso*BI	–	++	++	nd	nd	+	nd	nd
*Bso*FI	+	++	++	nd	nd	++	nd	nd
*Bsp*106I					nd	++	++	++
*Bsp*120I	–	+	–	nd	nd	++	nd	nd
*Bsp*1286I	+	++	+	–	nd	++	nd	nd
*Bsp*CI					nd	++	++	++
*Bsp*DI	+	++	+	nd	nd	++	nd	nd
*Bsp*EI	–	–	++	nd	nd	–	nd	nd
*Bsp*HI	+	++	++	+	+	++	++	+
*Bsp*LU11I	+	++	++	nd				
*Bsp*MI	+	+	++	++				
*Bsp*MII	+	+	++	++				
*Bsr*I	–	+	++	nd	nd	–	nd	++
*Bsr*BI	+	++	++	nd	nd	++	nd	nd
*Bsr*BRI	+	++	++	++				
*Bsr*DI	+	++	+	nd	nd	++	nd	nd
*Bsr*FI	+	++	++	nd	nd	++	nd	nd
*Bsr*GI	+	++	++	nd	nd	++	nd	nd
*Bsr*SI	–	+	++	nd	nd	–	nd	nd
*Bss*HII	++	++	++	++	++	++	++	++
*Bss*KI	–	+	++	nd	nd	+	nd	nd
*Bss*SI	–	+	++	nd	nd	+	nd	nd
*Bst*71I	+	+	nd	++				
*Bst*98I	–	+	nd	++				
*Bst*1107I	++	–	++	nd	nd	–	++	nd
*Bst*BI	++	++	+	–	nd	++	nd	nd
*Bst*EII	–	+	++	++	–	+	++	++
*Bst*NI	+	++	++	++	nd	++	++	++
*Bst*OI	+	++	nd	+				
*Bst*PI	–	++	++	nd				
*Bst*UI	++	++	+	–	nd	++	nd	nd
*Bst*XI	–	–	++	++	–	–	++	++
*Bst*YI	+	++	++	+	nd	++	nd	nd
*Bst*ZI	–	–	nd	++				
*Bsu*36I	–	+	++	+	nd	–	++	++
*Cac*8I	+	++	++	nd	nd	++	nd	nd
*Cel*II	+	+	++	nd				
*Cfo*I	++	++	+	+				
*Cfr*10I	–	++	++	++	–	++	++	++

Table 6. Effects of salt (continued)

Enzyme	Effects of NaCl[b] (mM)				Effects of KCl[b] (mM)			
	0	50	100	150	0	50	100	150
*Cfr*13I	++	++	++	+	++	++	++	+
*Cla*I	+	++	++	+	nd	++	nd	nd
*Csp*I	++	++	++	++	++	++	++	++
*Csp*45I	++	++	++	+	++	++	++	+
*Csp*6I	+	+	+	nd	nd	++	nd	nd
*Dde*I	+	+	++	++	+	+	++	++
*Dpn*I	++	++	++	++	nd	++	++	++
*Dra*I	++	++	++	+	++	++	++	+
*Dra*II	++	++	–	nd				
*Dra*III	+	++	++	+	nd	++	++	++
*Drd*I	+	+	–	nd	nd	++	nd	nd
*Eae*I	+	++	+	–	+	++	+	–
*Eag*I	+	+	++	++	nd	–	nd	nd
*Eam*1104I					nd	++	++	–
*Ear*I	++	++	+	nd	nd	++	nd	nd
*Ecl*136II	++	++	+	nd	nd	++	nd	nd
*Ecl*HKI	–	++	++	+				
*Eco*47I	–	+	++	++	–	+	++	++
*Eco*47III	+	++	++	+	+	++	++	++
*Eco*52I	+	++	++	+	+	++	++	+
*Eco*57I	+	+	+	nd	nd	++	nd	nd
*Eco*72I	+	++	nd	+	nd	++	++	++
*Eco*81I	++	+	–	–	++	++	–	–
*Eco*105I	++	++	–	–	++	++	–	–
*Eco*ICRI	+	++	nd	–				
*Eco*NI	++	++	++	++	nd	++	nd	nd
*Eco*O109I	++	++	–	–	++	++	+	–
*Eco*RI	++	++	++	–	++	++	++	–
*Eco*RII	+	++	++	++	+	+	++	++
*Eco*RV	–	+	++	++	–	+	++	++
*Eco*T22I	–	+	++	++	–	–	++	++
*Ehe*I	++	++	–	–	++	++	–	–
*Esp*I	–	++	++	nd				
*Fdi*II	++	++	–	nd				
*Fnu*4HI	–	+	+	–	nd	++	nd	nd
*Fnu*DII	++	–	–	–				
*Fok*I	++	++	++	++	++	++	++	++
*Fse*I	++	+	–	nd	nd	++	nd	nd
*Fsp*I	–	++	+	+	nd	++	nd	nd
*Hae*II	++	++	++	–	++	++	++	–
*Hae*III	+	++	++	++	+	++	++	++
*Hap*II	++	++	–	nd				
*Hga*I	++	++	+	–	nd	++	nd	nd
*Hgi*AI	–	–	+	++				
*Hha*I	–	+	++	++	–	+	++	++
*Hin*1I	++	++	+	–	+	+	+	–

Table 6. Effects of salt (continued)

Enzyme	Effects of NaCl[b] (mM)				Effects of KCl[b] (mM)			
	0	50	100	150	0	50	100	150
*Hinc*II	+	++	++	++	–	+	++	++
*Hind*II	+	++	++	nd				
*Hind*III	–	++	++	–	–	++	++	–
*Hinf*I	–	+	++	++	–	+	++	++
*Hin*P1I	++	++	++	nd	nd	++	nd	nd
*Hpa*I	++	++	+	–	++	++	++	–
*Hpa*II	++	++	+	–	++	++	+	+
*Hph*I	++	++	++	–	++	++	++	–
*Hsp*92I	–	++	++	+				
*Hsp*92II	+	+	nd	–	nd	nd	nd	++
*Ita*I	–	–	++	nd				
*Kas*I	++	++	++	nd	nd	++	nd	nd
*Kpn*I	++	+	–	–	++	++	–	–
*Ksp*I	++	–	–	nd				
*Ksp*632I	+	+	–	nd	nd	++	nd	nd
*Mae*I	–	–	++	nd				
*Mae*II	–	+	++	nd				
*Mae*III	–	–	++	nd				
*Mam*I	++	++	++	nd				
*Mbo*I	+	++	++	++	nd	++	++	++
*Mbo*II	++	++	++	+	++	++	++	+
*Mfe*I	++	+	–	nd	nd	++	nd	nd
*Mfl*I	++	++	–	nd				
*Mlu*I	–	–	++	++	–	–	++	++
*Mlu*NI	+	+	–	nd	nd	++	++	++
*Mnl*I	++	++	+	+	nd	++	++	++
*Mro*I	++	++	+	–	++	++	+	–
*Msc*I	+	++	++	+	++	++	++	+
*Mse*I	++	++	++	–	nd	++	nd	nd
*Msl*I	++	++	+	nd	nd	++	nd	nd
*Msp*I	++	++	++	–	++	++	++	–
*Msp*A1I	+	++	+	nd	nd	++	nd	nd
*Mst*I		++	++	++				
*Mst*II	–	+	++	++				
*Mun*I	++	++	–	nd	nd	++	++	–
*Mva*I	–	+	++	++	–	+	++	++
*Mvn*I	++	++	+	nd				
*Mwo*I	–	++	++	++	nd	++	nd	nd
*Nae*I	++	++	+	–	++	++	+	–
*Nar*I	++	+	–	–	++	+	–	–
*Nci*I	++	++	+	–	++	++	+	–
*Nco*I	+	++	++	++	+	++	++	++
*Nde*I	++	++	++	++	nd	++	++	++
*Nde*II	–	–	++	nd				
*Ngo*MI	++	+	–	nd	nd	++	nd	nd
*Nhe*I	++	++	+	–	++	++	+	–

Table 6. Effects of salt (continued)

Enzyme	**Effects of NaCl[b] (mM)**				**Effects of KCl[b] (mM)**			
	0	**50**	**100**	**150**	**0**	**50**	**100**	**150**
*Nla*III	+	+	+	nd	nd	++	nd	nd
*Nla*IV	–	–	–	nd	nd	++	nd	nd
*Not*I	–	+	++	++	–	+	++	++
*Nru*I	–	–	++	++	–	–	++	++
*Nsi*I	++	+	++	++	nd	++	++	++
*Nsp*I	++	–	–	nd				
*Nsp*V	++	++	++	–	++	++	+	–
*Pac*I	++	++	–	nd	nd	++	nd	nd
*Pae*R7I	+	++	–	–	nd	++	nd	nd
*Pal*I					nd	++	++	++
*Pfl*MI	–	++	++	+	nd	+	nd	nd
*Pin*AI	++	++	–	nd				
*Ple*I	++	++	++	–	nd	++	nd	nd
*Pma*CI	++	++	–	nd				
*Pme*I	–	+	–	nd	nd	++	nd	nd
*Pml*I	++	++	–	nd	nd	++	nd	nd
*Ppu*10I	–	+	–	nd	nd	++	nd	nd
*Ppu*MI	–	+	–	–	nd	++	nd	nd
*Psh*AI	+	++	–	nd	nd	++	nd	nd
*Psp*1406I	+	+	+	nd	nd	++	nd	nd
*Psp*AI					nd	+	++	+
*Pst*I	+	+	++	++	+	+	++	++
*Pvu*I	–	+	+	+	–	+	++	++
*Pvu*II	++	++	++	++	++	++	++	++
*Rca*I	+	++	+	nd				
*Rsa*I	++	++	+	–	++	++	+	–
*Rsr*I	+	++	–	nd	nd	++	nd	nd
*Rsr*II	++	+	–	–				
*Sac*I	++	+	+	–	++	+	+	–
*Sac*II	++	++	+	–	++	++	+	–
*Sal*I	–	–	+	++	–	–	+	++
*San*DI					nd	–	+	++
*Sap*I	++	+	–	nd	nd	++	nd	nd
*Sau*3AI	+	++	++	++	+	++	++	++
*Sau*96I	++	++	++	++	++	++	++	++
*Sca*I	–	++	++	+	–	++	++	+
*Scr*FI	++	++	++	++	nd	++	nd	nd
*Sex*AI	++	++	++	nd				
*Sfa*NI	–	++	++	++	nd	+	nd	nd
*Sfc*I	++	+	–	nd	nd	++	nd	nd
*Sfi*I	++	++	++	–	++	++	++	–
*Sfu*I	+	+	++	nd				
*Sgr*AI	++	+	–	nd	nd	++	nd	nd
*Sin*I	++	–	–	–				
*Sma*I	++	+	–	–	++	++	–	–
*Sna*BI	+	++	+	–	nd	++	++	+

Table 6. Effects of salt (continued)

Enzyme	**Effects of NaCl[b] (mM)**				**Effects of KCl[b] (mM)**			
	0	**50**	**100**	**150**	**0**	**50**	**100**	**150**
*Spe*I	+	++	++	+	+	++	++	+
*Sph*I	–	+	++	++	–	+	++	++
*Spl*I	–	+	++	++				
*Spo*I	–	++	++	+	–	++	++	+
*Srf*I					nd	++	++	+
*Sse*8387I	++	++	–	nd	nd	nd	++	nd
*Ssp*I	–	+	++	–	–	++	++	–
*Ssp*BI	+	++	++	nd				
*Sst*I	++	++	–	–				
*Stu*I	++	++	++	++	nd	++	++	++
*Sty*I	++	++	++	++	nd	++	++	+
*Swa*I	–	–	++	nd				
*Tai*I	+	++	++	nd	nd	++	++	nd
*Taq*I	–	++	++	++	–	++	++	++
*Tfi*I	++	++	++	nd	nd	++	nd	nd
*Tha*I	++	++	–	–				
*Tru*9I	++	++	++	+				
*Tse*I	++	++	++	nd	nd	++	nd	nd
*Tsp*45I	++	+	–	nd	nd	++	nd	nd
*Tsp*509I	++	++	++	nd	nd	–	nd	nd
*Tsp*RI	+	+	+	nd	nd	++	nd	nd
*Tth*111I	++	+	+	–	nd	++	nd	nd
*Tth*HB8I	++	++	++	nd				
*Van*91I	–	+	–	nd				
*Vsp*I	–	+	nd	++	nd	–	–	+
*Xba*I	–	++	++	+	–	++	++	+
*Xcm*I	–	++	+	nd	nd	+	nd	nd
*Xho*I	+	++	++	++	+	++	++	++
*Xho*II	+	++	nd	+	nd	++	++	+
*Xma*I	++	++	–	–	nd	++	nd	nd
*Xma*III	++	++	++	–				
*Xma*CI	++	++	–	–	nd	++	++	+
*Xmn*I	++	++	++	–	nd	++	++	++
*Xor*II	++	–	–	–				

[a]Data taken from various sources: personal experience, commercial catalogues, original descriptions of the enzymes.

[b]Symbols: ++, enzyme displays 50–100% of maximal activity with this salt concentration; +, enzyme displays 10–50% of maximal activity with this salt concentration; –, enzyme displays <10% of maximal activity with this salt concentration. The entries for an individual enzyme relate to a single 'maximal activity' for that enzyme, so the activities described for the two salts are directly comparable. The absence of an entry or the symbol 'nd' indicates that no data are available.

Table 7. Star activities of restriction enzymes[a, b]

Enzyme	Recognition sequence[c]		Conditions inducing star activity[d]
	Standard	Under star conditions	
*Apo*I	R↑AATTY		A B C D
*Ase*I	AT↑TAAT		A B C D
*Ava*I	C↑YCGRG		A B
*Ava*II	G↑GWCC		B E
*Bam*HI	G↑GATCC	G↑GNTCC G↑GANCC G↑RATCC	A B C F
*Ban*II	GRGCY↑C		B C E
*Bgl*II	A↑GATCT		E
*Bst*I	G↑GATCC	N↑GATCN	A B
*Bst*1107I	GTA↑TAC		C
*Bst*PI	G↑GTNACC		B C
*Dde*I	C↑TNAG		D E
*Eam*1105I	GACNNN↑NNGTC		B C D
*Eco*RI	G↑AATTC	N↑AATTN	A B C D F
*Eco*RV	GAT↑ATC	RAT↑ATC GNT↑ATC GAN↑ATC GAT↑NTC GAT↑ANC GAT↑ATY	E
*Eco*T22I	ATGCA↑T		C E
*Fba*I	T↑GATCA		B C D E
*Fsp*I	TGC↑GCA		B C
*Hae*III	GG↑CC		A B
*Hha*I	GCG↑C		A B E
*Hinc*II	GTY↑RAC		E
*Hind*III	A↑AGCTT	R↑AGCTT A↑NGCTT A↑AKCTT A↑AGCNT	E F
*Hpa*I	GTT↑AAC		A B
*Kpn*I	GGTAC↑C		A B D
*Mun*I	C↑AATTG		B C

Table 7. Star activities (continued)

Enzyme	Recognition sequence[c]		Conditions inducing star activity[d]
	Standard	Under star conditions	
*Nco*I	C↑CATGG		B E
*Nhe*I	G↑CTAGC		B C D E
*Pae*R7I	C↑TCGAG		B C
*Pst*I	CTGCA↑G		A B E
*Pvu*II	CAG↑CTG	CCG↑CTG CAT↑CTC CAG↑ATG CAG↑GTG CAG↑CGG	A B C D E
*Sac*I	GAGCT↑C		B E
*Sal*I	G↑TCGAC		A B E
*Sau*3AI	↑GATC	↑GAGC ↑CATC	A B
*Sca*I	AGT↑ACT		A
*Spe*I	A↑CTAGT		C E
*Sse*8387I	CCTGCA↑GG		C E
*Ssp*I	AAT↑ATT		B C D E
*Sst*I	GAGCT↑C		A B D
*Sst*II	CCGC↑GG		A B
*Tth*111I	GACN↑NNGTC	NACN↑NNGTC GACN↑NNNTC GACN↑NNGNC	C D F
*Tth*HB8I	T↑CGA		C D E
*Xba*I	T↑CTAGA		A B C
*Xmn*I	GAANN↑NNTTC		A B C D

[a]Information from various sources, in particular refs 5 and 6.
[b]Some additional enzymes may exhibit star properties but the activities have not been well characterized: *Apy*I, *Hin*fI, *Taq*I.
[c]For a description of the convention used in describing recognition sequences see *Table 1*, footnote b. For abbreviations see *Table 1*, footnote c.
[d]Key: A, high enzyme concentration (usually >100 units per µg DNA but can be less: 40 units μg^{-1} for *Pvu*II, 30 units μg^{-1} for *Sca*I); B, high glycerol content (>5%); C, low ionic strength (<25 mM); D, high pH (>8.0); E, presence of ethanol, dimethyl sulphoxide (DMSO), dimethylformamide (DMF) or some other organic solvents; F, replacement of Mg^{2+} with Mn^{2+}, Cu^{2+}, Zn^{2+} or Co^{2+}.

Table 8. Activities of restriction enzymes with single-stranded DNA[a, b]

Enzyme	Activity with single-stranded DNA[c]
*Bst*NI	<1%
*Dde*I	<1%
*Hae*III	10%
*Hga*I	<1%
*Hha*I	50%
*Hin*fI	<1%
*Hin*P1I	50%
*Mnl*I	50%
*Taq*I	<1%

[a]Taken from ref. 5.
[b]If an enzyme is not listed then presume it has no activity with single-stranded DNA.
[c]Expressed as the percentage of the enzyme's activity with double-stranded DNA.

3. EFFECTS OF DNA METHYLATION ON RESTRICTION

(Data kindly provided by Michael McClelland, California Institute of Biological Research, 11099 North Torrey Pines Road, La Jolla, CA 92037, USA. Updated by the Editor from ref. 7.)

Most restriction enzymes will not cut their recognition sequence if this region has been methylated. The methylation sensitivities of commonly used enzymes are shown in *Table 9*. This information, in conjunction with knowledge of the modification specificities of DNA methyltransferases (*Table 10*) and the differing methylation sensitivities of pairs of restriction enzymes recognizing identical sequences (*Table 11*), enables the restriction pattern of certain enzymes to be changed in predetermined ways (7). The methylation sensitivities of restriction enzymes also means that naturally methylated DNA may not be completely cut. This problem has been recognized for some time with reference to the Dam and Dcm methylases of *E. coli* and a *dam dcm* host should be used if cloned DNA is to be restricted with a sensitive enzyme (see Chapter 1, Section 1.3). In addition, both plant and mammalian DNAs are often methylated at specific positions, preventing complete restriction of genomic DNA preparations. *Table 12* shows the sensitivities of commonly used restriction enzymes to methylation by Dam, Dcm and the plant and mammalian methylases.

Table 9. Methylation sensitivity of restriction endonucleases[a]

Enzyme	Recognition sequence[b]	Sequences cut	Sequences not cut	Notes[c]
*Aat*I	AGGCCT	?	$AGG^{m5}CCT$ $AGGC^{m5}CT$ $AGGC^{m4}CT$	
*Aat*II	GACGTC	?	$GACGT^{m5}C$ $GA^{m5}CGTC$	
*Acc*I	GTMKAC	?	$GTMK^{m6}AC$ $GTMKA^{m5}C$	1
*Acc*II	CGCG	?	$^{m5}CGCG$	
*Acc*III	TCCGGA	$T^{m5}CCGGA$ $TC^{m5}CGGA$	$TCCGG^{m6}A$	2

Table 9. Methylation sensitivities (continued)

Enzyme	Recognition sequence[b]	Sequences cut	Sequences not cut	Notes[c]
*Acc*65I	GGTACC	?	GGTAC^{m5}C	
*Aci*I	CCGC	?	C^{m5}CGC	
*Afl*I	GGWCC	GGWC^{m5}C GGWC^{m4}C	?	3
*Afl*II	CTTAAG	?	^{m5}CTTAAG CTTA^{m6}AG	
*Afl*III	ACRYGT	?	A^{m5}CRYGT	
*Age*I	ACCGGT	?	A^{m5}CCGGT AC^{m5}CGGT	
*Aha*II	GRCGYC	?	GR^{m5}CGYC GRCGY^{m5}C	4
*Alu*I	AGCT	?	^{m6}AGCT AG^{m4}CT AG^{m5}CT AG^{hm5}CT	
*Alw*I	GGATC	?	GG^{m6}ATC GGAT^{m4}C	
*Alw*26I	GTCTC	?	GT^{m5}CTC GAG^{m6}AC	
*Alw*44I	GTGCAC	GTGC^{m6}AC	GTG^{m5}CAC	
*Alw*NI	CAGNNNCTG	?	CAGNNC^{m5}CTG	
*Ama*I	TCGCGA	TCGCG^{m6}A	?	
*Aos*II	GRCGYC	?	GR^{m5}CGYC	
*Apa*I	GGGCCC	?	GGG^{m5}CCC GGGCC^{m5}C	
*Apa*LI	GTGCAC	GTGC^{m6}AC GTG^{m5}CAC	GTGCA^{m5}C	
*Apy*I	CCWGG	C^{m5}CWGG	^{m5}CCWGG	
*Aqu*I	CYCGRG	?	^{m5}CYCGRG	
*Asc*I	GGCGCGCC	?	GG^{m5}CGCGCC GGCG^{m5}CGCC GGCGCG^{m5}CC GGCGCGC^{m5}C	
*Asp*700I	GAANNNNTTC	GA^{m6}ANNNNTTC GAANNNNTT^{m5}C	G^{m6}AANNNNTTC	
*Asp*718I	GGTACC	GGT^{m6}A^{m5}CC GGTA^{m5}CC	GGTAC^{m5}C GGTA^{m5}C^{m5}C	5
*Asp*MDI	GATC	G^{m6}ATC	?	
*Asu*I	GGNCC	GGNC^{m5}C	?	

Restriction/Methylation

Table 9. Methylation sensitivities (continued)

Enzyme	Recognition sequence[b]	Sequences cut	Sequences not cut	Notes[c]
*Asu*II	TTCGAA	TT^{m5}CGAA	?	
*Atu*CI	TGATCA	?	TG^{m6}ATCA	
*Ava*I	CYCGRG	C^{m5}CCGGG	^{m5}CYCGRG CY^{m5}CGRG CTCG^{m6}AG	6
*Ava*II	GGWCC	GGWC^{m4}C	GGW^{m5}CC GGWC^{m5}C GGW^{hm5}C^{hm5}C	7
*Avi*II	TGCGCA	?	TG^{m5}CGCA	
*Bae*I	ACNNNNGTAYC	?	ACNNNNGTYA^{m5}C	
*Bal*I	TGGCCA	?	TGG^{m5}CCA TGGC^{m5}CA	8
*Bam*FI	GGATCC	GG^{m6}ATCC	GGAT^{m4}CC	
*Bam*HI	GGATCC	GGATC^{m5}C GG^{m6}ATCC GG^{m6}ATC^{m5}C GGATC^{m4}C	GGAT^{m4}CC GGAT^{m5}CC GGAT^{hm5}C^{hm5}C GGA^{hm5}UCC	
*Bam*KI	GGATCC	GG^{m6}ATCC	GGAT^{m4}CC	
*Ban*I	GGYRCC	GG^{m5}CGCC GGYRC^{m4}C	?	9
*Ban*II	GRGCYC	GRGCY^{m5}C	GRG^{m5}CYC	
*Ban*III	ATCGAT	?	ATCG^{m6}AT	
*Bbe*I	GGCGCC	GGCG^{m5}CC GGCGC^{m5}C	GG^{m5}CGCC	
*Bbi*II	GRCGYC	?	GR^{m5}CGYC	
*Bbr*PI	CACGTG	?	CA^{m5}CGTG	
*Bbs*I	GAAGAC	GAAGA^{m5}C	?	
*Bbu*I	GCATGC	GCATG^{m5}C	GC^{m6}ATGC	
*Bbv*I	GCAGC	?	G^{m5}CAGC	
*Bca*77I	WCCGGW	WC^{m5}CGGW	W^{m5}CCGGW	
*Bcl*I	TGATCA	TGAT^{m5}CA	TG^{m6}ATCA TGAT^{hm5}CA	
*Bcn*I	CCSGG	^{m5}CCSGG	C^{m4}CSGG C^{m5}CSGG	
*Bep*I	CGCG	?	^{m5}CGCG	
*Bfr*I	CTTAAG	?	^{m5}CTTAAG	
*Bgl*I	GCCNNNNNGGC	GC^{m5}CNNNNNGGC	G^{m5}CCNNNNNGGC GCCNNNNNGG^{m5}C GC^{m4}CNNNNNGGC	10

Table 9. Methylation sensitivities (continued)

Enzyme	Recognition sequence[b]	Sequences cut	Sequences not cut	Notes[c]
*Bgl*II	AGATCT	AG^{m6}ATCT AGA^{hm5}UC^{hm5}U	AGAT^{m5}CT AGAT^{hm5}CT	
*Bin*I	GGATC	?	GG^{m6}ATC	
*Bma*DI	CGATCG	CG^{m6}ATCG	CGAT^{m6}CG	
*Bme*216I	GGWCC	?	GGWC^{m5}C	
*Bna*I	GGATCC	GG^{m6}ATCC	GGAT^{m4}CC GGAT^{m5}CC	
*Bsa*I	GGTCTC	GAGA^{m5}C^{m5}C	GGTCT^{m5}C	
*Bsa*AI	YACGTR	?	YA^{m5}CGTR	
*Bsa*BI	GATNNNNATC	?	GATNNNNAT^{m5}C G^{m6}ATNNN^{m6}ATC	
*Bsa*HI	GRCGYC	?	GR^{m5}CGYC	
*Bsa*JI	CCNNGG	C^{m5}CNNGG	?	
*Bsa*WI	WCCGGW	WC^{m5}GCCW	?	
*Bse*CI	ATCGAT	?	ATCG^{m6}AT	
*Bsg*I	CTGCAC	?	CTGCA^{m5}C	
*Bsh*1365I	GATNNNNATC	?	G^{m6}ATNNNNATC GATNNNN^{m6}ATC	
*Bsi*BI	GATNNNNATC	?	G^{m6}ATNNNNATC GATNNNN^{m6}ATC	
*Bsi*EI	CGRYCG	?	C^{m5}RY^{m5}CG	
*Bsi*LI	CCWGG	?	C^{m5}CWGG	
*Bsi*MI	TCCGGA	?	TCCGG^{m6}A	
*Bsi*WI	CGTACG	?	^{m5}CGTA^{m5}CG	
*Bsi*XI	ATCGAT	?	^{m6}ATCGAT ATCG^{m6}AT	
*Bsl*I	CCNNNNNNNGG	?	C^{m5}CNNNNNNNGG	
*Bsm*I	GAATGC	GAATG^{m5}C	G^{m6}AATGC	
*Bsm*AI	GTCTC	G^{m6}AGAC	GTCT^{m5}C	
*Bsp*106I	ATCGAT	?	ATCG^{m5}AT	
*Bsp*1286I	GDGCHC	GDGCH^{m5}C	GDG^{m5}CHC	
*Bsp*143I	GATC	?	G^{m6}ATC	
*Bsp*DI	ATCGAT	?	AT^{m5}CGAT ^{m6}ATCG^{m6}AT	
*Bsp*EI	TCCGGA	TC^{m5}CGGA	TCCGG^{m6}A	11
*Bsp*HI	TCATGA	?	TC^{m6}ATGA TCATG^{m6}A	
*Bsp*MI	ACCTGC	ACCTG^{m5}C	?	

Table 9. Methylation sensitivities (continued)

Enzyme	Recognition sequence[b]	Sequences cut	Sequences not cut	Notes[c]
*Bsp*MII	TCCGGA	TCCGG^{m6}A	T^{m5}CCGGA TC^{m5}CGGA	
*Bsp*XI	ATCGAT	?	ATCG^{m6}AT	
*Bsp*XII	TGATCA	?	TG^{m6}ATCA	
*Bsr*BI	GAGCGG	GAG^{m5}CGG	?	12
*Bsr*FI	RCCGGY	?	RC^{m5}CGGY	
*Bss*HII	GCGCGC	?	G^{m5}CGCGC	13
*Bst*I	GGATCC	GG^{m6}ATCC GGATC^{m5}C	GGAT^{m4}CC GGAT^{m5}CC GGATC^{m4}C	14
*Bst*1107I	GTATAC	?	GTATA^{m5}C	
*Bst*BI	TTCGAA	?	TTCG^{m6}AA TT^{m5}CGAA	
*Bst*EII	GGTNACC	GGTNA^{m5}C^{m5}C	GGTNA^{hm5}C^{hm5}C GGTNAC^{m4}C	15
*Bst*EIII	GATC	?	G^{m6}ATC	
*Bst*GI	TGATCA	?	TG^{m6}ATCA	
*Bst*NI	CCWGG	^{m5}CCWGG C^{m5}CWGG ^{m5}C^{m5}CWGG	^{hm5}C^{hm5}CWGG C^{m4}CWGG	16
*Bst*OI	CCWGG	C^{m5}CWGG	?	
*Bst*UI	CGCG	?	^{m5}CGCG	
*Bst*XI	CCANNNNNNTGG	C^{m5}CANNNNNNTGG	^{m5}CCANNNNNNTGG	
*Bst*YI	RGATCY	RG^{m6}ATCY	RGAT^{m4}CY RGAT^{m5}CY	
*Bsu*15I	ATCGAT	?	ATCG^{m6}AT	
*Bsu*36I	CCTNAGG	CCTN^{m6}AGG	^{m5}CCTNAGG	
*Bsu*BI	CTGCAG	?	CTGC^{m6}AG	
*Bsu*EII	CGCG	?	^{m5}CGCG	
*Bsu*FI	CCGG	?	^{m5}CCGG	
*Bsu*MI	CTCGAG	?	CT^{m5}CGAG	
*Bsu*RI	GGCC	?	GG^{m5}CC	17
*Cbi*I	TTCGAA	TTCG^{m6}AA	?	
*Ccr*I	CTCGAG	?	CTCG^{m6}AG	
*Cfo*I	GCGC	?	G^{m5}CGC GCG^{m5}C G^{hm5}CG^{hm5}C	
*Cfr*I	YGGCCR	?	YGG^{m5}CCR	
*Cfr*6I	CAGCTG	?	CAG^{m4}CTG CAG^{m5}CTG	

Table 9. Methylation sensitivities (continued)

Enzyme	Recognition sequence[b]	Sequences cut	Sequences not cut	Notes[c]
*Cfr*9I	CCCGGG	$C^{m5}CCGGG$ $CC^{m5}CGGG$	$^{m4}CCCGGG$ $^{m5}CCCGGG$ $C^{m4}CCGGG$ $CC^{m4}CGGG$	
*Cfr*10I	RCCGGY	?	$R^{m5}CCGGY$ $RC^{m5}CGGY$	
*Cfr*13I	GGNCC	?	$GGN^{m5}CC$	
*Cfr*AI	GCANNNNNNNNGTGG	?	$GC^{m6}ANNNNNNNNGTGG$	
*Cfr*BI	CCWWGG	?	$^{m4}CCWWGG$	
*Cla*I	ATCGAT	?	$^{m6}ATCGAT$ $AT^{m5}CGAT$ $ATCG^{m6}AT$	18
*Cpe*I	TGATCA	?	$TG^{m6}ATCA$	
*Csp*I	CGGWCCG	$CGGWC^{m5}CG$	$CGGW^{m5}CCG$ $^{m5}CGGWCCG$	
*Csp*45I	TTCGAA	?	$TTCG^{m6}AA$	
*Cty*I	GATC		$G^{m6}ATC$	
*Cvi*AI	GATC	$GAT^{m5}C$	$G^{m6}ATC$	
*Cvi*AII	CATG	$^{m5}CATG$	$C^{m6}ATG$	
*Cvi*BI	GANTC	?	$G^{m6}ANTC$	
*Cvi*JI	RGCY	?	$RG^{m5}CY$	
*Cvi*QI	GTAC	$GTA^{m5}C$	$GT^{m6}AC$	
*Cvi*RI	TGCA	?	$TGC^{m6}A$ $TG^{m5}CA$	
*Cvi*RII	GTAC	?	$GT^{m6}AC$	
*Dde*I	CTNAG	?	$^{m5}CTNAG$ $^{hm5}CTNAG$	
*Dpn*I	$G^{m6}ATC$	$G^{m6}ATC$ $G^{m6}AT^{m5}C$ $G^{m6}AT^{m4}C$	GATC $GAT^{m4}C$ $GAT^{m5}C$	19
*Dpn*II	GATC	?	$G^{m6}ATC$	
*Dra*I	TTTAAA	$TTTA^{m6}AA$	?	
*Dra*II	RGGNCCY	?	$RGGNC^{m5}CY$	
*Drd*I	GACNNNNNNGTC	?	$GA^{m5}CNNNNNNGT^{m5}C$	
*Dsa*V	CCNGG	?	$C^{m5}CNGG$	
*Eae*I	YGGCCR	?	$YGG^{m5}CCR$ $YGGC^{m5}CR$	

Restriction/Methylation

Table 9. Methylation sensitivities (continued)

Enzyme	Recognition sequence[b]	Sequences cut	Sequences not cut	Notes[c]
*Eag*I	CGGCCG	?	CGG^{m5}CCG ^{m5}CGGC^{m5}CG	
*Eam*1105I	GACNNNNNGTC	GA^{m5}CNNNNNGT^{m5}C	?	
*Ear*I	GAAGAG	CTCTT^{m5}C	G^{m6}AAGAG GAAG^{m6}AG ^{m5}CT^{m5}CTT^{m5}C	
*Eca*I	GGTNACC	?	GGTN^{m6}ACC	
*Ecl*136II	GAGCTC	?	GAGCT^{m5}C	
*Ecl*XI	CGGCCG	?	^{m5}CGGC^{m5}CG CGG^{m5}CCG	
*Eco*31I	GGTCTC	?	GGT^{m5}CTC G^{m6}AGACC	
*Eco*47I	GGWCC	?	GGWC^{m5}C	
*Eco*47II	GGNCC	?	GGNC^{m5}C	
*Eco*47III	AGCGCT	^{m6}AGCGCT	AG^{m5}CGCT	
*Eco*57I	CTGAAG	?	CTGA^{m6}AG CTTC^{m6}AG	
*Eco*AI	GAGNNNNNNNGTCA	?	G^{m6}AGNNNNNNNG^{m}TCA	20
*Eco*BI	TGANNNNNNNNTGCT	?	TG^{m6}ANNNNNNNN^{m}TGCT	20
*Eco*DI	TTANNNNNNNGTCY	?	TT^{m6}ANNNNNNNGTCY	
*Eco*DXXI	TCANNNNNNNATC	?	TCANNNNNNN^{m6}AA^{m}TC	20
*Eco*EI	GAGNNNNNNNATGC	?	G^{m6}AGNNNNNNNATGC	
*Eco*HI	CCSGG	?	C^{m5}CSGG	
*Eco*KI	AACNNNNNNGTGC	?	A^{m6}ACNNNNNNG^{m}TGC	20
*Eco*O109I	RGGNCCY	?	RGGNC^{m5}CY	
*Eco*PI	AGACC	AGA^{hm5}C^{hm5}C	AG^{m6}ACC	21
*Eco*P15I	CAGCAG	?	CAGC^{m6}AG	21
*Eco*RI	GAATTC	GAATT^{hm5}C GAA^{hm5}U^{hm5}UC	G^{m6}AATTC GA^{m6}ATTC GAATT^{m5}C	22
*Eco*RII	CCWGG	^{m5}CCWGG	^{m4}CCWGG C^{m4}CWGG C^{m5}CWGG CC^{m6}AGG ^{hm5}C^{hm5}CWGG	23
*Eco*RV	GATATC	GATAT^{m5}C GATAT^{hm5}C	G^{m6}ATATC GAT^{m6}ATC	24
*Eco*T22I	ATGCAT	?	ATG^{m6}CAT ATGC^{m6}AT	

Table 9. Methylation sensitivities (continued)

Enzyme	Recognition sequence[b]	Sequences cut	Sequences not cut	Notes[c]
*Ehe*I	GGCGCC	?	GG^{m5}CGCC GG^{m5}CGCC GGCG^{m5}CC	
*Esp*I	GCTNAGC	GCTNAG^{m5}C	G^{m5}CTNAGC	
*Esp*3I	CGTCTC	?	^{m5}CGTCTC CGT^{m5}CTC	
*Esp*1396I	CCANNNNNTGG	C^{m5}CANNNNNTGG	?	
*Fnu*4HI	GCNGC	?	G^{m5}CNGC GCNG^{m5}C	
*Fnu*DII	CGCG	?	^{m5}CGCG CG^{m5}CG	
*Fnu*EI	GATC	G^{m6}ATC	?	
*Fok*I	CATCC	CAT^{m5}CC CATC^{m5}C	GG^{m6}ATG C^{m6}ATCC CATC^{m4}C	25
*Fse*I	GGCCGGCC	?	GG^{m5}CCGG^{m5}CC GGC^{m5}CGGCC GG^{m5}CCGGCC	
*Fsp*I	TGCGCA	?	TG^{m5}CGCA	
*Hae*II	RGCGCY	?	RG^{m5}CGCY RGCG^{m5}CY RG^{hm5}CG^{hm5}CY	26
*Hae*III	GGCC	GGC^{m5}C	GG^{m5}CC GG^{hm5}C^{hm5}C	27
*Hap*II	CCGG	?	C^{m5}CGG	
*Hga*I	GACGC	?	GA^{m5}CGC G^{m5}CGTC GACG^{m5}C	
*Hgi*AI	GWGCWC	GWGCW^{m5}C	GWG^{m5}CWC	
*Hgi*BI	GGWCC	?	GGWCC (^{m5}C)	
*Hgi*CI	GGYRCC	?	GGYRC^{m5}C	
*Hgi*CII	GGWCC	?	GGWC^{m5}C	
*Hgi*DI	GRCGYC	?	GRCGYC	
*Hgi*EI	GGWCC	?	GGWC^{m5}C	
*Hgi*GI	GRCGYC	?	GR^{m5}CGY^{m5}C	
*Hha*I	GCGC	?	G^{m5}CGC GCG^{m5}C G^{hm5}CG^{hm5}C	
*Hha*II	GANTC	?	G^{m6}ANTC	

Restriction/Methylation

Table 9. Methylation sensitivities (continued)

Enzyme	Recognition sequence[b]	Sequences cut	Sequences not cut	Notes[c]
*Hinc*II	GTYRAC	GT^{m5}CRAC	GTYRA^{m5}C GTYR^{m6}AC GTYRA^{hm5}C	28
*Hind*II	GTYRAC	?	GTYR^{m6}AC GTYRA^{hm5}C	29
*Hind*III	AAGCTT	A^{m6}AGCTT AAGC^{hm5}U^{hm5}U	^{m6}AAGCTT AAG^{m5}CTT AAG^{hm5}CTT	
*Hin*fI	GANTC	GANT^{m5}C	G^{m6}ANTC GANT^{hm5}C	30
*Hin*P1I	GCGC	?	G^{m5}CGC	
*Hpa*I	GTTAAC	GTTAA^{m5}C	GTTA^{m6}AC GTTAA^{hm5}C G^{hm5}U^{hm5}UAAC	
*Hpa*II	CCGG	?	^{m4}CCGG ^{m5}CCGG C^{m4}CGG C^{m5}CGG ^{hm5}C^{hm5}CGG	31
*Hph*I	TCACC	TCAC^{m5}C	T^{m5}CACC TCA^{m5}CC GGTG^{m6}A	
*Kas*I	GGCGCC	?	GG^{m5}CGCC	
*Kpn*I	GGTACC	GGTA^{m5}CC GGTAC^{m5}C	GGT^{m6}ACC GGT^{m4}ACC GGTA^{m5}C^{m5}C GGT^{m6}ACC	32
*Kpn*2I	TCCGGA	TCCGG^{m6}A	T^{m5}CCGGA TC^{m5}CGGA	
*Ksp*I	CCGCGG	?	^{m5}CCGCGG C^{m5}CGCGG	
*Mae*II	ACGT	?	A^{m5}CGT	33
*Mam*I	GATNNNNATC	?	G^{m6}ATNNNN^{m6}ATC	
*Mbo*I	GATC	GAT^{m4}C GAT^{m5}C	G^{m6}ATC GAT^{hm5}C GA^{hm5}UC	34
*Mbo*II	GAAGA	T^{m5}CTT^{m5}C G^{m6}AAGA	GAAG^{m6}A GA^{m6}AGA	35
*Mfl*I	RGATCY	?	RG^{m6}ATCY RGAT^{m4}CY RGAT^{m5}CY	36
*Mlu*I	ACGCGT	^{m6}ACGCGT	A^{m5}CGCGT ACG^{m5}CGT	

Table 9. Methylation sensitivities (continued)

Enzyme	Recognition sequence[b]	Sequences cut	Sequences not cut	Notes[c]
*Mme*II	GATC	?	G^{m6}ATC	
*Mnl*I	CCTC	?	^{m5}CCTC ^{m5}C^{m5}CT^{m5}C	
*Mph*I	CCWGG	?	C^{m5}CWGG	
*Mro*I	TCCGGA	TCCGG^{m6}A	T^{m5}CCGGA TC^{m5}CGGA	
*Msc*I	TGGCCA	?	TGGC^{m5}CA	
*Msp*I	CCGG	^{m4}CCGG C^{m4}CGG C^{m5}CGG	^{m5}CCGG ^{hm5}C^{hm5}CGG	37
*Mst*II	CCTNAGG	^{m5}CCTNAGG	CCTN^{m6}AGG	
*Mth*TI	GGCC	?	GG^{m5}CC	
*Mth*ZI	CTAG	?	^{m4}CTAG	
*Mun*I	CAATTG	?	CA^{m6}ATTG	
*Mva*I	CCWGG	C^{m5}CWGG ^{m5}CCWGG	C^{m4}CWGG CC^{m6}AGG ^{m4}CCWGG ^{m5}C^{m5}CWGG	38
*Mvn*I	CGCG	?	^{m5}CGCG	
*Nae*I	GCCGGC	?	G^{m5}CCGGC GC^{m5}CGGC GCCGG^{m5}C	
*Nan*II	G^{m6}ATC	G^{m6}ATC	GATC G^{m6}AT^{m5}C GAT^{m5}C	39
*Nar*I	GGCGCC	GGCGC^{m5}C	GG^{m5}CGCC GGCGC^{m4}C GG^{hm5}CG^{hm5}C^{hm5}C	
*Nci*I	CCSGG	^{m5}CCSGG	C^{m4}CSGG C^{m5}CSGG	40
*Nco*I	CCATGG	CC^{m6}ATGG	^{m4}CCATGG ^{m5}CCATGG	41
*Ncr*I	AGATCT	AG^{m6}ATCT	?	
*Ncu*I	GAAGA	?	GAAG^{m6}A	
*Nde*I	CATATG	^{m5}CATATG	CAT^{m6}ATG	42
*Nde*II	GATC	GAT^{m5}C	G^{m6}ATC	42
*Ngo*I	RGCGCY	?	RG^{m5}CGCY	
*Ngo*II	GGCC	?	GG^{m5}CC GGC^{m5}C	43

Table 9. Methylation sensitivities (continued)

Enzyme	Recognition sequence[b]	Sequences cut	Sequences not cut	Notes[c]
*Ngo*BI	TCACC	?	$T^{m5}CACC$	
*Nhe*I	GCTAGC	?	$GCTAG^{m5}C$	
*Nla*III	CATG	?	$C^{m6}ATG$ $^{m5}CATG$	
*Nla*IV	GGNNCC	?	$GGNN^{m4}CC$	
*Nmu*DI	$G^{m6}ATC$	$G^{m6}ATC$	GATC	44
*Nmu*EI	$G^{m6}ATC$	$G^{m6}ATC$	GATC	44
*Not*I	GCGGCCGC	$GCGGCCG^{m5}C$	$GCGG^{m5}CCGC$ $GCGGC^{m5}CGC$	
*Nru*I	TCGCGA	$TCG^{m5}CGA$	$T^{m5}CGCGA$ $TCGCG^{m6}A$	
*Nsi*I	ATGCAT	?	$ATGC^{m6}AT$ $ATG^{m5}CAT$	
*Nsp*I	RCATGY	?	$R^{m5}CATGY$ $RC^{m6}ATGY$	
*Nsp*BII	CMGCKG	$C^{m5}CGCKG$	?	
*Pae*R7I	CTCGAG	?	$CTCG^{m6}AG$ $CT^{m5}CGAG$	45
*Pfa*I	GATC	$G^{m6}ATC$	?	
*Pfl*MI	CCANNNNNTGG	?	$C^{m4}CANNNNNTGG$ $C^{m5}CANNNNNTGG$	
*Pfu*I	CGTACG	?	$CGTA^{m5}CG$	
*Pme*I	GTTTAAAC	$GTTAAA^{m5}C$	?	
*Pml*I	CACGTG	?	$CA^{m5}CGTG$	
*Ppu*AI	CGTACG	?	$CGTA^{m5}CG$	
*Ppu*MI	RGGWCCY	?	$RGGWC^{m5}CT$	
*Pst*I	CTGCAG	?	$^{m5}CTGCAG$ $C^{hm5}UGCAG$ $CTG^{m5}CAG$ $CTGC^{m6}AG$	46
*Pvu*I	CGATCG	$CG^{m6}ATCG$	$CGAT^{m4}CG$ $CGAT^{m5}CG$	
*Pvu*II	CAGCTG	?	$CAG^{m4}CTG$ $CAG^{m5}CTG$	47
*Rfl*FI	GTCGAC	?	$GTCG^{m6}AC$	
*Rfl*FII	AGTACT	?	$AGT^{m6}ACT$	
*Rrh*4273I	GTCGAC	?	$GTCG^{m6}AC$	
*Rsa*I	GTAC	$GTA^{m5}C$	$GT^{m6}AC$ $GTA^{m4}C$	48
*Rsh*I	CGATCG	$CG^{m6}ATCG$	?	

Table 9. Methylation sensitivities (continued)

Enzyme	Recognition sequence[b]	Sequences cut	Sequences not cut	Notes[c]
*Rsp*XI	TCATGA	?	TC^{m6}ATGA TCATG^{m6}A	
*Rsr*I	GAATTC	?	G^{m6}AATTC GA^{m6}ATTC	49
*Rsr*II	CGGWCCG	?	^{m5}CGGWCCG CGGW^{m5}CCG CGGWC^{m5}CG	
*Sac*I	GAGCTC	G^{m6}AGCTC GAGCT^{m5}C	GAG^{m5}CTC	
*Sac*II	CCGCGG	?	^{m5}CCGCGG C^{m5}CGCGG	
*Sal*I	GTCGAC	GTCGA^{m5}C	GT^{m5}CGAC GTCG^{m6}AC G^{hm5}UCGAC	50
*Sal*DI	TCGCGA	TCGCG^{m6}A	T^{m5}CGCGA	
*Sau*3AI	GATC	G^{m6}ATC GA^{hm5}UC	GAT^{m5}C GAT^{m4}C GAT^{hm5}C	51
*Sau*96I	GGNCC	?	GGN^{m5}CC GGNC^{m5}C GGN^{hm5}C^{hm5}C	
*Sau*3239I	CTCGAG	?	CTCG^{m6}AG	
*Sau*LPI	GCCGGC	?	G^{m5}CCGGC	
*Sbo*13I	TCGCGA	TCGCG^{m6}A	T^{m5}CGCGA	
*Sca*I	AGTACT	AGTA^{m5}CT	?	
*Scr*FI	CCNGG	^{m5}CCNGG	C^{m5}CNGG C^{m4}CNGG	
*Sfa*NI	GATGC	GATG^{m5}C	G^{m6}ATGC	
*Sfi*I	GGCCNNNNNGGCC	GG^{m5}CCNNNNNGG^{m5}CC	GGC^{m5}CNNNNNGGCC GGCCNNNNNGGC^{m5}C	
*Sfl*I	CTGCAG	?	CTGC^{m6}AG	
*Sfu*I	TTCGAA	TT^{m5}CGAA	TTCG^{m6}AA	
*Sgr*AI	CRCCGGYG	?	CRC^{m5}CGGYG	
*Sin*I	GGWCC	?	GGW^{m5}CC	
*Sma*I	CCCGGG	C^{m5}CCGGG	^{m4}CCCGGG ^{m5}CCCGGG C^{m4}CCGGG CC^{m4}CGGG CC^{m5}CGGG	52
*Sna*BI	TACGTA	?	TA^{m5}CGTA T^{m6}ACGT^{m6}A	
*Sno*I	GTGCAC	?	GTG^{m5}CA^{m5}C	

Table 9. Methylation sensitivities (continued)

Enzyme	Recognition sequence[b]	Sequences cut	Sequences not cut	Notes[c]
*Spe*I	ACTAGT	?	^{m6}ACTAGT A^{m5}CTAGT	53
*Sph*I	GCATGC	GCATG^{m5}C	GC^{m6}ATGC G^{hm5}CATG^{hm5}C	
*Spl*I	CGTACG	CGT^{m6}ACG	CGTA^{m5}CG CGTA^{m4}CG	54
*Spo*I	TCGCGA	TCGCG^{m6}A	T^{m5}CGCGA TCG^{m5}CGA	
*Srf*I	GCCCGGGC	?	G^{m5}CCCGGGC GC^{m5}CCGGGC GCC^{m5}CGGGC GCCCGGG^{m5}C	
*Sso*I	GAATTC	?	G^{m6}AATTC	
*Sso*II	CCNGG	?	C^{m5}CNGG ^{m5}CCNGG	
*Ssp*RFI	TTCGAA	?	TTCG^{m6}AA	
*Sst*I	GAGCTC	?	GAG^{m5}CTC GAG^{hm5}CT^{hm5}C	
*Sst*II	CCGCGG	?	^{m5}CCGCGG C^{m5}CGCGG	
*Sts*I	GGATG	?	GG^{m6}ATG C^{m6}ATCC	
*Stu*I	AGGCCT	?	AGG^{m5}CCT AGGC^{m5}CT AGGC^{m4}CT	
*Sty*LTI	CAGAG	?	CAG^{m6}AG	
*Sty*SJ	GAGNNNNNNGTRC	?	GAGNNNNNNG^{m}TRC	55
*Sty*SPI	AACNNNNNNGTRC	?	A^{m6}ACNNNNNNG^{m}TRC	55
*Taq*I	TCGA	T^{m5}CGA	TCG^{m6}A T^{hm5}CGA	56
*Taq*II	GACCGA CACCCA	?	G^{m6}ACCGA	
*Taq*XI	CCWGG	^{m5}CCWGG C^{m5}CWGG	?	
*Tfi*I	GAWTC	GAWT^{m5}C	?	
*Tfl*I	TCGA	?	TCG^{m6}A	
*Tha*I	CGCG	?	^{m5}CGCG ^{hm5}CG^{hm5}CG	
*Tsp*509I	AATT	?	^{m6}AATT	

Table 9. Methylation sensitivities (continued)

Enzyme	Recognition sequence[b]	Sequences cut	Sequences not cut	Notes[c]
*Tth*H111I	GACNNNGTC	GA^{m5}CNNNGTC GACNNNGT^{m5}C	?	
*Tth*HB8I	TCGA	T^{m5}CGA	TCG^{m6}A	
*Van*91I	CCANNNNNTGG	?	C^{m5}CANNNNNTGG	
*Xba*I	TCTAGA	?	TCTAG^{m6}A T^{m5}CTAGA T^{hm5}CTAGA	57
*Xcy*I	CCCGGG	?	C^{m4}CCCGGG	
*Xho*I	CTCGAG	? CTCG^{m6}AG	CT^{m5}CGAG ^{m5}CTCGAG	58
*Xho*II	RGATCY	RG^{m6}ATCY	RGAT^{m5}CY	59
*Xma*I	CCCGGG	CC^{m5}CGGG	^{m4}CCCGGG ^{m5}CCCGGG C^{m4}CCGGG CC^{m4}CGGG	60
*Xma*III	CGGCCG	?	CGG^{m5}CCG	
*Xmn*I	GAANNNNTTC	GA^{m6}ANNNNTTC	G^{m6}AANNNNTTC GAANNNNTT^{m5}C	61
*Xor*II	CGATCG	?	CG^{m6}ATCG CGAT^{m5}CG ^{hm5}CGAT^{hm5}CG	62

[a]For the primary sources of data see ref. 7.

[b]For a description of the convention used in describing recognition sequences see *Table 1*, footnote b. For abbreviations see *Table 1*, footnote c. Additional abbreviations: ^{hm5}C, hydroxymethylcytosine; ^{hm5}U, hydroxymethyuracil; ^{m}C, N^{4}- or C^{5}-methylcytosine; ^{m4}C, N^{4}-methylcytosine; ^{m5}C, C^{5}-methylcytosine; ^{m6}A, N^{6}-methyladenine.

[c]Additional notes:

1. *Acc*I nicking occurs slowly in the unmethylated strand of the hemimethylated sequence GTMKA^{m5}C.
2. *Acc*III cuts slowly at T^{m5}CCGGA and TC^{m5}CGGA.
3. *Afl*I cuts slowly at GGWC^{m4}C.
4. *Aha*II (GRCGYC) will cut GRCGCC faster if these sites are methylated at GRCG^{m5}CC, but will not cut GRCGY^{m5}C sites.
5. *Asp*718I cuts GT^{m6}AC- and ^{m5}CC-modified *Chlorella* virus NY2A DNA. *Asp*718I does not cut GGTAC^{m5}CWGG overlapping *dcm* sites or ^{m5}C-substituted phage XP12 DNA, whereas *Kpn*I cuts these sites readily.
6. *Ava*I nicking occurs slowly in the unmethylated strand of the hemimethylated sequence CTCG^{m6}AG/CTCGAG.
7. *Ava*II cuts slowly at GGWC^{m4}C.
8. *Bal*I sites overlapping *dcm* sites (TGGC^{m5}CAGG) are cut 50-fold slower than unmethylated sites.
9. *Ban*I gives various rate effects when its recognition sequence is ^{m4}C- or ^{m5}C-methylated at different positions.
10. The *Bgl*I cleavage rate at certain GC^{m5}CNNNNNGGC, GC^{m4}CNNNNNGGC, and GCCNNNNNGG^{m5}C hemimethylated sites is extremely slow. However, ^{m5}C bi-methylated M·*Hae*III–*Bgl*I sites are completely refractory to *Bgl*I.

Table 9. Methylation sensitivities (continued)

11. *Bsp*EI cleavage is slowed by TC^{m5}CGGA.
12. *Bsr*BI cleavage is slowed by GAG^{m5}CGG.
13. *Bss*HII does not cut M·*Hha*I-modified DNA, in which two different cytosine positions are hemimethylated, G^{m5}CGCGC/GCG^{m5}CGC.
14. M·*Bst*I modifies the internal cytosine GGAT^{m}CC, but it is not known whether this modification is ^{m5}C or ^{m4}C.
15. *Bst*EII cuts the fully ^{m5}C-substituted phage XP12 DNA.
16. *Bst*NI isoschizomers that are insensitive to C^{m5}CWGG include *Aor*I, *Apy*I, *Bsp*NI, *Mva*I and *Taq*XI.
17. *Bsu*RI nicking occurs in the unmethylated strand of the hemimethylated sequence GG^{m5}CC/GGCC.
18. *Cla*I cuts slowly at hemimethylated AT^{m5}CGAT.
19. *Dpn*I requires adenine methylation on both DNA strands. Isoschizomers of *Dpn*I include *Cfu*I, *Nan*II, *Nmu*EI, *Nmu*DI and *Nsu*DI. *Dpn*I cuts *dam*-modified XP12 DNA.
20. *Eco*AI, *Eco*BI, *Eco*DXXI and *Eco*KI are Type I restriction endonucleases. ^{m}T represents a 6-methyladenine in the complementary strand.
21. *Eco*PI and *Eco*P15I are Type III restriction endonucleases.
22. *Eco*RI cannot cut hemimethylated G^{m6}AATTC/GAATTC sites. Bimethylated GA^{m6}ATTC/GA^{m6}ATTC sites are not cut by *Eco*RI or *Rsr*I. *Eco*RI shows a reduced rate of cleavage at hemimethylated GAATT^{m5}C and does not cut an oligonucleotide that contains GAATT^{m5}C in both strands.
23. *Eco*RII does not cleave some DNA molecules that carry only a single site. However, oligonucleotides containing the *Eco*RII site can be used to transactivate sites that are resistant to cleavage. *Eco*RII isoschizomers that are sensitive to C^{m5}CWGG include *Atu*BI, *Atu*II, *Bst*GII, *Bin*SI, *Ecl*II, *Eca*II and *Mph*I. *Eco*RII shows a reduced rate of cleavage at hemimethylated ^{m5}CCWGG/CCWGG sites.
24. *Eco*RV cuts the fully ^{m5}C-substituted phage XP12 DNA.
25. *Fok*I cuts about twofold to fourfold more slowly at CATC^{m5}C than at unmodified sites.
26. *Hae*II shows a reduction in rate of cleavage when its recognition sequence is modified at RGCG^{m5}CY.
27. *Hae*III nicking occurs in the unmethylated strand of the hemimethylated sequence GG^{m5}CC/GGCC.
28. There are conflicting data regarding cleavage of GTYRA^{m5}C by *Hinc*II.
29. *Hin*dIII cuts slowly at hemimethylated AAG^{m5}CTT.
30. *Hin*fI cuts GANT^{m5}C; however, detectable rate differences are observed between unmethylated, hemimethylated (GANT^{m5}C/GANTC) and bimethylated (GANT^{m5}C/GANT^{m5}C) target sequences. *Hin*fI does cut ^{m5}C-substituted phage XP12 DNA, although at a reduced rate. However, the rate difference for unmethylated and fully methylated *Hin*fI sites is only about 10-fold.
31. *Hpa*II nicking in the unmethylated strand of the hemimethylated sequence ^{m5}CCGG/CCGG is in dispute. *Hpa*II cuts 50-fold slower at hemimethylated ^{m}CCGG and 3000-fold slower at fully methylated ^{m}CCGG compared with unmethylated DNA.
32. *Kpn*I cuts ^{m5}C-substituted phage XP12 DNA, but cuts slowly at hemimethylated GGTA^{m5}C^{m5}C, GGTA^{m5}CC and GGTAC^{m5}C.
33. *Mae*II nicks slowly in the unmethylated strand of hemimethylated A^{m5}CGT/ACGT.
34. *Mbo*I isoschizomers that are sensitive to G^{m6}ATC include *Bsa*PI, *Bsp*74I, *Bsp*76I, *Bsp*105I, *Bss*GII, *Bst*EIII, *Bst*XII, *Cpa*I, *Cty*I, *Cvi*AI, *Cvi*HI, *Dpn*II, *Fnu*AII, *Fnu*CI, *Hac*I, *Meu*I, *Mkr*AI, *Mme*II, *Mno*III, *Mos*I, *Msp*67II, *Mth*I, *Mth*AI, *Nde*II, *Nfl*AII, *Nfl*BI, *Nfl*I, *Nla*DI *Nla*II, *Nme*CI, *Nph*I, *Nsi*AI, *Nsp*AI, *Nsu*I, *Pfa*I, *Rlu*1I, *Sal*AI, *Sal*HI, *Sau*6782I, *Sin*MI and *Tru*II.
35. *Mbo*II cuts the fully ^{m5}C-substituted phage XP12 DNA, although certain hemimethylated ^{m5}C-containing substrates are reported not to be cut.
36. *Mfl*I cuts slowly at ^{m6}AGATCY sites.
37. *Msp*I cuts the hemimethylated sequence C^{m5}CGG/CCGG and C^{m4}CGG/CCGG duplexes. *Msp*I cuts very slowly at GGC^{m5}CGG. An M·*Msp*I clone methylates ^{m5}CCGG. However, there is a report that *Moraxella* sp. chromosomal DNA is methylated at ^{m5}C^{m5}CGG.
38. *Mva*I nicking occurs in the unmethylated strand of the hemimethylated sequence ^{m4}CCWGG/CCWGG and CC^{m6}AGG/CCTGG. *Mva*I cuts ^{m5}C-substituted XP12 DNA very slowly at ^{m5}C^{m5}CWGG.
39. *Nan*II requires adenine methylation on both DNA strands. *Nan*II cuts *dam*-modified XP12 DNA.
40. The ability of *Nci*I to cut C^{m5}CGGG is in dispute.

Table 9. Methylation sensitivities (continued)

41. *Nco*I is blocked by M·*Sec*I (CCNNGG).
42. *Nde*I and *Nde*II cut the fully ^{m5}C-substituted phage XP12 DNA.
43. *Ngo*II does not cut overlapping *dcm* sites.
44. *Nmu*DI, *Nmu*EI require adenine methylation on both DNA strands.
45. *Pae*R7I cuts hemimethylated CT^{m5}CGAG/CTCGAG sites 100-fold slower and cuts fully methylated CT^{m5}CGAG/CT^{m5}CGAG sites 2900-fold slower than unmethylated sites. Hemi- or full methylation at ^{m6}A completely protects against *Pae*R7I cleavage.
46. *Pst*I cuts slowly at hemimethylated ^{m5}CTG^{m5}CAG.
47. *Pvu*II cuts slowly at hemimethylated ^{m5}CAG^{m5}CTG.
48. *Rsa*I cuts the fully ^{m5}C-substituted phage XP12 DNA according to some sources but not others.
49. *Rsr*I cannot cut hemimethylated G^{m6}AATTC/GAATTC sites.
50. *Sal*I cuts slowly at hemimethylated GT^{m5}CGAC.
51. *Sau*3AI nicking occurs in the unmethylated strand of the hemimethylated sequence GAT^{m5}C/GATC. *Sau*3AI cuts at a reduced rate at ^{m6}AGATC. *Sau*3AI isoschizomers that are insensitive to G^{m6}ATC include *Bce*243I, *Bsp*49I, *Bsp*51I, *Bsp*52I, *Bsp*54I, *Bsp*57I, *Bsp*58I, *Bsp*59I, *Bsp*60I, *Bsp*61I, *Bsp*64I, *Bsp*65I, *Bsp*66I, *Bsp*67I, *Bsp*72I, *Bsp*91I, *Bsp*AI, *Bsr*PII, *Cpf*I, *Csp*5I, *Fnu*EI, *Msp*BI, *Sau*CI, *Sau*DI, *Sau*EI, *Sau*FI, *Sau*GI and *Sau*MI.
52. *Sma*I nicking occurs in the unmethylated strand of the hemimethylated sequence CC^{m5}CGGG/CCCGGG. *Sma*I may cut C^{m5}C^{m5}CGGG methylated DNA. Possibly the second methylation negates the effect of CC^{m5}CGGG. There are conflicting results regarding *Sma*I: some results suggest that ^{m5}CCCGGG is not cut when modified by M·*Aqu*I methyltransferase or when present in an overlapping M·*Hae*III–*Sma*I site (GG^{m5}CCCGGG), another report indicates that *Sma*I cuts at a reduced rate at hemimethylated ^{m5}CCCGGG sites.
53. *Spe*I cuts slowly at hemimethylated A^{m5}CTAGT.
54. *Spl*I cuts GT^{m6}AC-modified *Chlorella* virus NY2A DNA, but does not cut *Kpn*I-digested XP12 DNA.
55. *Sty*SJ and *Sty*SPI are Type I restriction endonucleases. ^{m}T represents a 6-methyladenine in the complementary strand.
56. *Taq*I cuts very slowly at T^{hm5}CGA. *Taq*I cuts the fully ^{m5}C-substituted phage XP12 DNA. M·*Taq*I methylates T^{m5}CGA at least 20-fold slower than unmodified TCGA.
57. *Xba*I will cut T^{m5}CTAGA/TCTAGA hemimethylated DNA at high enzyme levels (>100 units μg^{-1}), but will not cut this sequence in 20- to 40-fold overdigestions.
58. *Xho*I may cut CT^{m5}CGAG according to New England Biolabs.
59. *Xho*II nicking occurs slowly in the unmethylated strand of the hemimethylated sequence RGAT^{m5}CY/RGATCY.
60. *Xma*I is claimed not to cut CC^{m5}CGGG in one report.
61. *Xmn*I cuts the fully ^{m5}C-substituted phage XP12 DNA. *Xmn*I cuts slowly at some sites in DNA methylated on both strands at GAANNNNTT^{m5}C.
62. *Xor*II, according to the Life Technologies catalogue, may cut CG^{m6}ATCG.

Table 10. DNA methyltransferases and their modification specificities[a]

Enzyme	Specificity[b]	Enzyme	Specificity[b]
M·*Acc*I	GTMK^{m6}AC	M·*Apa*I	GGG^{m5}CCC
M·*Ala*K21	GAT^{m5}C	M·*Aqu*I	^{m5}CYCGRG
M·*Alu*I	AG^{m5}CT	M·*Bal*I	TGG^{m5}CCA
M·*Alw*26I	GT^{m5}CTC G^{m6}AGAC	M·*Bam*HI	GGAT^{m4}CC
		M·*Bam*HII	G^{m}CWGC?

Table 10. Methyltransferases (continued)

Enzyme	Specificity[b]
M·*Ban*III	ATCG^{m6}AT
M·*Bbv*I	G^{m5}CWGC
M·*Bbv*SI	G^{m}CWGC
M·*Bbv*SII	G^{m6}AT
M·*Bbv*SIII	A^{m6}AG
M·*Bcg*I	CG^{m6}ANNNNNNTGC
M·*Bcn*I	C^{m4}CSGG
M·*Bep*I	^{m5}CGCG
M.*Bme*216I	GGWC^{m}C
M·*Bna*I	GGAT^{m5}CC
M·*Bse*CI	ATCG^{m6}AT
M·*Bsp*RI	GG^{m5}CC
M·*Bsp*106I	ATCG^{m6}AT
M·*Bsp*RI	GG^{m5}CC
M·*Bst*I	GGAT^{m}CC
M·*Bst*NI	C^{m4}CWGG
M·*Bst*VI	CTCG^{m6}AG
M·*Bst*YI	RGAT^{m}CY
M·*Bsu*15I	ATCG^{m6}AT
M·*Bsu*BI	CTGC^{m6}AG
M·*Bsu*EII	^{m5}CGCG
M·*Bsu*FI	^{m5}CCGG
M·*Bsu*H2	GG^{m5}CC G^{m5}CNGC
M·*Bsu*MI	CT^{m5}CGAG
M·*Bsu*QI	^{m}CCGG
M·*Bsu*RI	GG^{m5}CC
M·*Bsu*SPb	GG^{m5}CC G^{m5}CNGC
M·*Bsu*SPRI	GG^{m5}CC ^{m5}C^{m5}CGG
M·*Bsu*SPR191	^{m5}C^{m5}CGG C^{m}CWGG

Enzyme	Specificity[b]
M·*Bsu*SPR83I	GG^{m5}CC C^{m5}CWGG
M·*Bsu*σ3T	GG^{m5}CC G^{m5}CNGC
M·*Bsu*ρ11I	GG^{m5}CC G^{m5}CNGC
M·*Cfr*I	YGG^{m5}CCR
M·*Cfr*6I	CAG^{m4}CTG
M·*Cfr*9I	C^{m4}CCGGG
M·*Cfr*10I	R^{m5}CCGGY
M·*Cfr*13I	GGN^{m5}CC
M·*Cla*I	ATCG^{m6}AT
M·*Cre*I	T^{m5}CR
M·*Cty*I	G^{m6}ATC
M·*Cvi*AI	G^{m6}ATC
M·*Cvi*AII	C^{m6}ATG
M·*Cvi*BI	G^{m6}ANTC
M·*Cvi*BII	G^{m6}ATC
M·*Cvi*BIII	TCG^{m6}A
M·*Cvi*JI	RG^{m5}CB
M·*Cvi*JII	G^{m6}ANTC
M·*Cvi*PI	^{m5}CC
M·*Cvi*QI	GT^{m6}AC
M·*Cvi*QII	G^{m6}ANTC
M·*Cvi*QIII	C^{m6}ATG
M·*Cvi*QIV	R^{m6}AR
M·*Cvi*RI	TGC^{m6}A
M·*Cvi*RII	GT^{m6}AC
M·*Cvi*SI	TGC^{m6}A
M·*Cvi*SII	C^{m6}ATG
M·*Cvi*SIII	TCG^{m6}A
M·*Cvi*SIV	G^{m6}ATC
M·*Cvi*TI	RG^{m5}CB

Table 10. Methyltransferases (continued)

Enzyme	Specificity[b]
M·*Dde*I	^{m5}CTNAG
M·*Dpn*II	G^{m6}ATC
M·*Dpn*A	G^{m6}ATC
M·*Eae*I	YGG^{m5}CCR
M·*Eca*I	GGTN^{m6}ACC
M·*Eco dam*	G^{m6}ATC
M·*Eco dcm*I	C^{m5}CWGG
M·*Eco dcm*II	R^{m}CCGG
M·*Eco dcm*III	^{m}CCWGG
M·*Eco dcm*IV	GGWC^{m}C
M·*Eco*57I	CTGA^{m6}AG CTTC^{m6}AG
M·*Eco*31I	GGT^{m5}CTC G^{m6}AGACC
M·*Eco*A	G^{m6}AGNNNNNNNG^{m}TCA
M·*Eco*BI	TG^{m6}ANNNNNNNN^{m}TGCT
M·*Eco*KI	A^{m6}ACNNNNNNG^{m}TGC
M·*Eco*PI	AG^{m6}ACC
M·*Eco P1 dam*	G^{m6}ATC
M·*Eco*P15I	CAGC^{m6}AG
M·*Eco*RI	GA^{m6}ATTC
M·*Eco*RII	C^{m5}CWGG
M·*Eco*RV	G^{m6}ATATC
M·*Eco*R124	GA^{m6}AANNNNNNRTCG
M·*Eco*R124/3	GA^{m6}ANNNNNNNRTCG
M·*Eco*T1 *dam*	G^{m6}ATC
M·*Eco*T2 *dam*	G^{m6}AT
M·*Eco*T4 *dam*	G^{m6}ATC
M·*Erh*I	G^{m6}ATC
M·*Esp*3I	GGT^{m6}CTC GAG^{m6}ACC
M·*Fnu*DI	GG^{m5}CC
M·*Fnu*DII	^{m5}CGCG
M·*Fok*I	GG^{m6}ATG C^{m6}ATCC
M·*Hae*III	GG^{m5}CC
M·*Hap*II	C^{m5}CGG
M·*Hgi*CI	GGYRC^{m5}C
M·*Hha*I	G^{m5}CGC
M·*Hha*II	G^{m6}ANTC
M·*Hinc*II	GTYR^{m6}AC
M·*Hind*II	GTYR^{m6}AC
M·*Hind*III	^{m6}AAGCTT
M·*Hinf*I	G^{m6}ANTC
M·*Hpa*I	GTTA^{m6}AC
M·*Hpa*II	C^{m5}CGG
M·*Hph*I	T^{m5}CACC
M·*Kpn*I	GGT^{m6}ACC
M·*Mbo*I	G^{m6}ATC
M·*Mbo*II	GAAG^{m6}A
M·*Msp*I	^{m5}CCGG
M·*Mth*TI	GG^{m5}CC
M·*Mth*ZI	^{m4}CTAG
M·*Mun*I	CA^{m6}ATTG
M·*Mva*I	C^{m4}CWGG
M·*Nde*I	CAT^{m6}ATG
M·*Ngo*I	RG^{m5}CGCY
M·*Ngo*II	GG^{m5}CC
M·*Ngo*IV	G^{m5}CCGGC
M·*Ngo*V	GGNN^{m5}CC
M·*Ngo*VI	G^{m6}ATC
M·*Ngo*VII	G^{m5}CWGC
M·*Ngo*IX	GTANNNNN^{m5}CCTC
M·*Ngo*AI	GG^{m5}CC
M·*Ngo*BI	T^{m5}CACC

Table 10. Methyltransferases (continued)

Enzyme	Specificity[b]	Enzyme	Specificity[b]
M·*Ngo*BII	GTANNNNN^{m5}CTC	M·*Sin*I	GGW^{m5}CC
M·*Ngo*MI	G^{m5}CCGGC	M·*Sma*I	CC^{m4}CGGG
M·*Ngo*MVI	GGNN^{m5}CC	M·*Sso*I	G^{m6}AATTC
M·*Nla*III	C^{m6}ATG	M·*Sso*II	C^{m5}CNGG
M·*Nla*IV	G^{m5}CCGGC	M·*Ssp*MQI	^{m5}CG
M·*Pae*R7I	CTCG^{m6}AG	M·*Sts*I	GG^{m6}ATG C^{m6}ATCC
M·*Pgi*I	G^{m6}ATC		
M·*Pst*I	CTGC^{m6}AG	M·*Sty*LTI	CAG^{m6}AG
M·*Pvu*II	CAG^{m4}CTG	M·*Sty*R124	GAANNNNNNRTCG (^{m6}A)
M·*Rrh*4273I	GTCG^{m6}AC	M·*Sty*R124/3	GAANNNNNNNRTCG (^{m6}A)
M·*Rsa*I	GT^{m6}AC	M·*Sty*SBI	G^{m6}AGNNNNNNR^{m}TYG
M·*Rsr*I	GA^{m6}ATTC	M·*Sty*SJ	G^{m6}AGNNNNNNG^{m}TRC
M·*Sal*I	GTCG^{m6}AC	M·*Sty*SPI	A^{m6}ACNNNNNNG^{m}TRC
M·*Sau*3A	GAT^{m5}C	M·*Sty*SQ	A^{m6}ACNNNNNNR^{m}TAYG
M·*Sau*96I	GGN^{m5}CC	M·*Taq*I	TCG^{m6}A
M·*Sau*3239I	CTCG^{m6}AG	M·*Tfl*I	TCG^{m6}A
M·*Sau*LPI	G^{m5}CCGGC	M·*Tth*HB8I	TCG^{m6}A
M·*Scr*FI	C^{m}CNGG	M·*Xba*I	TCTAG^{m6}A
M·*Sfi*I	GG^{m4}CCNNNNNGGCC	M·*Xcy*I	C^{m4}CCGGG
		M·*Xma*III	CGG^{m4}CCG

[a]For the primary sources of data see ref. 7.
[b]For a description of the convention used in describing recognition sequences see *Table 1*, footnote b. For abbreviations see *Table 1*, footnote c. Additional abbreviations: ^{hm5}C, hydroxymethylcytosine; ^{m}C, N^4- or C^5-methylcytosine; ^{m4}C, N^4-methylcytosine; ^{m5}C, C^5-methylcytosine; ^{m6}A, N^6-methyladenine.

Table 11. Isoschizomer pairs that differ in their sensitivity to sequence-specific methylation[a]

Methylated sequence[b]	Cut by[c]	Not cut by[c]
^{m5}CATG	*Cvi*AII	*Nla*III
C^{m5}CANNNNNTGG	*Esp*1396I	*Pfl*MI *Van*91I
^{m4}CCGG	*Msp*I	*Hpa*II
C^{m5}CGG	*Msp*I	*Hpa*II *Hap*II
C^{m4}CGG	*Msp*I	*Hpa*II

Table 11. Isoschizomer pairs (continued)

Methylated sequence[b]	Cut by[c]	Not cut by[c]
CC^{m5}CGGG	*Xma*I *Cfr*9I	*Sma*I
^{m5}CCTNAGG	*Mst*II	*Bsu*36I
C^{m5}CSGG	*Bcn*I	*Eco*HI *Eco*1831I
C^{m5}CWGG	*Bst*NI *Mva*I *Apy*I	*Eco*RII
^{m5}CCWGG	*Bst*NI *Eco*RII *Mva*I	*Apy*I
CG^{m6}ATCG	*Pvu*I	*Xor*II
GAANNNNTT^{m5}C	*Asp*700I	*Xmn*I
GAGCT^{m5}C	*Sac*I	*Ecl*136II
G^{m6}ATC	*Sau*3AI *Fnu*EI	*Mbo*I *Nde*II
GAT^{m5}C	*Mbo*I	*Sau*3AI
GAT^{m4}C	*Mbo*I	*Sau*3AI
GGC^{m5}C	*Hae*III	*Ngo*II
GGNC^{m5}C	*Asu*I	*Sau*96I
GTG^{m5}CAC	*Apa*LI	*Alw*44I
GGTAC^{m5}C	*Kpn*I	*Asp*718I *Acc*65I
GGTA^{m5}C^{m5}C	*Kpn*I	*Asp*718I
GGWC^{m5}C	*Afl*I	*Ava*II *Eco*47I
RG^{m6}ATCY	*Xho*II *Bst*YI	*Mfl*I
RGAT^{m5}CY	*Bst*YI	*Xho*II *Mfl*I
T^{m5}CCGGA	*Acc*III	*Bsp*MII *Mro*I *Kpn*2I
TC^{m5}CGGA	*Acc*III *Bsp*EI	*Bsp*MII *Mro*I *Kpn*2I
TCCGG^{m6}A	*Bsp*MII *Kpn*2I *Mro*I	*Acc*III

Table 11. Isoschizomer pairs (continued)

Methylated sequence[b]	Cut by[c]	Not cut by[c]
TCGCG^{m6}A	*Sbo*13I *Ama*I *Sal*DI *Spo*I	*Nru*I
TCG^{m5}CGA	*Nru*I	*Spo*I
TT^{m5}CGAA	*Asu*II *Sfu*I	*Bst*BI
TTCG^{m6}AA	*Cbi*I	*Bst*BI *Csp*45I *Ssp*RFI
CGGWC^{m5}CG	*Csp*I	*Rsr*II

[a]For the primary sources of data see ref. 7.
[b]For a description of the convention used in describing recognition sequences see *Table 1*, footnote b. For abbreviations see *Table 1*, footnote c. Additional abbreviations: ^{hm5}C, hydroxymethylcytosine; ^{m}C, N^{4}- or C^{5}-methylcytosine; ^{m4}C, N^{4}-methylcytosine; ^{m5}C, C^{5}-methylcytosine; ^{m6}A, N^{6}-methyladenine.
[c]An enzyme is classified as insensitive to methylation if it cuts the methylated sequence at a rate that is at least one tenth the rate at which it cuts the unmethylated sequence. An enzyme is classified as sensitive to methylation if it is inhibited at least 20-fold by methylation relative to the unmethylated sequence.

Table 12. Abilities of restriction enzymes to cut DNAs methylated by common methylases[a]

Enzyme	Cuts methylated DNA?[b]			
	Dam[c]	Dcm[d]	Plant[e]	Mammalian[f]
*Aat*I		✘		
*Aat*II	✔	✔	✘	✘
*Acc*III	✘	✔	✔	✔
*Acc*65I		✘		
*Afl*I		✔		
*Afl*II			✔	
*Aha*II		✘		
*Aha*III			✔	
*Alw*I	✘	✘		
*Alw*NI		✘		
*Apa*I	✔	✘	✘	✘
*Apy*I		✔		
*Asc*I				✘
*Ase*I			✔	
*Asp*700I			✔	
*Asp*718I		✘		

Table 12. Cutting methylated DNA (continued)

Enzyme	Cuts methylated DNA?[b]			
	Dam[c]	Dcm[d]	Plant[e]	Mammalian[f]
*Asu*II			✔	✔
*Ava*I	✔	✔	✘	✘
*Ava*II	✔	✘	✘	✘
*Avi*II				✘
*Bal*I	✔	✘	✘	✔
*Bam*HI	✔	✔	✘	✔
*Ban*II		✔		
*Ban*III	✘			
*Bbe*I		✔	✘	✘
*Bbu*I	✔	✔	✔	✔
*Bcl*I	✘		✔	
*Bgl*I		✔		
*Bgl*II	✔	✔	✘	✔
*Bma*DI				✘
*Bpm*I		✘		
*Bsa*BI	✘			
*Bsa*JI		✔		
*Bsl*I		✘		
*Bsp*106I	✘			
*Bsp*120I		✘		
*Bsp*1286I		✔		
*Bsp*DI	✘			✔
*Bsp*EI	✘			
*Bsp*HI	✘		✔	
*Bsp*MII	✔		✘	✘
*Bsp*NI			✔	
*Bsr*BI				✘
*Bss*HII			✘	✘
*Bss*KI		✘		
*Bst*I	✔	✔		
*Bst*BI			✘	✘
*Bst*EII	✘	✔	✔	
*Bst*NI		✔	✔	
*Bst*XI		✘		
*Bst*YI	✔			
*Cfr*9I				✔
*Cla*I	✘	✔	✘	✘
*Csp*I	✔	✔	✘	✘
*Csp*45I	✔	✔	✘	✘
*Cvi*QI			✔	
*Dpn*I	see footnote g		✔	
*Dpn*II	✘			
*Dra*I			✔	
*Dra*II		✘		
*Eae*I		✘		
*Eag*I			✘	✘
*Ecl*XI				✘

Restriction/Methylation

Table 12. Cutting methylated DNA (continued)

Enzyme	Cuts methylated DNA?[b]			
	Dam[c]	Dcm[d]	Plant[e]	Mammalian[f]
*Eco*47I		✘		
*Eco*47III	✔	✔	✘	✘
*Eco*O109I		✘		
*Eco*RI	✔	✔	✘	✘
*Eco*RII		✘		
*Eco*RV			✔	
*Ehe*I		✔		
*Fse*I			✘	✘
*Fsp*I			✘	✘
*Hae*III	✔	✔	✘	✔
*Hin*1I		✘		
*Hinc*II	✔	✔	✔	✔
*Hind*III	✔	✔	✔	✔
*Hpa*I			✔	
*Hpa*II	✔		✘	✘
*Hph*I	✘			
*Kpn*I	✔	✔	✔	✔
*Kpn*2I			✘	✘
*Mbo*I	✘	✔	✔	✔
*Mbo*II	✘	✔	✔	✔
*Mlu*I		✔	✔	✘
*Mlu*9273I				✘
*Mlu*9273II				✘
*Mro*I	✔		✘	✘
*Msc*I		✘		
*Mse*I			✔	
*Msp*I	✔	✔	✘	✔
*Mva*I		✔		
*Nae*I			✘	✘
*Nar*I	✔	✔	✘	✘
*Nde*I			✔	
*Nde*II	✘		✔	
*Not*I	✔	✔	✘	✘
*Nru*I	✘	✔	✘	✘
*Pac*I			✔	✔
*Pal*I		✔		✘
*Pfl*MI		✘		
*Pfu*I				✘
*Pme*I				✔
*Pml*I			✘	✘
*Ppu*AI				✘
*Ppu*MI		✘		
*Pst*I	✔	✔	✘	✔
*Pvu*I	✔		✘	✘
*Pvu*II	✔	✔	✘	✘
*Rsa*I		✔	✔	
*Rsp*XI	✘		✔	

Table 12. Cutting methylated DNA (continued)

Enzyme	Cuts methylated DNA?[b]			
	Dam[c]	Dcm[d]	Plant[e]	Mammalian[f]
*Rsr*II			✗	✗
*Sac*I	✔	✔		✔
*Sac*II	✔	✔	✗	✗
*Sal*I	✔	✔	✗	✗
*Sal*DI				✗
*Sau*3AI	✔	✔	✗	✗
*Sau*96I	✔	✗	✗	✗
*Sbo*13I				✗
*Scr*FI		✗		
*Sfi*I	✔		✗	✗
*Sfu*I			✔	✔
*Sin*I	✔	✔	✗	✔
*Sma*I	✔	✔	✗	✗
*Sna*BI	✔	✔	✗	✗
*Spe*I			✔	
*Sph*I	✔	✔	✔	✔
*Spl*I				✗
*Spo*I	✔			✗
*Srf*I				✗
*Sse*8387I				✔
*Ssp*I			✔	
*Stu*I	✔	✗	✗	✔
*Swa*I			✔	✔
*Taq*I	✗	✔	✔	✔
*Tsp*509I			✔	
*Tth*HBI	✗		✔	
*Xba*I	✗	✔	✔	✔
*Xho*I	✔	✔	✔	✗
*Xho*II	✔	✔	✗	✔
*Xma*I	✔	✔		✔
*Xmn*I			✔	
*Xor*II	✔			✗

[a]Data taken from various sources, in particular ref. 7.
[b]✔ = the enzyme activity is not affected by this type of methylation, all sites will be cut; ✗ = the enzyme activity is partially or completely blocked by this type of activity, some or all sites will not be cut. No entry means status is uncertain.
[c]The DNA adenine methylase (Dam) of *E. coli* modifies the N^6 position of the adenine present in the target sequence 5′-GATC-3′. See Chapter 1, Section 1.3.
[d]The DNA cytosine methylase (Dcm) of *E. coli* modifies the C^5 positions of the cytosines in the target sequences 5′-CCAGG-3′ and 5′-CCTGG-3′. See Chapter 1, Section 1.3.
[e]Plant DNA often displays C^5 methylation of the cytosines in 5′-CG-3′ and 5′-CNG-3′.
[f]Mammalian DNA often displays C^5 methylation of the cytosine in 5′-CG-3′.
[g]*Dpn*I does not cut unmethylated DNA, the recognition sequence must be Dam-methylated for cleavage to occur.

4. REFERENCES

1. http://www.neb.com/rebase/rebase.html

2. Roberts, R.J. and Macelis, D. (1997) *Nucl. Acids Res.*, **25**, 248.

3. Williams, R., Kline, M. and Smith, R. (1996) *Promega Notes*, **59**, 46.

4. Lepinske, M. (1996) *Promega Notes*, **60**, 28.

5. Anon (1996) *Catalog*. New England Biolabs, Beverly.

6. Anon (1994) *Molecular Biology Reagents Catalogue*. United States Biochemical Corp., Cleveland.

7. McClelland, M., Nelson, M. and Raschke, E. (1994) *Nucl. Acids Res.*, **22**, 3640.

CHAPTER 3
DNA AND RNA MODIFYING ENZYMES

This chapter provides information on the enzymes and proteins commonly used to modify DNA and/or RNA, with the exception of restriction endonucleases and DNA methylases, which were described in Chapter 2. The chapter is divided into two sections. Section 1 provides an overview of the important features of these enzymes and proteins, grouped for convenience into the following categories:

(i) DNA polymerases, *including thermostable DNA polymerases and the reverse transcriptases*

(ii) RNA polymerases

(iii) Nucleases, *including some endonucleases with sequence specificity, but excluding the restriction endonucleases (Chapter 2)*

(iv) End-modification enzymes

(v) Ligases

(vi) Topoisomerases

(vii) Glycosylases

(viii) DNA-binding proteins

Section 2 gives details of the properties, applications and reaction conditions for each individual enzyme or protein.

Modifying Enzymes

1. OVERVIEW

1.1 DNA polymerases

DNA polymerases are enzymes that synthesize DNA molecules from deoxyribonucleotide subunits. Broadly speaking, there are two types of DNA polymerase:

(i) Template-dependent DNA polymerases, which synthesize a new strand of DNA complementary to an existing DNA or RNA template (DNA- and RNA-dependent DNA polymerases, respectively).

(ii) Template-independent DNA polymerases, which synthesize DNA without the aid of a template.

This section deals only with template-dependent enzymes as the only template-independent DNA polymerase relevant to recombinant DNA technology is terminal deoxynucleotidyl transferase, which is usually categorized as an end-modification enzyme.

Polymerase and exonuclease activities of DNA-dependent DNA polymerases
All template-dependent DNA polymerases synthesize DNA in the 5′ to 3′ direction and require a short DNA or RNA primer to initiate the reaction (*Figure 1a*). The length of new strand that is synthesized before the enzyme dissociates from the template varies but is usually between several hundred and a few thousand nucleotides under test tube conditions. This is referred to as the processivity of the polymerase. In general, DNA polymerases with high processivity (e.g. T7 DNA polymerase and its derivatives) are more useful than enzymes with low processivity (e.g. *E. coli* DNA polymerase and its derivatives) since the highly processive enzymes enable longer DNA strands to be synthesized.

Template-dependent DNA polymerases may also possess zero, one or both of the following exonuclease activities:

(i) A 5′ to 3′ exonuclease, removing individual nucleotides or short oligonucleotides from the 5′ end of a DNA molecule. In the cell this activity enables the polymerase to replace an existing DNA strand with a new polynucleotide (*Figure 1b*), as occurs during DNA repair and replication of the lagging strand. This activity is utilized in nick translation labelling, but causes problems in those applications (e.g. DNA sequencing) where 5′ end degradation of newly synthesized strands is undesirable.

(ii) A 3′ to 5′ exonuclease, removing individual nucleotides from the 3′ end of a DNA molecule. This gives the polymerase a 'proof reading' activity, enabling it to reverse along a newly synthesized strand and replace nucleotides that are not complementary with the template (*Figure 1c*). This activity increases the enzyme's accuracy when copying the template strand, but at low dNTP concentrations (as often used in DNA sequencing) the extent of the exonuclease activity can approach that of the polymerase, limiting the efficiency of strand synthesis.

The exonuclease properties of individual DNA-dependent DNA polymerases are summarized in *Table 1*. Note that the activities vary in degree between different enzymes and can be suppressed or enhanced by the reaction conditions. For applications such as PCR and DNA sequencing the reaction conditions recommended for a particular enzyme should always be used and it should not be assumed that a set of conditions that gives good results with one polymerase will be equally successful with a second enzyme.

Thermostable DNA polymerases
The growing importance of techniques that require thermostable DNA polymerases has led to a wide range of these enzymes being marketed by different companies. No attempt is made here to describe all the enzymes that are available. As well as the exonuclease activities (*Table 1*), the thermostabilities of the individual enzymes (*Table 2*) are important since these determine the length of time that an enzyme will remain functional during high-temperature DNA polymerization. A few suppliers are prepared to provide real data on the thermostabilities of the enzymes they sell, others are more reticent and prefer non-quantifiable statements such as 'stable at 95°C'. In fact, the most useful information would be the amount of DNA synthesized per half-life, because in practical terms a highly thermostable enzyme with a low rate of strand synthesis might be less desirable than an enzyme that degrades more quickly but makes DNA more rapidly. This information does not appear in *Table 2* as it is virtually impossible to obtain for any thermostable enzyme.

Figure 1. The polymerase and exonuclease activities of a template-dependent DNA polymerase.

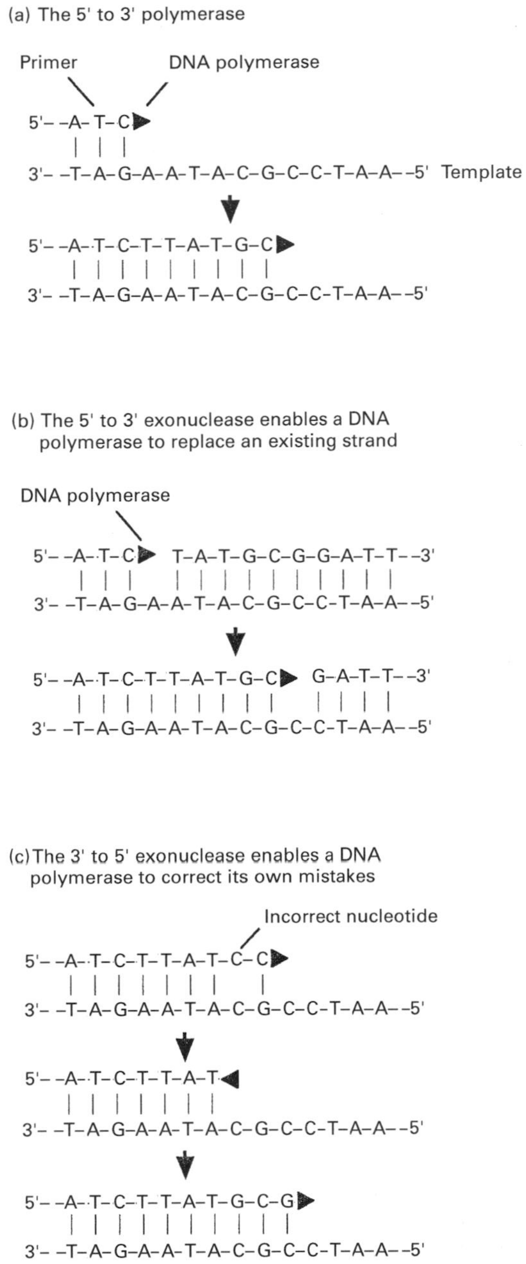

Reverse transcriptases

RNA-dependent DNA polymerases, or reverse transcriptases, are used to make cDNA copies of RNA molecules, usually as a preliminary to cloning or PCR amplification. The key features when comparing the available enzymes (*Table 3*) are the presence or absence of an

RNase H exoribonuclease activity (which can reduce the efficiency of cDNA synthesis by degrading the RNA template), the ability of the enzyme to use DNA as well as RNA as the template (which extends the applications of the enzyme) and the preferred incubation temperature.

Table 1. Exonuclease activities of DNA-dependent DNA polymerases[a]

Enzyme	3′ to 5′ exonuclease	5′ to 3′ exonuclease
Alpha DNA polymerase	No	No
Bst DNA polymerase	No	Yes
Bst DNA polymerase, large fragment	No	No
Deep Vent DNA polymerase	Yes	No
Deep Vent (exo$^-$) DNA polymerase	No	No
DNA polymerase I, Klenow fragment	Yes	No
DNA polymerase I, Klenow fragment (exo$^-$)	No	No
DNA polymerase I (Kornberg polymerase)	Yes	Yes
Hot *Tub* DNA polymerase	?	?
Micrococcal DNA polymerase	Yes	Yes
Pfu DNA polymerase	Yes	?
Pfu (exo$^-$) DNA polymerase	No	?
Pwo DNA polymerase	Yes	No
Sequenase	No	No
T4 DNA polymerase	Yes	No
T7 DNA polymerase	Yes	No
Taq DNA polymerase	No	Yes
Tfl DNA polymerase	?	?
Tli DNA polymerase	Yes	No
Tth DNA polymerase	No	?
Vent DNA polymerase	Yes	No
Vent (exo$^-$) DNA polymerase	No	No

[a]The degree of each exonuclease activity varies between different enzymes and can be suppressed or enhanced by the reaction conditions.

Table 2. Temperature stability of thermophilic DNA polymerases[a]

Enzyme	Half-life at 95°C	Half-life at 100°C
Bst DNA polymerase[b]	Inactivated after 10 min at 80°C	
Deep Vent DNA polymerase[c]	23 h	8 h
Hot *Tub* DNA polymerase	?[d]	?
Pfu DNA polymerase[c]	?[e]	?[e]
Pwo DNA polymerase	?	>2 h
Taq DNA polymerase	1.6 h	<5 min
Tfl DNA polymerase	?[f]	?
Tli DNA polymerase	?	?[g]
Tth DNA polymerase	Half-life of 3–6 h at 85°C	?[h]
Vent DNA polymerase[c]	6.7 h	1.8 h

[a]Data taken from suppliers' product literature.
[b]Includes *Bst* polymerase, large fragment.
[c]Includes the exo$^-$ versions.
[d]'Can withstand temperatures of up to 95°C' (Amersham).
[e]'95% active after 1-hour incubation at 98°C' (Stratagene).
[f]'Stability to prolonged incubation at 95°C' (Promega).
[g]'Remains functional even after incubation at 100°C' (Promega).
[h]'Resistant to prolonged incubations at high temperatures (95°C)' (Boehringer–Mannheim).

Table 3. Properties of reverse transcriptases

Enzyme	Exoribonuclease[a]	DNA-dependent DNA polymerase	Incubation temperature (°C)
AMV reverse transcriptase	Yes	Yes	42
HIV-1 reverse transcriptase	Yes	No	37
M-MuLV reverse transcriptase	Reduced	Yes	37
RAV2 reverse transcriptase	Yes	Yes	42
Superscript reverse transcriptase	No	Yes	37
Tth DNA polymerase	No	Yes	70–75

[a]'RNase H' activity, specific for the RNA component of a DNA–RNA hybrid.

1.2 RNA polymerases

As with DNA polymerases, RNA polymerases include enzymes that do or do not require a template. The template-independent RNA polymerases used in recombinant DNA experiments [poly(A) polymerase and polynucleotide phosphorylase] are both categorized as end-modification enzymes.

Two template-dependent RNA polymerases – *E. coli* RNA polymerase and RNA polymerase II from wheat germ – have been available for many years but their broad promoter specificities and low RNA yields have limited their applications. Much more useful are the three phage RNA polymerases, from SP6, T3 and T7, as these recognize only their individual promoter sequences and provide large quantities of RNA. More recently, a thermostable enzyme has become available (*Thermus* RNA polymerase) but this is a broad specificity, low yield enzyme. The properties of template-dependent RNA polymerases are summarized in *Table 4*.

Table 4. Properties of RNA polymerases

Enzyme	**Promoter specificity**[a]	**Yield of RNA**[b]	**Incubation temperature (°C)**
E. coli RNA polymerase	Broad	Low	37
RNA polymerase II	Broad	Low	25
SP6 RNA polymerase	Specific for SP6 promoter sequence	High	37
T3 RNA polymerase	Specific for T3 promoter sequence	High	37
T7 RNA polymerase	Specific for T7 promoter sequence	High	37
Thermus RNA polymerase	Broad	Low	65

[a]RNA polymerases with broad promoter specificity recognize a range of sequences related to a consensus. Each of the phage RNA polymerases are highly specific for a single sequence.
[b]The highly active phage RNA polymerases can synthesize up to 30 µg of RNA from 1 µg of template DNA in 30 min.

1.3 Nucleases

Nucleases are enzymes that degrade nucleic acids. Individual enzymes display a range of activities (*Table 5*). Some are specifically deoxyribonucleases and degrade only DNA, others are ribonucleases, active only with RNA, others degrade both DNA and RNA. Some are specific for single-stranded molecules, others for double-stranded molecules and some show no specificity. Some are exonucleases, removing nucleotides from one or both ends of a polynucleotide, while others are endonucleases that cut within polynucleotides. Endonucleases may cut at random positions or may display sequence specificity. The special category of sequence-specific endonucleases called restriction enzymes are described in Chapter 2. Other types of sequence-specific nuclease, described in this chapter, are listed in *Table 6*.

Table 5. Properties of nucleases

Enzyme	**Type**	**Activity on**			
		ssDNA	**dsDNA**	**ssRNA**	**dsRNA**
Bal31 nuclease	3′ to 5′exonuclease	Yes	Yes	Inefficient	
	Endonuclease	Yes	No	Inefficient	
Deoxyribonuclease I	Endonuclease	Yes	Yes	No	No
Exonuclease I	3′ to 5′ exonuclease	Yes	No	No	No
Exonuclease III	3′ to 5′ exonuclease	No	Yes	No	No
	Endonuclease	Baseless sites[a]		DNA–RNA hybrid[b]	

Table 5. Nucleases (continued)

Enzyme	Type	Activity on			
		ssDNA	dsDNA	ssRNA	dsRNA
Exonuclease V	3′ to 5′, 5′ to 3′ exonuclease	Yes	Yes	No	No
Exonuclease VII	3′ to 5′, 5′ to 3′ exonuclease	Yes	No	No	No
Lambda exonuclease	5′ to 3′ exonuclease	No	Yes	No	No
Mung bean nuclease	Endonuclease	Yes	No	Yes	No
N. crassa nuclease	Exonuclease Endonuclease	Yes Yes	Yes No	No Yes	No No
P1 nuclease	Exonuclease Endonuclease	Yes Yes	Yes Yes	Yes Yes	Yes Yes
Phosphodiesterase I	Exonuclease	Yes	Yes	Yes	Yes
Phosphodiesterase II	Exonuclease	Yes	Yes	Yes	Yes
Ribonuclease A	Endonuclease	No	No	Yes	Yes
Ribonuclease H	Endonuclease	No	No	DNA–RNA hybrid[b]	
Ribonuclease ONE	Endonuclease	No	No	Yes	No
S1 nuclease	Endonuclease	Yes	No	Yes	No
S7 nuclease	Endonuclease	Yes	Yes	Yes	Yes
T7 endonuclease	Endonuclease	Yes	Low[c]	No	No
T7 gene 6 exonuclease	5′ to 3′ exonuclease	No	Yes	No	No

[a]Endodeoxyribonuclease activity is specific for nucleotides from which the purine or pyrimidine base has been cleaved.
[b]Specific for the RNA component of a DNA–RNA hybrid.
[c]The activity of T7 endonuclease with dsDNA is approximately 1% that with ssDNA.

Table 6. Sequence specific nucleases (other than restriction endonucleases)

Enzyme	Source	Target sequence[a]
Deoxyribonucleases		
I-*Ceu* I	*Chlamydomonas eugametos* LrRNA gene intron	5′-TAACTATAACGGTC CTAA↑GGTAGCGA-3′ 3′-ATTGATATTGCCAG↑GATT CCATCGCT-5′
Lambda terminase	Phage λ	5′- CCCGCCGCTGGA↑-3′[b] 3′-↑GGGCGGCGACCT -5′
Meganuclease I-*Sce* I	*Saccharomyces cerevisiae* LrRNA gene intron	5′-TAGGG ATAA↑CAGGGTAAT-3′ 3′-ATCCC↑TATT GTCCCATTA-5′
PI-*Psp* I	*Pyrococcus* GB-D	5′-TGGCAAACAGCTA TTAT↑GGGTATTATGGGT-3′ 3′-ACCGTTTGTCGAT↑AATA CCCATAATACCCA-5′

Table 6. Sequence specific nucleases (continued)

Enzyme	Source	Target sequence[a]
Ribonucleases		
B. cereus RNAase	*Bacillus cereus*	5′-Up↑N-3′, 5′-Cp↑N-3′
RNase CL3	Chicken liver	5′-Cp↑N-3′
RNase M1	*Cucumis melo*	5′-N↑pA-3′, 5′-N↑pU-3′, 5′-N↑pG-3′
RNase Phy1	*Physarum polycephalum*	5′-Gp↑N-3′, 5′-Ap↑N-3′, 5′-Up↑N-3′
RNase PhyM	*Physarum polycephalum*	5′-Up↑N-3′, 5′-Ap↑N-3′
RNase T1	*Aspergillus oryzae*	5′-Gp↑N-3′
RNase T2	*Aspergillus oryzae*	5′-Ap↑N-3′ > all others
RNase U2	*Ustilago sphaerogena*	5′-Ap↑N-3′ > 5′-Gp↑N-3′
RNAzyme RCH 1.0, 1.1	Hammerhead ribozyme	5′-ACGGUCUC↑ACGAGC-3′
RNAzyme TET 1.0	*Tetrahymena* intron	5′-CUCU↑-3′
S. aureus RNase	*Staphylococcus aureus*	pH 7.5: 5′-A↑pN-3′, 5′-U↑pN-3′ pH 3.5, no Ca^{2+}: 5′-C↑pN-3′, 5′-U↑pN-3′

[a]Arrows indicate the cleavage site. The endodeoxyribonucleases leave staggered ends as shown; the endoribonucleases cut ssRNA.

[b]Lambda terminase sometimes leaves overhangs of 4–5 nucleotides.

1.4 End-modification enzymes

This category includes a group of enzymes displaying a range of activities that are useful in DNA and RNA manipulation (*Table 7*). The properties and specific applications of each enzyme are described in Section 2 of this chapter. The four alkaline phosphatases (*Table 8*) differ in their stabilities, the search for alternatives to bacterial alkaline phosphatase being prompted by the extreme stability of this enzyme, which makes it difficult to avoid carry over of active phosphatase to later stages of an experiment.

Table 7. Properties of end-modification enzymes

Enzyme	Property
Acid pyrophosphatase	Removes the cap structure from mRNA
Bacterial alkaline phosphatase	Removes the 5′-P group from DNA and RNA
Calf intestine alkaline phosphatase	Removes the 5′-P group from DNA and RNA
HK alkaline phosphatase	Removes the 5′-P group from DNA and RNA
mRNA guanyltransferase	Adds a 5′ cap structure to RNA
Poly(A) polymerase	Adds a poly(A) tail to the 3′-OH terminus of ssRNA
Polynucleotide phosphorylase	Adds ribonucleotides to the 3′-OH terminus of RNA
Shrimp alkaline phosphatase	Removes the 5′-P group from DNA and RNA
T4 polynucleotide kinase	Adds a 5′-P group to DNA and RNA
Terminal deoxynucleotidyl transferase	Adds deoxyribonucleotides to the 3′-OH terminus of DNA

Table 8. Properties of alkaline phosphatases

Alkaline phosphatase	Source	Stability
Bacterial (BAP)	*E. coli*	Stable at 68°C, resistant to phenol
Calf intestine (CIP, CIAP)	Calf	Inactivated by 10 min at 70°C
HK	Arctic bacterium	Inactivated by 15 min at 65°C
Shrimp	Arctic shrimp	Inactivated by 15 min at 65°C

1.5 Ligases

Ligases are ATP- or NAD-dependent enzymes that join DNA and/or RNA molecules together by synthesizing phosphodiester bonds between 5′-P and 3′-OH termini. A range of enzymes are now available (*Table 9*) but these have specific applications. Only T4 DNA ligase is able to carry out the full range of activities on DNA: synthesis of missing phosphodiester bonds in a duplex molecule ('nick repair'), ligation of cohesive ends and ligation of blunt ends. The enzyme has specific optimal reaction conditions for different applications, as described in its entry in Section 2. *E. coli* DNA ligase appears to be rarely used but has the advantage over T4 DNA ligase of greater discrimination for correctly matched cohesive ends and so creates fewer spurious ligation products. T4 RNA ligase has several specialist applications, including stimulation of the activity of T4 DNA ligase towards blunt ends.

Several thermostable DNA ligases have recently been introduced. These have the advantage of enabling ligations to be carried out at elevated temperatures at which mismatches between hybridized molecules are minimized, for example in gene synthesis from single stranded oligonucleotides and in the ligase chain reaction (1–5).

Table 9. Properties of ligases

Ligase	Cofactor	Substrate[a]	Incubation temperature (°C)
Ampligase DNA ligase	NAD	DNA	45–80
E. coli DNA ligase	NAD	DNA	4–25
Pfu DNA ligase	ATP	DNA	45–80
T4 DNA ligase	ATP	DNA	4–25
T4 RNA ligase	ATP	DNA, RNA	4–25
Taq DNA ligase	NAD	DNA	45–65

[a]Only T4 DNA ligase is able to ligate blunt ends.

1.6 Topoisomerases

Two topoisomerases – DNA gyrase and DNA topoisomerase I – are occasionally used in recombinant DNA research for applications that require control over supercoiling and other aspects of DNA tertiary structure.

1.7 Glycosylases

Glycosylases cleave the *N*-glycosidic bond linking the base to the sugar component of a nucleotide. Uracil DNA–glycosylase, which removes uracil from single- or double-stranded

Modifying Enzymes

DNA, has several applications, most importantly as part of a procedure for reducing carry-over of DNA in PCR reactions (6).

1.8 DNA-binding proteins

RecA and the single strand-binding protein of *E. coli*, as well as the T4 gene 32 protein, are predominantly ssDNA binding proteins (RecA also has ATPase activity) and are used for a variety of applications including mutagenesis techniques (7–9) and electron microscopy of DNA (10–14).

2. DETAILS OF INDIVIDUAL ENZYMES AND PROTEINS

The following pages are an alphabetical listing of the 90 enzymes and proteins mentioned in Section 1, with each entry giving source, description, properties, applications and reaction conditions.

ACID PYROPHOSPHATASE

Source
Tobacco.

Description
A 'decapping' enzyme able to remove the cap structure at the 5′ termini of most eukaryotic mRNAs. Do not confuse with standard pyrophosphatases which hydrolyse inorganic pyrophosphates.

Properties
Pyrophosphatase hydrolysing the phosphoric acid anhydride bonds in the triphosphate linkage of the mRNA cap structure, removing the cap nucleoside and leaving a 5′-P terminus:

Applications
Pretreatment of mRNA prior to:

(i) 3′ sequence analysis (15).

(ii) 5′ end mapping (16–18).

(iii) Studies of mRNA degradation (19).

(iv) 5′ end labelling.

Reaction conditions
Use at 37°C or room temperature.

10 × TAP Buffer: 500 mM NaAc (pH 6.0), 10 mM Na_2·EDTA, 1% β-mercaptoethanol, 0.1% Triton X-100.

ALPHA DNA POLYMERASE

Source
Calf thymus.

Description
Alpha DNA polymerase is the most active eukaryotic DNA polymerase and the one most clearly linked to chromosomal DNA replication (20). The enzyme lacks the exonuclease activities of prokaryotic DNA polymerases.

Properties
5′ to 3′ DNA-dependent DNA polymerase, requiring a ssDNA template and a DNA or RNA primer with a 3′-OH terminus:

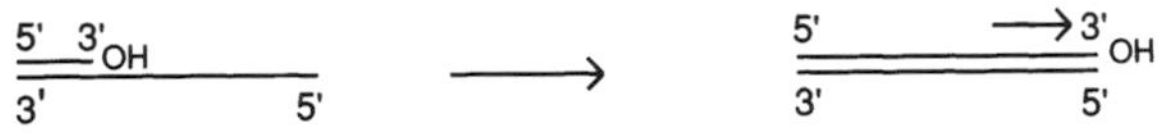

Applications
As a standard eukaryotic DNA polymerase.

Reaction conditions
Requirements are similar to prokaryotic DNA polymerases; 37°C.

10 × Alpha Buffer: 600 mM Tris-HCl (pH 7.5), 60 mM $MgCl_2$, 50 mM β-mercaptoethanol, 5 mg ml^{-1} BSA.

Thermostable ligase

AMPLIGASE DNA LIGASE

Marketed by Epicentre Technologies

Source
Thermophilic bacterium.

Description
An NAD-requiring DNA ligase, apparently functionally similar to the *E. coli* enzyme but active at high temperatures.

Properties
DNA ligase, catalysing the formation of a phosphodiester bond between adjacent 5′-P and 3′-OH termini hybridized to a complementary target DNA:

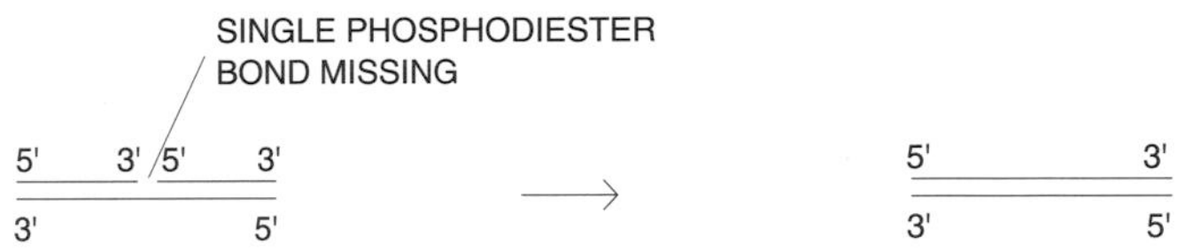

Applications
Because of the instability of base-paired structures at elevated temperatures, ampligase DNA ligase cannot be used as a general purpose ligation enzyme. It has specialist applications:

(i) Ligase chain reaction (1–5).

(ii) Repeat expansion detection (21).

(iii) High-fidelity gene synthesis.

(iv) *In vitro* mutagenesis procedures (22, 23).

(v) Padlock probe hybridization system (24).

Reaction conditions
Requires Mg^{2+} and NAD. Reactions conditions should be optimized for different substrates.

10 × Basic AmL Buffer: 200 mM Tris-HCl (pH 8.3), 250 mM KCl, 100 mM $MgCl_2$, 5 mM NAD, 0.1% Triton X-100.

AMV REVERSE TRANSCRIPTASE

Source
Chicks infected with avian myeloblastosis virus (AMV), or a recombinant *E. coli* expressing the AMV *pol* gene.

Description
A dimer of 62 kd and 94 kd subunits, with the small subunit being a proteolytic product of the large one (25). The enzyme possesses a complex array of activities (26, 27).

Properties
(i) 5′ to 3′ DNA- or RNA-dependent DNA polymerase, requiring a ssDNA or ssRNA template and a DNA or RNA primer with a 3′-OH terminus (26, 27):

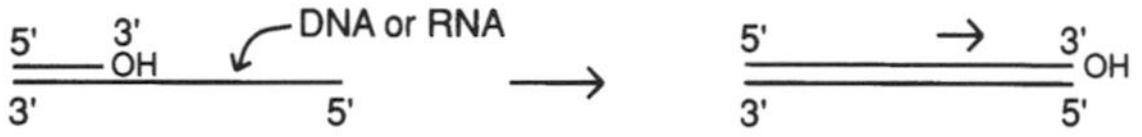

(ii) 5′ to 3′ and 3′ to 5′ exoribonuclease, specific for the RNA component of a DNA–RNA hybrid (28):

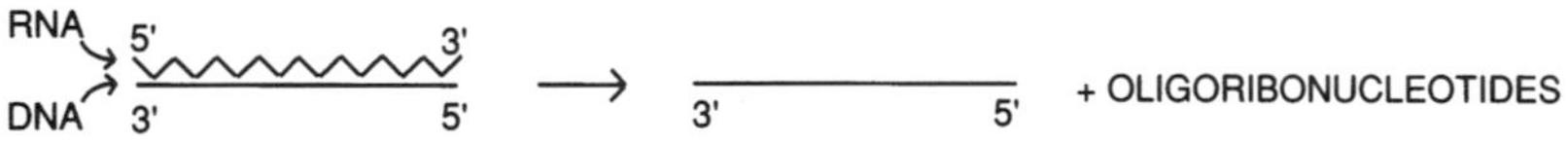

(iii) DNA endonuclease (29).

(iv) DNA unwinding activity.

Applications
(i) First strand cDNA synthesis (30–32).

(ii) Polymerase chain reaction with RNA substrates (RT-PCR) (33).

(iii) Chain termination RNA sequencing (34).

(iv) Modified chain termination DNA sequencing, especially to read GC-rich regions (35).

(v) Probing RNA secondary structure (36).

(vi) 3′ end labelling DNA fragments (37).

Reaction conditions
Requires Mg^{2+}. AMV reverse transcriptase can be incubated at 42°C. Actinomycin D (50 µg ml^{-1}) can be added to the reaction mix to inhibit the DNA-dependent DNA polymerase activity (38).

10 × AMV RT Buffer: 500 mM Tris-HCl (pH 8.3), 400 mM KCl, 100 mM $MgCl_2$, 50 mM DTT.

End-modification enzyme

BACTERIAL ALKALINE PHOSPHATASE (BAP)

Source

Escherchia coli (39).

Description

Dimeric, contains zinc and magnesium atoms. Stable at 68°C and resistant to phenol extraction. Preparations of bacterial alkaline phosphatase frequently contain a contaminating exodeoxyribonuclease. BAP is active on 5′ overhangs, recessed 5′ ends and blunt ends.

Properties

5′ phosphatase, removing 5′ phosphate groups from ssDNA, dsDNA, ssRNA and dsRNA:

E.g.

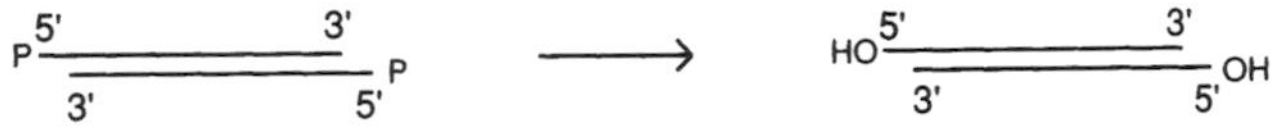

Applications

(i) Dephosphorylation of dsDNA to prevent self-ligation (40, 41).

(ii) Dephosphorylation of DNA and RNA to allow 5′ end labelling with T4 polynucleotide kinase (42).

(iii) As a component of a conjugated antibody immunodetection system (43, 44).

(iv) Component of a non-isotopic labelling method (45).

Reaction conditions

Alkaline phosphatases do not have particularly stringent buffer requirements. Incubate at 65°C for DNA (to suppress the contaminating exonuclease) and 45°C for RNA. If your enzyme has no significant exonuclease activity then with dsDNA use 37°C for blunt ends or 5′ overhangs, and 60°C for 3′ overhangs.

10 × BAP Buffer: 500 mM Tris-HCl (pH 8.0).

Bal31 NUCLEASE

Source
Alteromonas espejiana Bal31.

Description
An extracellular nuclease. Bal31 is highly sequence dependent, so the rate at which it shortens a DNA molecule can vary considerably. At extreme dilutions only the ssDNA endonuclease activity is evident.

Properties (see refs 46–52)
(i) 3′ to 5′ exonuclease that progressively removes nucleotides from the 3′-OH termini of ssDNA and dsDNA molecules having either blunt or sticky ends:

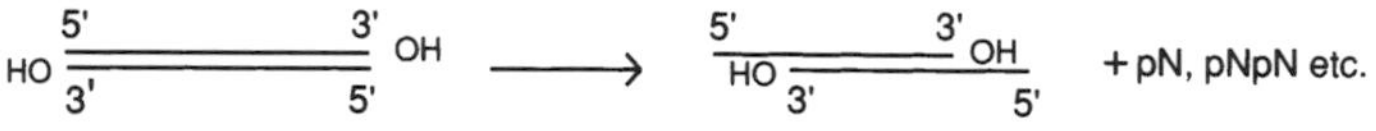

(ii) Endodeoxyribonuclease, with high specificity for ssDNA:

(iii) The combination of (i) and (ii) enables Bal31 to shorten dsDNA molecules progressively:

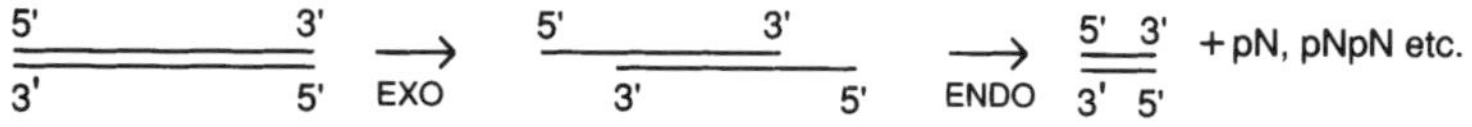

(iv) Inefficient ribonuclease (53).

Applications
(i) Controlled trimming of dsDNA molecules to remove unwanted sequences prior to cloning (54).

(ii) Generating overlapping subclones for DNA sequencing.

(iii) Generating nested deletions for linker-scanning mutagenesis (55).

(iv) Restriction mapping (49).

(v) Mapping DNA secondary structures (47, 52).

Reaction conditions
Bal31 has an absolute requirement for Mg^{2+} and Ca^{2+}. Reactions can therefore be stopped with EGTA. Incubate at 30°C. Some suppliers warn against freezing stocks of Bal31.

2 × Bal31 Buffer: 40 mM Tris-HCl (pH 7.4), 600 mM NaCl, 25 mM $MgCl_2$, 25 mM $CaCl_2$, 2 mM Na_2·EDTA.

B. CEREUS RIBONUCLEASE

Source
Bacillus cereus.

Description
One of several endoribonucleases displaying partial sequence specificity and used in enzymatic sequencing of RNA (56). *B. cereus* ribonuclease has a relatively low degree of specificity.

Properties
Endoribonuclease cutting primarily at the arrow in the sequences Up↑N and Cp↑N.

Applications
Enzymatic sequencing of RNA (57).

Reaction conditions
For RNA sequencing, use a reaction mixture containing 8 M urea and incubate at 50°C to reduce RNA secondary structure. Use the buffer recommended by the supplier.

Bst DNA POLYMERASE

Source
Recombinant *E. coli* expressing a cloned gene from *Bacillus stearothermophilus* (58).

Description
Bst DNA polymerase is equivalent to the Kornberg polymerase of *E. coli* but has a temperature optimum of 65°C (59). It is inactivated by incubation at 80°C for 10 min and has no 3′ to 5′ exonuclease activity.

Properties
(i) 5′ to 3′ DNA-dependent DNA polymerase, requiring a ssDNA template and a DNA or RNA primer with a 3′-OH terminus:

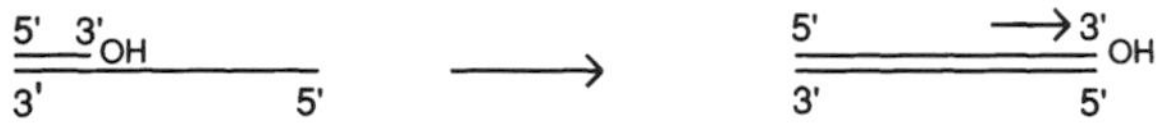

(ii) 5′ to 3′ exonuclease, degrading dsDNA or a DNA–RNA hybrid (including the RNA component) from a 5′-P terminus:

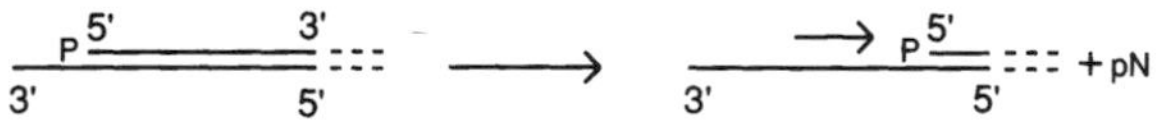

Applications
As permitted by the supplier.

Reaction conditions
Use the conditions recommended by the supplier.

Bst DNA POLYMERASE, LARGE FRAGMENT

Source
Recombinant *E. coli* expressing a modified version of a cloned gene from *Bacillus stearothermophilus*.

Description
This enzyme is an engineered version of *Bst* DNA polymerase lacking the 5′ to 3′ exonuclease activity. It is inactivated by incubation at 80°C for 10 min.

Properties
5′ to 3′ DNA-dependent DNA polymerase, requiring a ssDNA template and a DNA or RNA primer with a 3′-OH terminus:

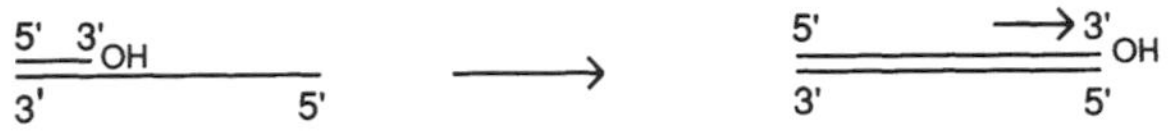

Applications
As permitted by the supplier.

Reaction conditions
Use the conditions recommended by the supplier.

CALF INTESTINE ALKALINE PHOSPHATASE (CIAP)

Source
Calf intestine (60).

Description
Dimeric glycoprotein, two identical 69 kd subunits. Contains four atoms of zinc per molecule (40). Inactivated by heating to 68°C. CIAP is active on 5′ overhangs, recessed 5′ ends and blunt ends.

Properties
5′ phosphatase, removing 5′ phosphate groups from ssDNA, dsDNA, ssRNA and dsRNA:

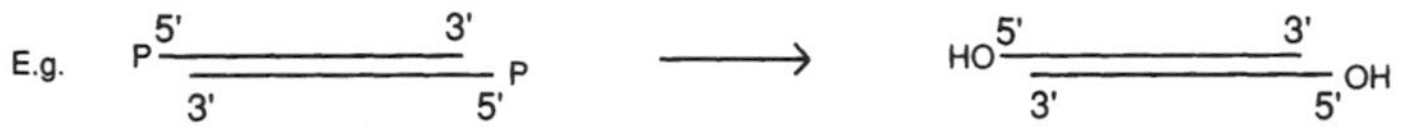

Applications
(i) Dephosphorylation of dsDNA to prevent self-ligation (40, 41).

(ii) Dephosphorylation of DNA and RNA to allow 5′ end labelling with T4 polynucleotide kinase (42).

(iii) As a component of a conjugated antibody immunodetection system (43, 44).

(iv) Component of a non-isotopic labelling method (45).

Reaction conditions
Alkaline phosphatases do not have particularly stringent buffer requirements. Incubate at 37°C for RNA and dsDNA with 5′ overhangs, or 56°C for blunt-ended dsDNA and dsDNA with 3′overhangs.

5 × CIAP Buffer: 250 mM Tris-HCl (pH 9.0), 50 mM $MgCl_2$, 5 mM $ZnCl_2$, 50 mM spermidine.

DEEP VENT DNA POLYMERASE

Marketed by New England Biolabs

Source

Recombinant *E. coli* expressing a cloned gene from *Pyrococcus* GB-D (61).

Description

A thermostable enzyme with greater temperature stability than either *Taq* polymerase or Vent DNA polymerase. It possesses a 3′ to 5′ exonuclease but no 5′ to 3′ exonuclease (62) and is highly processive.

Properties

(i) 5′ to 3′ DNA-dependent DNA polymerase, requiring a ssDNA template and a DNA or RNA primer with a 3′-OH terminus:

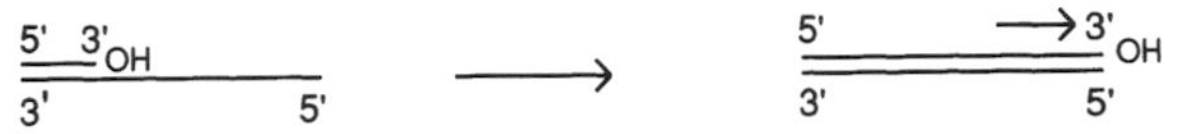

(ii) 3′ to 5′ exonuclease, degrading ssDNA or dsDNA from a 3′-OH terminus:

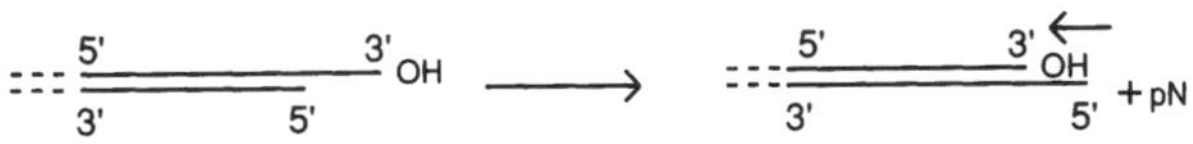

Applications

As permitted by the supplier.

Reaction conditions

The optimal Mg^{2+} concentration should be determined empirically for each experiment since it depends on the dNTP concentration and the nature of the template and primers. The provision of adequate buffering is critical and for this reason the supplier's buffer should be used, with the Mg^{2+} concentration adjusted as required.

DEEP VENT (exo⁻) DNA POLYMERASE

Marketed by New England Biolabs

Source

Recombinant *E. coli* expressing a modified version of a cloned gene from *Pyrococcus* GB-D (61).

Description

A thermostable enzyme with greater temperature stability than either *Taq* polymerase or Vent DNA polymerase. Deep Vent (exo^-) DNA polymerase is an engineered form of the Deep Vent enzyme lacking the 3′ to 5′ exonuclease. It is highly processive.

Properties

5′ to 3′ DNA-dependent DNA polymerase, requiring a ssDNA template and a DNA or RNA primer with a 3′-OH terminus:

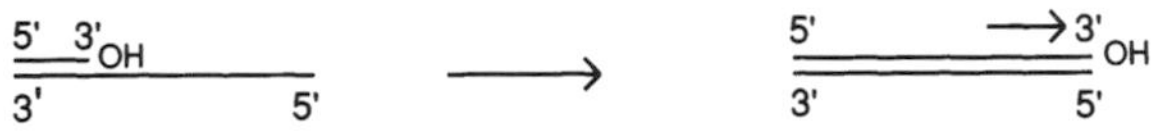

Applications

As permitted by the supplier.

Reaction conditions

The optimal Mg^{2+} concentration should be determined empirically for each experiment since it depends on the dNTP concentration and the nature of the template and primers. The provision of adequate buffering is critical and for this reason the supplier's buffer should be used, with the Mg^{2+} concentration adjusted as required.

DEOXYRIBONUCLEASE I

Source
Bovine pancreas.

Description
A glycoprotein, 31 kd, usually obtained as a mixture of several isoenzymes. Most suppliers market RNase-free versions.

Properties
Endodeoxyribonuclease, degrading ssDNA and dsDNA by cutting preferentially 5′ to pyrimidines, resulting in mono- and oligonucleotides with an average length of 4 nt (63, 64).

In the absence of Mn^{2+}:

In the presence of Mn^{2+}:

Applications
(i) Introducing nicks into dsDNA prior to labelling by nick translation (65).

(ii) DNase I protection and footprinting to locate proteins bound to DNA (66, 67).

(iii) Removal of DNA during RNA preparation.

(iv) Removal of the DNA template after *in vitro* transcription (68).

(v) Detection of transcriptionally active regions of chromatin (69).

(vi) Producing overlapping subclones for DNA sequencing (70).

(vii) Nicking circular DNA prior to bisulphite mutagenesis (71).

Reaction conditions
Requires Mg^{2+}, but this can be replaced by Mn^{2+} to cause dsDNA breaks (72). 37°C. For maximum activity include 20 mM $CaCl_2$ to the reaction mixture. Note that DNase I is very sensitive to physical denaturation by shaking, so mix solutions carefully.

10 × DNase Buffer: 500 mM Tris-HCl (pH 7.5), 10 mM $MgCl_2$, 1 mg ml^{-1} BSA.

RNase-free DNase I
Pancreatic DNase I is often contaminated with RNases. Commercial RNase-free preparations are available but the enzyme should be used with a suitable RNase inhibitor if RNase activity will be a problem. DNase I can be rendered RNase-free by dissolving at a concentration of 1 mg ml^{-1} in 0.1 M iodoacetic acid, 0.15 M NaAc (pH 5.2), heating to 55°C for 45 min, cooling on ice and then adding $CaCl_2$ to a concentration of 5 mM. Aliquots can be stored at –20°C for months.

DNA GYRASE

Source
Micrococcus luteus (73).

Description
Multimer probably of identical subunits. ATP is needed as a cofactor.

Properties
Type II DNA topoisomerase, able to introduce negative superhelical turns into cccDNA by double-stranded breakage followed by rejoining of the phosphodiester bonds (74, 75):

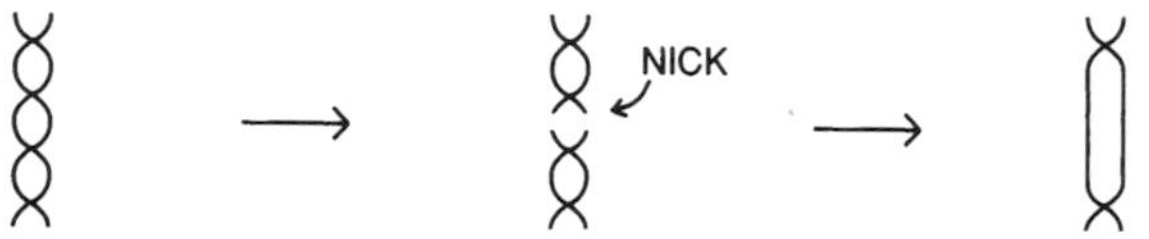

This activity enables DNA gyrase to reversibly knot and catenate cccDNA.

Applications
Introducing supercoils into plasmids and other cccDNA.

Reaction conditions
Requires Mg^{2+} and ATP. Incubate at 37°C. The addition of 10% glycerol to the reaction mixture improves DNA gyrase activity with some substrates.

5 × Gyrase Buffer: 175 mM Tris-HCl (pH 7.5), 100 mM KCl, 100 mM $MgCl_2$, 0.5 mM Na_2·EDTA, 10 mM spermidine, 5 mM ATP, 50 mM β-mercaptoethanol, 200 μg ml^{-1} BSA.

DNA POLYMERASE I, KLENOW FRAGMENT

Source
Escherichia coli, originally by subtilisin proteolysis of the Kornberg polymerase (76), now usually from a recombinant *E. coli* expressing a truncated *polA* gene (77).

Description
The large, or Klenow, fragment of DNA polymerase I has a molecular mass of 75 kd and lacks the 5′ to 3′ exonuclease activity of the Kornberg enzyme (76, 78–81). It has low processivity.

Properties

(i) 5′ to 3′ DNA-dependent DNA polymerase, requiring a ssDNA template and a DNA or RNA primer with a 3′-OH terminus:

(ii) 3′ to 5′ exonuclease, degrading ssDNA or dsDNA from a 3′-OH terminus:

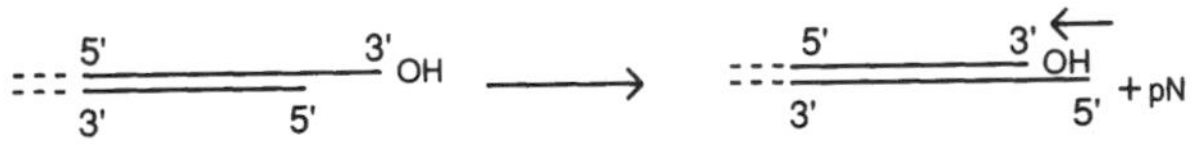

Applications
(i) End-filling and end-labelling dsDNA with a 5′ overhang (82–85)

(ii) Labelling ssDNA by random priming (86, 87).

(iii) Chain termination DNA sequencing (88).

(iv) Second strand synthesis of cDNA (89–92).

(v) Second strand synthesis in site-directed mutagenesis (93–97).

(vi) Production of ssDNA probes by primer extension (98–100).

Reaction conditions
The pH optimum is around 7.4 (101). Requires Mg^{2+}. Reactions are usually carried out at room temperature or 25°C, using different buffers for each application.

10 × End-Labelling Buffer: 500 mM Tris-HCl (pH 7.5), 100 mM $MgCl_2$, 10 mM DTT, 500 μg ml^{-1} BSA.

10 × Random Priming Buffer: 500 mM Tris-HCl (pH 7.5), 100 mM $MgCl_2$, 10 mM DTT, 500 μg ml^{-1} BSA.

10 × TM buffer (DNA sequencing): 100 mM Tris-HCl (pH 7.6), 600 mM NaCl, 66 mM $MgCl_2$.

2 × Second Strand cDNA Buffer: 200 mM HEPES–NaOH (pH 6.9), 140 mM KCl, 20 mM $MgCl_2$, 5 mM DTT.

DNA POLYMERASE I, KLENOW FRAGMENT (exo$^-$)

Source

A recombinant *E. coli* expressing a truncated *polA* gene (77) containing a double mutation that eliminates the 3′ to 5′ exonuclease activity (102).

Description

A genetically engineered version of the Klenow fragment of DNA polymerase I, molecular mass of 75 kd, lacking both the 5′ to 3′ and 3′ to 5′ exonuclease activities of the Kornberg enzyme. It has low processivity.

Properties

5′ to 3′ DNA-dependent DNA polymerase, requiring a ssDNA template and a DNA or RNA primer with a 3′-OH terminus:

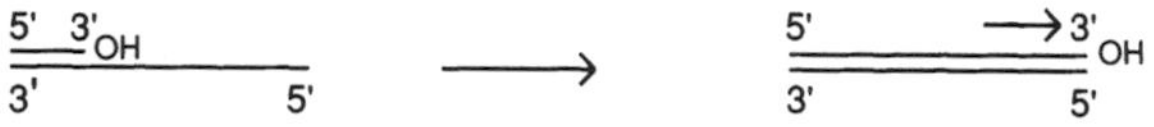

Applications

(i) Labelling ssDNA by random priming (86, 87).

(ii) Chain termination DNA sequencing (88).

(iii) Strand displacement amplification (103).

Reaction conditions

The pH optimum is around 7.4 (101). Requires Mg^{2+}. Reactions are usually carried out at room temperature or 25°C, using different buffers for each application.

10 × Random Priming Buffer: 500 mM Tris-HCl (pH 7.5), 100 mM $MgCl_2$, 10 mM DTT, 500 μg ml^{-1} BSA.

10 × TM buffer (DNA sequencing): 100 mM Tris-HCl (pH 7.6), 600 mM NaCl, 66 mM $MgCl_2$.

DNA POLYMERASE I (Kornberg polymerase)

Source

Escherichia coli. Most commercial preparations are obtained from an *E. coli* lysogen carrying a λ*polA* transducing phage, e.g. *E. coli* NM964 (104) or CM5199 (105).

Description

A single polypeptide chain, 109 kd, with one polymerase and two exonuclease activities (101, 106, 107). With dsDNA and excess dNTPs the 3′ to 5′ exonuclease is usually masked by the 5′ to 3′ polymerase. It has low processivity.

Properties

(i) 5′ to 3′ DNA-dependent DNA polymerase, requiring a ssDNA template and a DNA or RNA primer with a 3′-OH terminus:

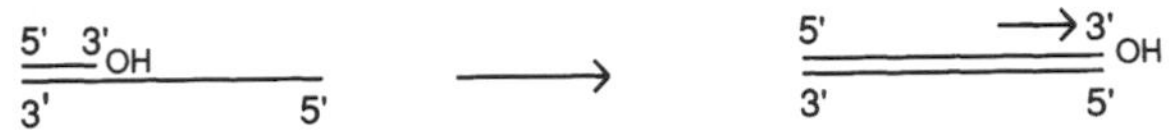

(ii) 5′ to 3′ exonuclease, degrading dsDNA or a DNA–RNA hybrid (including the RNA component) from a 5′-P terminus:

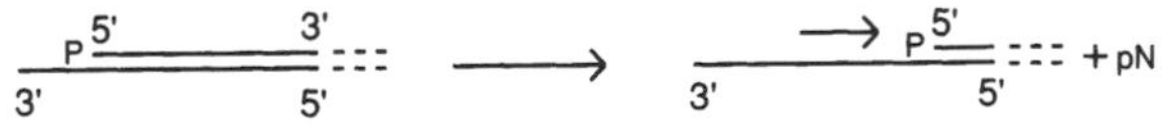

(iii) 3′ to 5′ exonuclease, degrading ssDNA or dsDNA from a 3′-OH terminus:

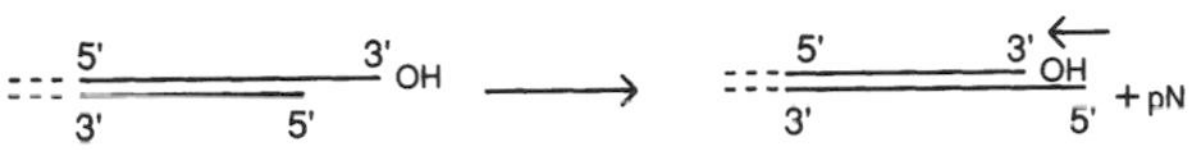

Applications

(i) Labelling DNA by nick translation (33, 65, 98, 104).

(ii) Second strand cDNA synthesis in conjunction with RNase H (30).

Reaction conditions

The pH optimum is around 7.4 (101). Requires Mg^{2+}. Reactions are usually carried out at room temperature or 25°C.

10 × Nick Translation Buffer: 500 mM Tris-HCl (pH 7.4), 100 mM $MgCl_2$, 10 mM DTT, 500 μg ml^{-1} BSA.

DNA TOPOISOMERASE I

Source
Calf thymus or wheat germ.

Description
Single polypeptide, 105 kd (108).

Properties
Type I DNA topoisomerase, able to relax supercoiled DNA by introducing transient single-stranded breaks in the sugar–phosphate backbone. After the intact strand has been passed through the break, the free polynucleotide ends are rejoined (75, 109, 110):

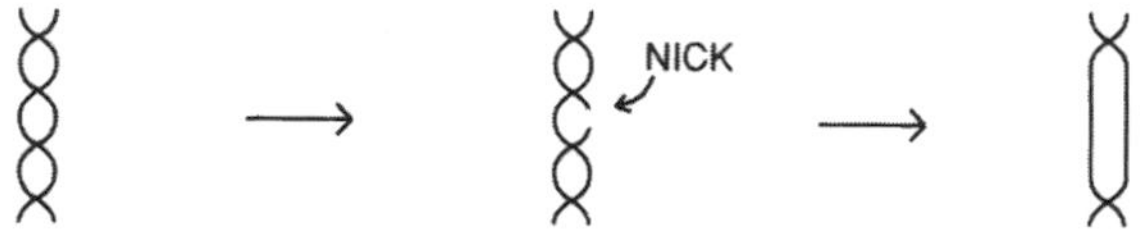

This activity enables topoisomerase I to:

(i) Catenate and decatenate nicked circular dsDNA.

(ii) Attach a ssDNA molecule to a 5′-OH terminus on a second ssDNA or dsDNA molecule (111–113).

Applications
(i) Studying nucleosome assembly (114, 115).

(ii) Studying DNA tertiary structure (116, 117).

(iii) Studying effects of DNA supercoiling on transcription *in vitro* (118, 119).

(iv) Assaying for mutant plasmids that differ in length by as little as 1 bp (120).

Reaction conditions
Requires Mg^{2+}. The pH optimum is 7.5 and the enzyme is active in 200 mM NaCl. Incubate at 37°C.

10 × TopoI Buffer: 500 mM Tris-HCl (pH 7.5), 500 mM KCl, 100 mM $MgCl_2$, 5 mM DTT, 1 mM Na_2·EDTA, 300 µg ml^{-1} BSA.

E. COLI DNA LIGASE

Source

Escherichia coli 594 (*su*⁻) lysogenic for λgt4-*lop*-11 *lig* *S*am7 (121, 122).

Description

Monomer, 74 kd (123), functionally similar to T4 DNA ligase but requires NAD instead of ATP as the cofactor and is more stringent in its activity: for instance, being inactive with RNA-primed DNA (124, 125), producing fewer spurious ligation products with cohesive ends and having no activity in blunt-end ligation.

Properties

DNA ligase, catalysing the formation of a phosphodiester bond between adjacent 5′-P and 3′-OH termini (126), e.g. with cohesive-ended molecules:

Applications

Joining DNA molecules, though generally T4 DNA ligase is preferred.

Reaction conditions

Requires Mg^{2+} and NAD. Reactions conditions should be optimized for different substrates.

10 × Basic EcL Buffer: 300 mM Tris-HCl (pH 8.0), 40 mM $MgCl_2$, 12 mM Na_2·EDTA, 10 mM DTT, 0.26 mM NAD, 500 μg ml^{-1} BSA.

E. COLI RNA POLYMERASE

Source
Escherichia coli.

Description
Both the core enzyme (subunit structure $\alpha_2\beta\beta'$) and the holoenzyme ($\alpha_2\beta\beta'\sigma$) are available commercially. The core enzyme can transcribe efficiently but requires the σ subunit in order to recognize the *E. coli* consensus promoter sequence (127).

Properties
DNA-dependent RNA polymerase; the holoenzyme is able to recognize a variety of promoter sequences related to the *E. coli* consensus:

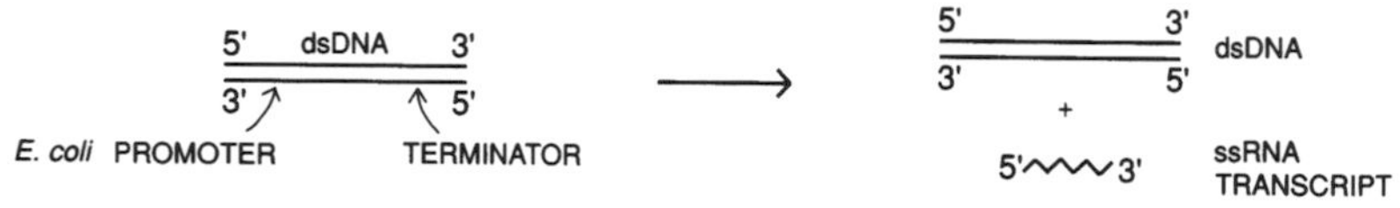

Applications
(i) *In vitro* transcription of genes with suitable promoters.

(ii) Synthesis of labelled RNA (128).

Reaction conditions
Requires Mg^{2+}; 37°C. EDTA is recommended.

10 × RNA Pol Buffer: 400 mM Tris-HCl (pH 7.9), 1500 mM KCl, 100 mM $MgCl_2$, 1 mM Na_2·EDTA, 1 mM DTT, 5 mg ml^{-1} BSA.

EXONUCLEASE I

Source
Escherichia coli.

Description
Monomer, 55 kd.

Properties
3′ to 5′ exodeoxyribonuclease, specific for ssDNA:

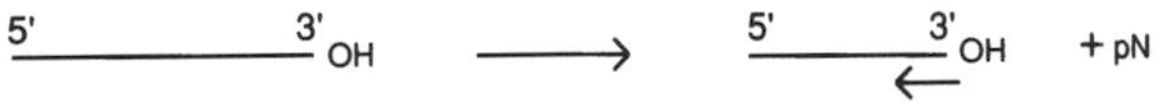

Applications
(i) Removing residual primers after PCR.

(ii) Creating blunt ends on dsDNA fragments with 3′ overhangs.

(iii) Studying DNA helicase activity (129).

(iv) Identifying regions of ssDNA in partially dsDNA molecules (130).

Reaction conditions
Incubate at 37°C.

10× Exo I Buffer: 670 mM glycine–KOH (pH 9.5), 67 mM $MgCl_2$, 100 mM β-mercaptoethanol.

EXONUCLEASE III

Source

E. coli BE247 or SR80, both of which contain a thermoinducible gene on the plasmid pSGR3 (131, 132).

Description

Monomer, 28 kd (133), with a number of activities. The exonuclease is inactive on ssDNA or dsDNA with a 3′ overhang of 4 nt or more.

Properties

(i) 3′ to 5′ exodeoxyribonuclease, specific for dsDNA with 3′-OH termini (133):

(ii) 3′ phosphatase, removing phosphate groups from the 3′ termini of ssDNA or dsDNA (132, 134):

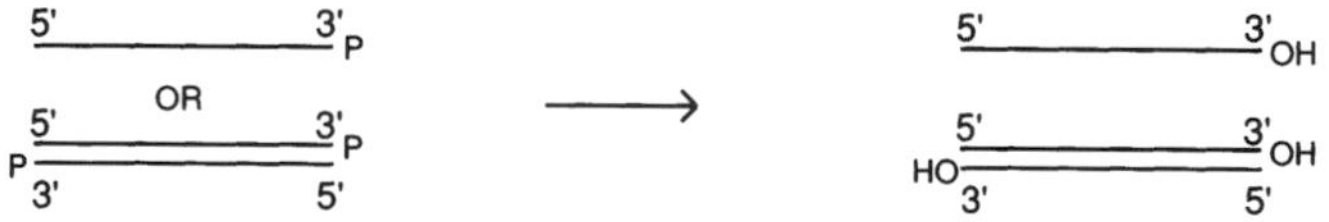

(iii) Endodeoxyribonuclease, specific for nucleotides from which the purine or pyrimidine base has been cleaved (132, 134, 135).

(iv) Endoribonuclease, specific for the RNA component of a DNA–RNA hybrid (132, 134):

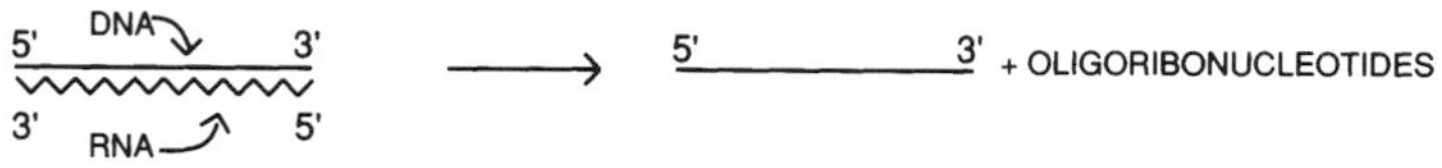

Applications

(i) Generation of overlapping subclones for DNA sequencing (135, 136–139).

(ii) End-labelling of blunt-ended dsDNA, in conjunction with Klenow polymerase (140, 141).

(iii) Deletion of sequences at the termini of dsDNA fragments, in conjunction with a single-strand specific endonuclease (135, 142).

(iv) Mutagenesis techniques (143).

(v) Protein protection studies (144–146).

(vi) Generating nested deletions for linker-scanning mutagenesis (55).

Reaction conditions

The exo- and endonucleases have pH optima of 7.6–8.5. The optimum for the phosphatase is pH 6.8–7.4. Requires Mg^{2+}. Salt may be added to the reaction mixture for certain applications.

10 × Basic Exo III Buffer: 500 mM Tris-HCl (pH 8.0), 50 mM $MgCl_2$, 100 mM β-mercaptoethanol.

EXONUCLEASE V

Source
Micrococcus luteus.

Description
An ATP-requiring deoxyribonuclease.

Properties
3′ to 5′ and 5′ to 3′ exodeoxyribonuclease, specific for ssDNA and dsDNA:

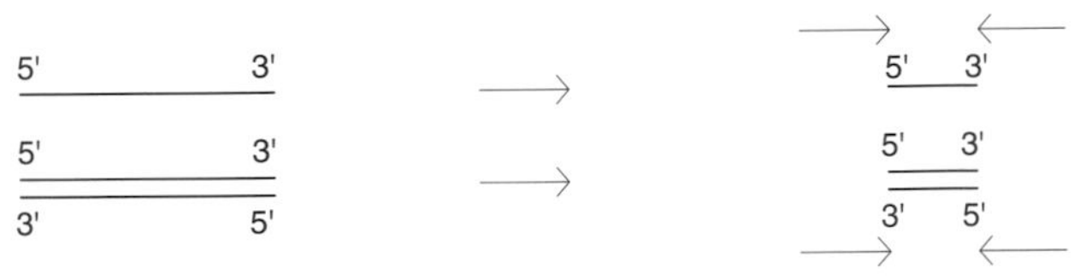

Applications
Apparently none reported other than the removal of DNA from RNA preparations.

Reaction conditions
Requires ATP. Incubate at 37°C. Use the buffer conditions recommended by the supplier.

EXONUCLEASE VII

Source
Escherichia coli HMS137.

Description
An exonuclease that can attack both the 5′ and 3′ termini of ssDNA. Exonuclease VII produces oligonucleotides, which means it may not leave blunt ends when used with dsDNA with overhangs.

Properties
3′ to 5′ and 5′ to 3′exodeoxyribonuclease, specific for ssDNA (147–149):

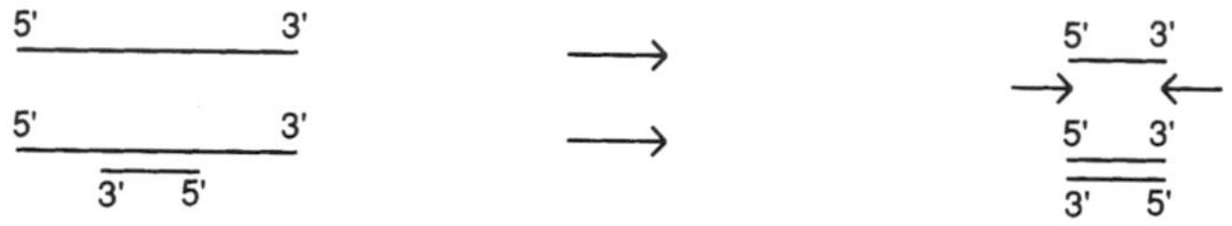

Applications
(i) Exon mapping (150, 151). Exonuclease VII has no endonucleolytic activity, so can be used in transcript mapping to distinguish ssDNA loops from ssDNA overhangs in DNA–RNA hybrids.

(ii) Excision of segments of DNA ligated into plasmid vectors by poly(dA–dT) tailing (152).

Note that exonuclease VII is not suitable for blunt-ending dsDNA as its mode of action may result in single nucleotide overhangs remaining after treatment.

Reaction conditions
Does not require Mg^{2+} and is active in 8 mM EDTA. The optimum temperature is 37°C, but exonuclease VII retains most of its activity at 42°C. The higher temperature plus low salt (up to 30 mM KCl) may be used to destabilize secondary structures in ssDNA.

10 × Exo VII Buffer: 670 mM K phosphate (pH 7.9), 83 mM Na_2·EDTA, 100 mM β-mercaptoethanol.

HIV-1 REVERSE TRANSCRIPTASE

Source
Recombinant *Escherichia coli* expressing the 66 kd and 51 kd subunits.

Description
Structurally and functionally similar to AMV reverse transcriptase.

Properties
(i) 5′ to 3′ RNA-dependent DNA polymerase, requiring a ssRNA template and a DNA or RNA primer with a 3′-OH terminus:

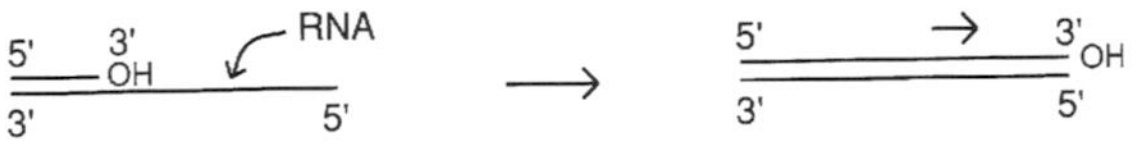

(ii) 5′ to 3′ and 3′ to 5′ exoribonuclease, specific for the RNA component of a DNA–RNA hybrid:

(iii) DNA endonuclease.

(iv) DNA unwinding activity.

Applications
Used primarily in studies of HIV-1 replication and in screening potential reverse transcriptase inhibitors. Can be used in recombinant DNA experiments in the same way as AMV reverse transcriptase.

Reaction conditions
Requires Mg^{2+}, usually incubated at 37°C.

10 × HIV-1 RT Buffer: 500 mM Tris-HCl (pH 8.3), 500 mM KCl, 100 mM $MgCl_2$, 30 mM DTT.

HK ALKALINE PHOSPHATASE

Marketed by Epicentre Technologies

Source
Antarctic bacterium.

Description
A bacterial alkaline phosphatase that is completed inactivated by incubation at 65°C for 15 min. HK alkaline phosphatase is not active on recessed 5′ ends and has only low activity with blunt ends. Unlike other bacterial alkaline phosphatases, preparations of this enzyme are free from contaminating nucleases.

Properties
5′ phosphatase:

Applications
(i) Dephosphorylation of dsDNA to prevent self-ligation (153, 154).

(ii) Dephosphorylation of DNA and RNA to allow 5′ end labelling with T4 polynucleotide kinase.

Reaction conditions
Alkaline phosphatases do not have particularly stringent buffer requirements. The enzyme has maximal activity in the presence of 5 mM $CaCl_2$ but this is usually omitted from the reaction buffer because Ca^{2+} inhibits the restriction enzymes used in subsequent steps.

10 × HKP Buffer: 330 mM Tris-acetate (pH 7.8), 660 mM KAc, 100 mM MgAc, 5 mM DTT.

HOT *Tub* DNA POLYMERASE

Marketed by Amersham International

Source

Described by the supplier as '*Thermus ubiquitous*'.

Description

A thermostable enzyme. There is no information on the presence or absence of exonuclease activities.

Properties

5′ to 3′ DNA-dependent DNA polymerase, requiring a ssDNA template and a DNA or RNA primer with a 3′-OH terminus:

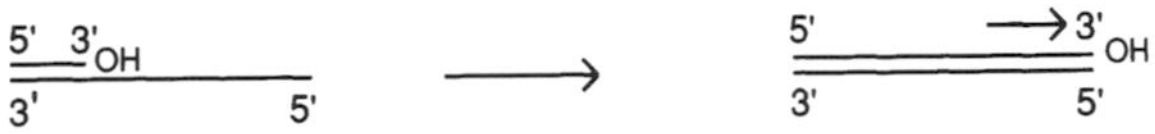

Applications

As permitted by the supplier.

Reaction conditions

The optimal Mg^{2+} concentration should be determined empirically for each experiment since it depends on the dNTP concentration and the nature of the template and primers. The provision of adequate buffering is critical and for this reason the supplier's buffer should be used, with the Mg^{2+} concentration adjusted as required.

I-*Ceu* I

Source
Recombinant *E. coli* containing a cloned sequence from *Chlamydomonas eugametos* mitochondrial DNA (155, 156).

Description
This is the endonuclease encoded by the group I intron from the *Chlamydomonas eugametos* mitochondrial LrRNA gene.

Properties
Endodeoxyribonuclease, specific for dsDNA and leaving a 3′ overhang of 4 nucleotides after cutting at the recognition sequence (with some variations) (156, 157):

```
5′-TAACTATAACGGTC CTAA↑GGTAGCGA-3′
3′-ATTGATATTGCCAG↑GATT CCATCGCT-5′
```

Applications
(i) Removal from clones of inserts flanked by the appropriate linker sequences.

(ii) As a rare cutting enzyme for physical mapping and cloning of chromosomal DNA.

Reaction conditions
Incubate at 37°C.

10 × Ceu Buffer: 100 mM Tris-HCl (pH 8.6), 100 mM $MgCl_2$, 10 mM DTT, 1 mg ml^{-1} BSA.

LAMBDA EXONUCLEASE

Source

Escherichia coli λ lysogen SG5519.

Description

This exonuclease is 10–100 times more active with dsDNA (blunt-ended or with a 3′ overhang) than with ssDNA (158, 159), but is inefficient with dsDNA with a 5′ overhang. Nicks and gaps are not usually recognized as starting points.

Properties

5′ to 3′ exodeoxyribonuclease, specific for dsDNA preferably with a terminal 5′-P (160):

Applications

(i) Removing 5′ overhangs on dsDNA, especially prior to labelling with terminal transferase (161).

(ii) Restriction mapping.

(iii) Selective digestion of one strand of a PCR product prior to direct sequencing (162).

Reaction conditions

Lambda exonuclease requires Mg^{2+} and a relatively high pH. 37°C is recommended, though the enzyme is active at room temperature.

10 × LE Buffer: 670 mM glycine–KOH (pH 9.4), 25 mM $MgCl_2$, 500 μg ml^{-1} BSA.

LAMBDA TERMINASE

Source
Recombinant *E. coli* JM109 containing a plasmid carrying the λ *A* and *Nu* genes.

Description
Lambda terminase is the gene *A* product responsible for cleaving replicated λ DNA concatamers into individual genomes. It recognizes a region of about 100 bp and cleaves specifically within the *cos* site.

Properties
Endodeoxyribonuclease, specific for dsDNA containing sequences with high similarity to the λ cos region (*c*. 100 bp) and resulting in 5′ overhangs, usually of 12 nucleotides but sometimes of 4–5 nucleotides (163), within the recognition sequence:

```
5′- CCCGCCGCTGGA↑-3′
3′-↑GGGCGGCGACCT -5′
```

Applications
(i) Linearization of recombinant cosmids prior to transformation.

(ii) End-labelling linearized cosmid clones for restriction mapping of inserts (164).

(iii) As a rare cutting enzyme for physical mapping of chromosomal DNA (165).

Reaction conditions
High concentrations of enzyme (up to 20 units per μg DNA in a 1 h digestion) are recommended for complete cleavage. ATP is needed as a cofactor. Spermidine, putrescine and Triton-X100 are usually added to increase the reactivity. Requires Mg^{2+}; 30°C.

10 × LT Buffer: 130 mM Tris-HCl (pH 8.0), 500 mM KAc, 30 mM $MgCl_2$, 5 mM Na_2·EDTA, 20 mM spermidine, 40 mM putrescine, 10 mM ATP, 50 mM DTT, 0.1% Triton X-100.

MEGANUCLEASE I-*Sce* I

Source
Recombinant *E. coli* TG1 containing a cloned sequence from *Saccharomyces cerevisiae* mitochondrial DNA.

Description
This is the endonuclease encoded by the group I intron from the yeast mitochondrial 21S rRNA gene.

Properties
Endodeoxyribonuclease, specific for dsDNA and leaving a 3′ overhang of 4 nucleotides after cutting at the recognition sequence (or variants with single base pair changes):

```
5′-TAGGG ATAA↑CAGGGTAAT-3′
3′-ATCCC↑TATT GTCCCATTA-5′
```

Applications
(i) Removal from clones of inserts flanked by the appropriate linker sequences.

(ii) As a rare cutting enzyme for physical mapping and cloning of chromosomal DNA, though target sites are apparently absent from human DNA.

Reaction conditions
Use the conditions recommended by the supplier.

MICROCOCCAL DNA POLYMERASE

Source
Micrococcus luteus.

Description
Functionally identical to the Kornberg DNA polymerase I of *E. coli* but has been preferred for certain specific applications.

Properties
(i) 5′ to 3′ DNA-dependent DNA polymerase, requiring a ssDNA template and a DNA or RNA primer with a 3′-OH terminus:

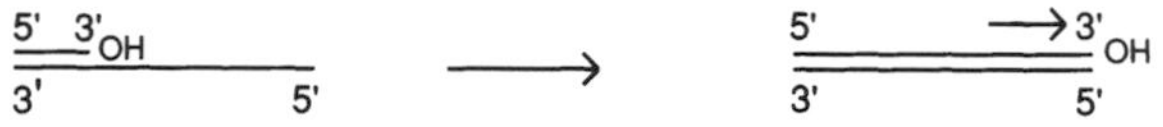

(ii) 5′ to 3′ exonuclease, degrading dsDNA:

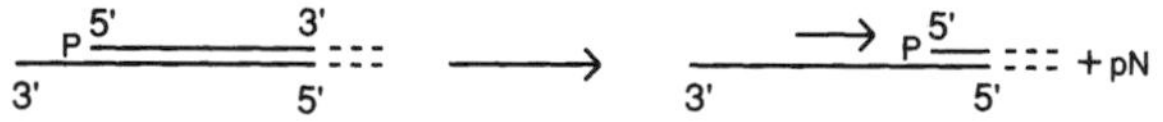

(iii) 3′ to 5′ exonuclease, degrading ssDNA or dsDNA from a 3′-OH terminus:

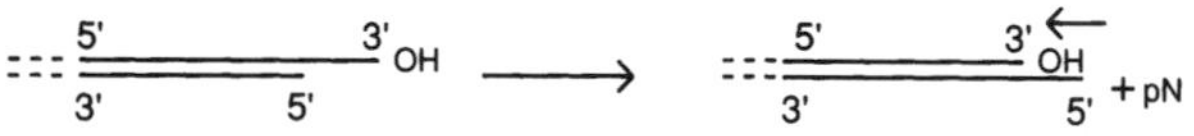

Applications
(i) Mutagenesis procedures (7, 166).

(ii) Synthesis of artificial copolymers.

Reaction conditions
Activity is reduced in Tris buffers. Requires Mg^{2+}; 37°C.

10 × MDP Buffer: 400 mM K phosphate (pH 7.0), 80 mM $MgCl_2$, 10 mM β-mercaptoethanol.

M-MuLV REVERSE TRANSCRIPTASE

Source

Recombinant *E. coli* containing plasmid pB6B15.23, which carries the *pol* gene of Moloney murine leukaemia virus (167).

Description

The enzyme is obtained as a fusion protein comprising the first six amino acids of *trpE* followed by the M-MuLV *pol* gene, lacking the last 48 amino acids (167). Total molecular mass is 71 kd. Less heat stable than AMV and RAV2 reverse transcriptases.

Properties

(i) 5′ to 3′ DNA- or RNA-dependent DNA polymerase, requiring a ssDNA or ssRNA template and a DNA or RNA primer with a 3′-OH terminus (26, 27):

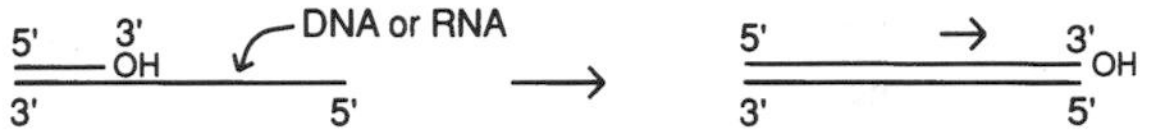

(ii) 5′ to 3′ and 3′ to 5′ exoribonuclease, specific for the RNA component of a DNA–RNA hybrid, much reduced compared with AMV reverse transcriptase (168):

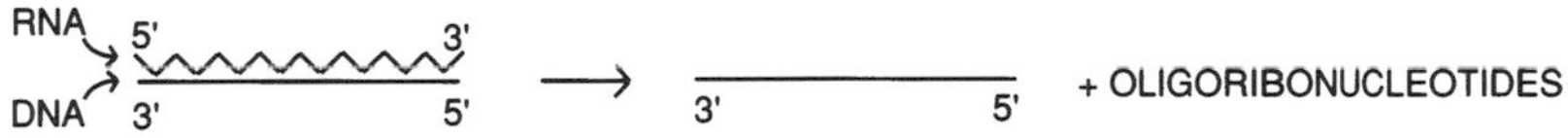

(iii) DNA unwinding activity.

Applications

For comparisons of AMV and M-MuLV reverse transcriptases in cDNA synthesis see refs 74 and 169.

(i) First strand cDNA synthesis (167, 170–172).

(ii) Polymerase chain reaction with RNA substrates (RT-PCR) (173).

(iii) Chain termination RNA sequencing (34).

(iv) Modified chain termination DNA sequencing, especially to read GC-rich regions (35).

(v) Probing RNA secondary structure (36).

(vi) 3′ end labelling DNA fragments (37).

Reaction conditions

Requires Mg^{2+}. M-MuLV reverse transcriptase can be incubated at 37°C. Actinomycin D (50 μg ml^{-1}) can be added to the reaction mix to inhibit the DNA-dependent DNA polymerase activity. Some suppliers recommend including 4 mM $MnCl_2$ in the reaction mixture.

10 × M-MuLV RT Buffer: 500 mM Tris-HCl (pH 8.3), 750 mM KCl, 30 mM $MgCl_2$, 100 mM DTT, 1 mg ml^{-1} BSA.

mRNA GUANYLTRANSFERASE

Source
Vaccinia virus or *Saccharomyces cerevisiae pep4*.

Description
An enzyme complex with a very high specificity for RNA molecules with a 5′ triphosphate terminus.

Properties
A three-stage enzymatic activity that attaches the 5′ cap structure to a suitable acceptor RNA.

(i) RNA triphosphatase:

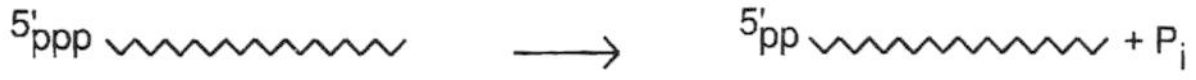

(ii) Guanyltransferase:

(iii) RNA (guanine-7) methyltransferase:

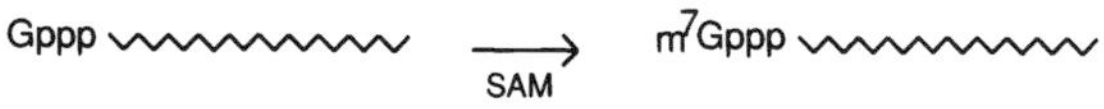

Applications
In transcript mapping, to distinguish a 5′ terminus created by transcription initiation from one resulting from RNA processing (174–178).

Reaction conditions
Requires Mg^{2+}; 37°C.

10 × GT Buffer: 500 mM Tris-HCl (pH 7.9), 60 mM KCl, 12.5 mM $MgCl_2$, 25 mM DTT, 400 μM GTP.

MUNG BEAN NUCLEASE

Source

Mung bean (*Vigna radiata*) sprouts.

Description

Functionally identical to S1 nuclease (179–183) but less likely to cause nicks in dsDNA, and also less likely to cut the intact strand opposite a nick.

Properties

Single strand-specific endonuclease, more active on ssDNA than ssRNA:

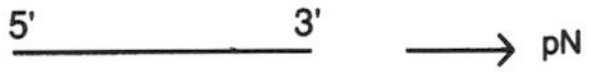

Applications

Useful as an alternative to S1 nuclease when cleavage of double-stranded molecules is a problem (182, 184–189).

Reaction conditions

Requirements are similar to S1 nuclease, except that the optimal pH is a little higher (pH 5.0) and the optimal temperature is 30°C. Mung bean nuclease is very 'sticky' and glycerol is added to prevent the enzyme adhering to the reaction tube.

10 × Mung Bean Buffer: 300 mM NaAc (pH 5.0), 500 mM NaCl, 10 mM $ZnSO_4$, 50% glycerol.

N. CRASSA NUCLEASE

Source
Neurospora crassa.

Description
Acts as an endonuclease on ssRNA and ssDNA, and as an exonuclease on ssDNA and dsDNA (190). The substrate can therefore be either linear or circular. At low enzyme concentration nicks can be made in supercoiled DNA (191). This is the only enzyme known to produce RNA fragments and ribonucleotides with 5′ terminal phosphates (192, 193).

Properties
(i) Endonuclease active on ssRNA and ssDNA:

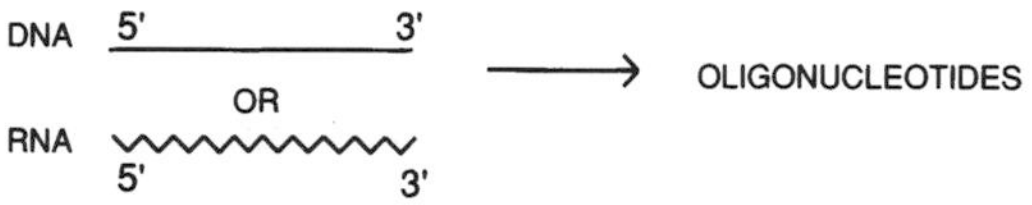

(ii) Exonuclease active on ssDNA and dsDNA:

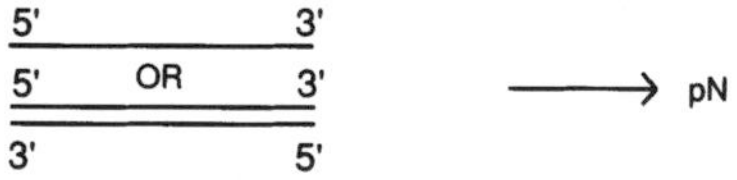

Applications
Degrading non-paired single-stranded regions in DNA molecules, for example during hybridization experiments (194).

Reaction conditions
The optimum 1 × buffer is 100 mM Tris-HCl (pH 7.5–8.5) with the salt and Mg^{2+} concentrations dependent on the substrate. Mg^{2+} is not essential but 10 mM $MgCl_2$ stimulates the activity on DNA and represses that on RNA. The exonuclease is inhibited by Zn^{2+}.

P1 NUCLEASE

Source

Penicillium citrinum.

Description

A Zn metalloprotein, highly glycosylated, 24 kd.

Properties

Non-specific phosphodiesterase that acts both as an endonuclease and an exonuclease. P1 is most active with ssDNA and ssRNA but also has substantial activity with dsDNA and dsRNA (195):

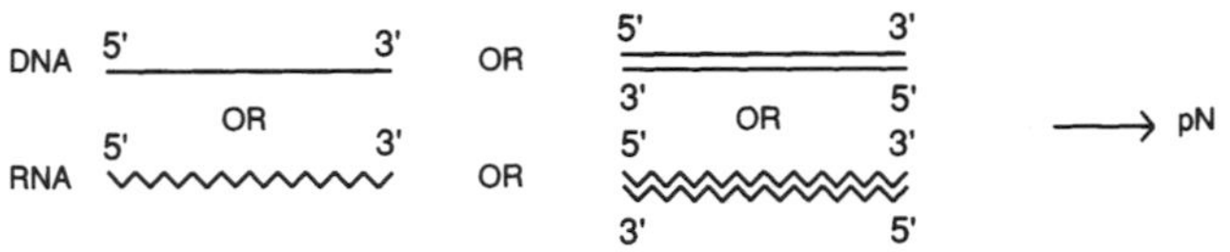

Applications

(i) Mobility shift analysis in RNA sequencing.

(ii) Possibly as an alternative to S1 and mung bean nucleases for ssDNA or ssRNA cleavage, though the high dsDNA and dsRNA activities make P1 nuclease a poor choice.

Reaction conditions

The optimal pH is between 4.5 and 8.0, depending on the substrate (195, 196). Zn^{2+} is not essential. The optimal temperature is 70°C, but 37°C is generally used. A suitable 10 × buffer is 500 mM NaAc pH 5.5.

Pfu DNA LIGASE

Marketed by Stratagene

Source

Pyrococcus furiosus.

Description

A thermostable ATP-requiring DNA ligase.

Properties

DNA ligase, catalysing the formation of a phosphodiester bond between adjacent 5′-P and 3′-OH termini hybridized to a complementary target DNA:

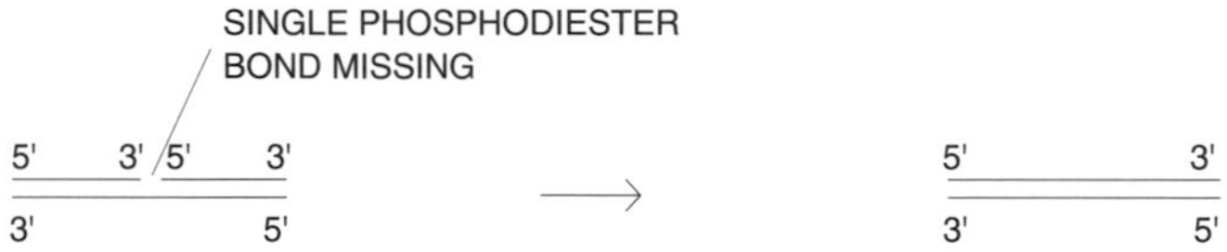

Applications

Because of the instability of base-paired structures at elevated temperatures, *Pfu* DNA ligase cannot be used as a general purpose ligation enzyme. It has specialist applications:

(i) Ligase chain reaction (1–5).

(ii) Repeat expansion detection (21).

(iii) High-fidelity gene synthesis.

(iv) *In vitro* mutagenesis procedures (22, 23).

(v) Padlock probe hybridization system (24).

Reaction conditions

Requires Mg^{2+} and ATP. Reactions conditions should be optimized for different substrates.

10 × Basic *Pfu* Ligase Buffer: 200 mM Tris-HCl (pH 7.5), 200 mM KCl, 100 mM $MgCl_2$, 0.1 mM ATP, 10 mM DTT, 1% Nonidet P40.

Pfu DNA POLYMERASE

Marketed by Stratagene

Source
Pyrococcus furiosus.

Description
A thermostable enzyme, with high temperature stability and claimed by the suppliers as 'the most accurate thermostable enzyme during DNA synthesis' (197–199). It possesses a 3′ to 5′ exonuclease. It is not clear if it also has a 5′ to 3′ exonuclease.

Properties
(i) 5′ to 3′ DNA-dependent DNA polymerase, requiring a ssDNA template and a DNA or RNA primer with a 3′-OH terminus:

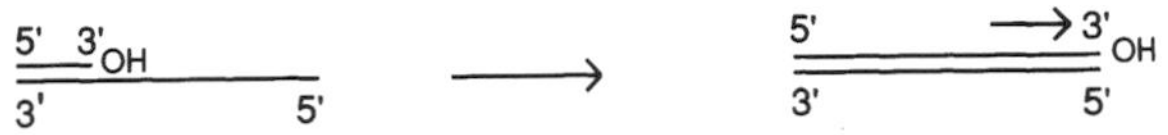

(ii) 3′ to 5′ exonuclease, degrading ssDNA or dsDNA from a 3′-OH terminus:

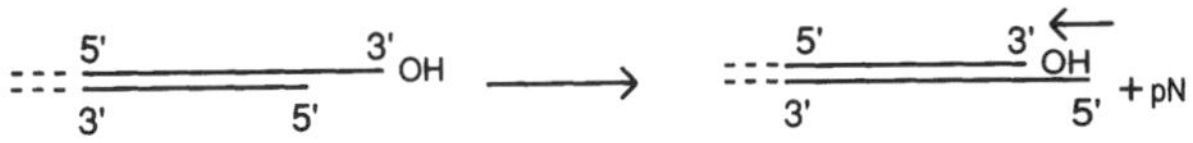

Applications
As permitted by the supplier.

Reaction conditions
The optimal Mg^{2+} concentration should be determined empirically for each experiment since it depends on the dNTP concentration and the nature of the template and primers. The provision of adequate buffering is critical and for this reason the supplier's buffer should be used, with the Mg^{2+} concentration adjusted as required.

Pfu (exo⁻) DNA POLYMERASE

Marketed by Stratagene

Source

Pyrococcus furiosus.

Description

A modified version of *Pfu* DNA polymerase, lacking the 3′ to 5′ exonuclease.

Properties

5′ to 3′ DNA-dependent DNA polymerase, requiring a ssDNA template and a DNA or RNA primer with a 3′-OH terminus:

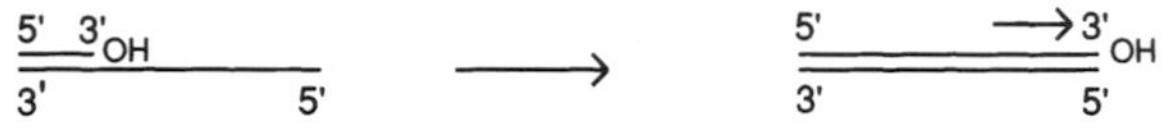

Applications

As permitted by the supplier.

Reaction conditions

The optimal Mg^{2+} concentration should be determined empirically for each experiment since it depends on the dNTP concentration and the nature of the template and primers. The provision of adequate buffering is critical and for this reason the supplier's buffer should be used, with the Mg^{2+} concentration adjusted as required.

PHOSPHODIESTERASE I

Source
Crotalus durissus venom (200).

Description
Glycoprotein, 115 kd (201). Acts on both DNA and RNA with the appropriate termini.

Properties
Exonuclease, specific for DNA and RNA with a 3′-OH terminus, yielding 5′-dNMPs and NMPs:

E.g.

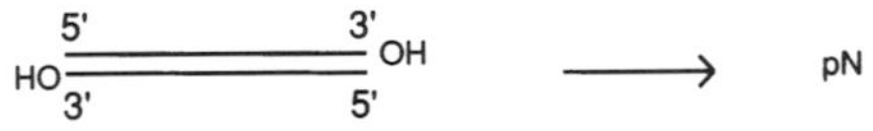

Applications
DNA and RNA structural studies.

Reaction conditions
Requires Mg^{2+}; 25°C.

10 × PI Buffer: 1000 mM Tris-HCl (pH 8.9), 1000 mM NaCl, 140 mM $MgCl_2$.

PHOSPHODIESTERASE II

Source
Calf spleen.

Description
Acts on both DNA and RNA with the appropriate termini.

Properties
Exonuclease, specific for DNA and RNA with a 5′-OH terminus, yielding 3′-dNMPs and NMPs:

E.g.

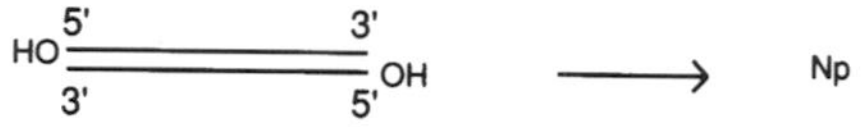

Applications
DNA and RNA structural studies.

Reaction conditions
Does not require Mg^{2+}; 37°C.

2 × PII Buffer: 250 mM succinate–HCl (pH 6.5).

PI-*Psp* I

Source
Recombinant *E. coli* containing a cloned sequence from *Pyrococcus* GB-D (202).

Description
This is an endonuclease that forms part of the DNA polymerase polypeptide of *Pyroccocus* GB-D (202). The two enzyme activities are separated by proteolytic cleavage of the pre-protein.

Properties
Endodeoxyribonuclease, specific for dsDNA and leaving a 3′ overhang of 4 nucleotides after cutting at the recognition sequence (with some variations):

```
5′-TGGCAAACAGCTA TTAT↑GGGTATTATGGGT-3′
3′-ACCGTTTGTCGAT↑AATA CCCATAATACCCA-5′
```

Applications
(i) Removal from clones of inserts flanked by the appropriate linker sequences.

(ii) As a rare cutting enzyme for physical mapping and cloning of chromosomal DNA.

Reaction conditions
Incubate at 65°C.

10 × *Psp* Buffer: 100 mM Tris-HCl (pH 8.6), 1500 mM KCl, 100 mM $MgCl_2$, 10 mM DTT, 1 mg ml^{-1} BSA.

POLY(A) POLYMERASE

Source

Escherichia coli.

Description

Monomer, 58 kd. As well as polymerizing ATP the enzyme uses CTP and UTP at 5% maximal activity, but does not work with GTP. It is specific for ssRNA, having no activity with ssDNA or dsDNA, and negligible activity with dsRNA. Short ssRNAs (e.g. trinucleotides) are poor acceptors. Up to 2000 residues can be added to a good acceptor.

Properties

Template-independent 3′ poly(A) polymerase, specific for ssRNA with a 3′-OH terminus (203):

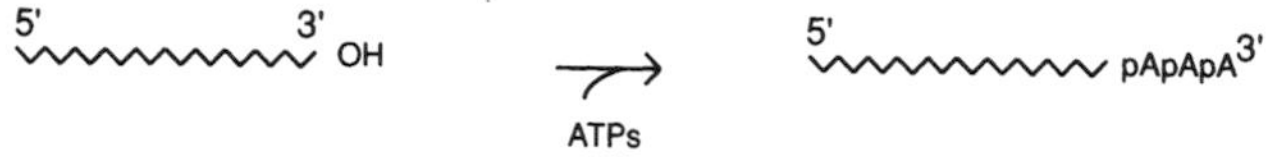

Applications

(i) Labelling RNA with a tail of labelled nucleotides (203) or a single cordycepin monophosphate (204).

(ii) Providing a poly(A) tail for priming cDNA synthesis with oligo(dT) (205).

Reaction conditions

Requires Mg^{2+} and a high concentration of monovalent cations. Stimulated by Mn^{2+}. Incubate at 37°C.

10 × Poly(A) Buffer: 500 mM Tris-HCl (pH 7.9), 2500 mM NaCl, 100 mM $MgCl_2$, 25 mM $MnCl_2$, 500 mg ml^{-1} BSA. Use 0.25 mM ATP in the final mix for a complete reaction with 250 µg RNA.

POLYNUCLEOTIDE PHOSPHORYLASE

Source

Micrococcus luteus or *Escherichia coli* B.

Description

Tetramer, total molecular mass 260 kd. The polymerization reaction is reversible with excess inorganic phosphate. The enzyme will use modified NDPs as substrates, especially if Mg^{2+} is replaced by Mn^{2+}.

Properties

Template-independent ribonucleotide polymerase, able to add ribonucleotides to the 3′-OH terminus of an RNA acceptor:

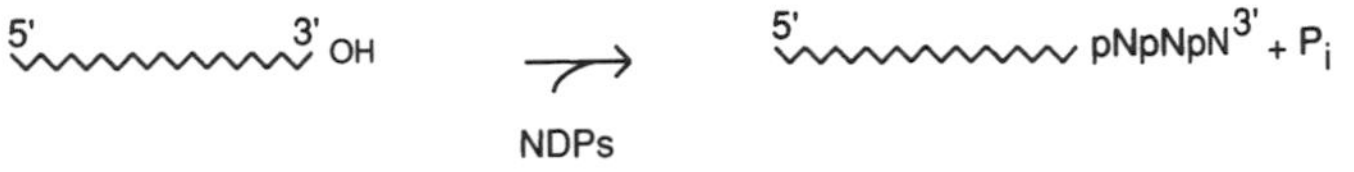

The enzyme also catalyses an exchange reaction between inorganic phosphate and the β-phosphate of an NDP:

$$^{*}P_i + N\text{-}P\text{-}P \longrightarrow P_i + N\text{-}P\text{-}P^{*}$$

Applications

Synthesis of artificial RNA molecules.

Reaction conditions

Requires Mg^{2+}; 37°C.

10 × PP Buffer: 1000 mM Tris-HCl (pH 9.0), 500 mM $MgCl_2$, 4 mM Na_2·EDTA, 200 mM NDP.

Pwo DNA POLYMERASE

Source

Pyrococcus woesei or recombinant *E. coli*.

Description

A thermostable enzyme, 90 kd. It lacks a 5′ to 3′ exonuclease but has a 3′ to 5′ exonuclease.

Properties

(i) 5′ to 3′ DNA-dependent DNA polymerase, requiring a ssDNA template and a DNA or RNA primer with a 3′-OH terminus:

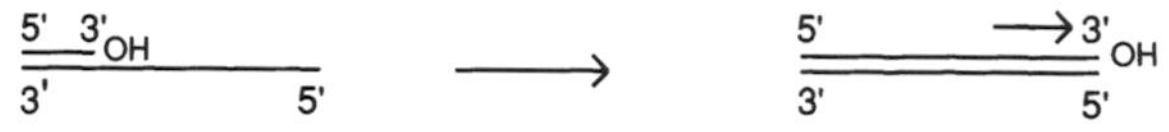

(ii) 3′ to 5′ exonuclease, degrading ssDNA or dsDNA from a 3′-OH terminus:

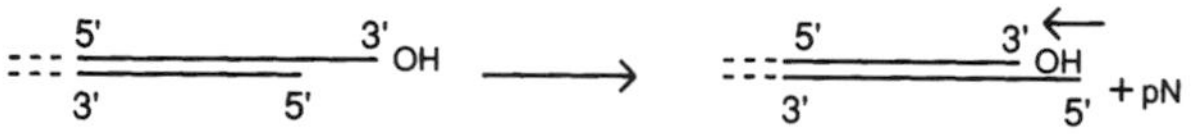

Applications

As permitted by the supplier.

Reaction conditions

The optimal Mg^{2+} concentration should be determined empirically for each experiment since it depends on the dNTP concentration and the nature of the template and primers. The provision of adequate buffering is critical and for this reason the supplier's buffer should be used, with the Mg^{2+} concentration adjusted as required.

RAV2 REVERSE TRANSCRIPTASE

Source
Rous associated virus 2.

Description
Thought to be structurally and functionally identical to AMV reverse transcriptase.

Properties
(i) 5′ to 3′ DNA- or RNA-dependent DNA polymerase, requiring a ssDNA or ssRNA template and a DNA or RNA primer with a 3′-OH terminus:

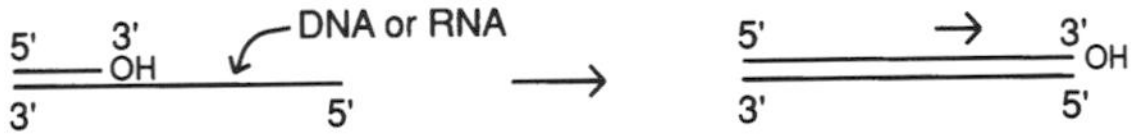

(ii) 5′ to 3′ and 3′ to 5′ exoribonuclease, specific for the RNA component of a DNA–RNA hybrid:

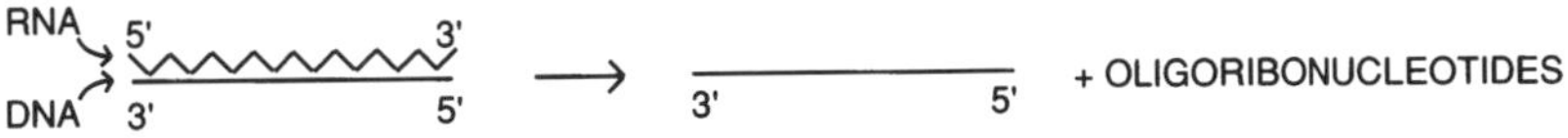

(iii) DNA endonuclease.

(iv) DNA unwinding activity.

Applications
Applications are as for AMV reverse transcriptase:

(i) First strand cDNA synthesis.

(ii) Polymerase chain reaction with RNA substrates (RT-PCR).

(iii) Chain termination RNA sequencing.

(iv) Modified chain termination DNA sequencing, especially to read GC-rich regions.

(v) Probing RNA secondary structure.

(vi) 3′ end labelling DNA fragments.

Reaction conditions
Requires Mg^{2+} and can be incubated at 42°C. The salt optimum for the RNA-dependent DNA polymerase is 50–100 mM, and 10–20 mM for the DNA-dependent DNA polymerase.

10 × RAV2 RT Buffer: 500 mM Tris-HCl (pH 8.3), 750 mM KCl, 1500 mM $MgCl_2$, 100 mM DTT.

RecA PROTEIN

Source
Escherichia coli, including recombinant forms.

Description
Required for homologous genetic recombination and DNA repair (206, 207). The protein promotes DNA strand exchange *in vitro* (208).

Properties
(i) Single strand DNA binding protein.

(ii) DNA-dependent ATPase.

Applications
(i) D-loop mutagenesis procedure (7, 8).

(ii) Visualizing ssDNA by electron microscopy (14).

(iii) Enrichment method for genomic cloning (209–211).

(iv) Sequence-specific cleavage of large DNA fragments (212, 213).

Reaction conditions
These depend on the application (7, 14). A suitable 10 × buffer for D-loop formation is 200 mM Tris-HCl (pH 8.0), 100 mM $MgCl_2$, 10 mM DTT, 13 mM ATP.

RIBONUCLEASE A

Source
Bovine pancreas.

Description
A very active and stable enzyme, 13.7 kd (214). Most preparations are contaminated with a separate DNase that can be inactivated by heating to 100°C. RNase A is the non-glycosylated form of the standard pancreatic ribonuclease.

Properties
Endoribonuclease, cleaving phosphodiester bonds 3′ to a pyrimidine (215). No activity with DNA:

5'
3'
OLIGONUCLEOTIDES WITH A TERMINAL
PYRIMIDINE 2', 3' - CYCLIC PHOSPHATE

Applications
(i) Removing RNA from DNA preparations.

(ii) Mapping single bp mutations in DNA and RNA (216, 217).

Reaction conditions
Ribonuclease A is active over a wide pH range (though the optimum is pH 7.0–7.5) and has no cofactor requirements. Normally, a DNA solution is not adjusted specifically to optimize conditions for treatment with RNase A.

DNase-free RNase A
Commercial preparations of RNase A are often contaminated with DNases. To prepare DNase-free RNase A the solid enzyme is dissolved in 10 mM Tris-HCl (pH 7.5), 15 mM NaCl, heated to 100°C for 15 min and then cooled slowly to room temperature. Aliquots can be stored at –20°C for months. A stock solution of 10 mg ml^{-1} is suitable for most applications.

RIBONUCLEASE CL3

Source
Chicken liver.

Description
One of several endoribonucleases displaying partial sequence specificity and used in enzymatic sequencing of RNA (218). Ribonuclease CL3 has a relatively low degree of specificity.

Properties
Endoribonuclease cutting ssRNA at the arrow in the sequence Cp↑N, with less activity for Ap↑N and Gp↑N and minimal activity for Up↑N.

Applications
Enzymatic sequencing of RNA (57).

Reaction conditions
For RNA sequencing, use a reaction mixture containing 8 M urea and incubate at 50°C to reduce RNA secondary structure. Use the buffer recommended by the supplier.

RIBONUCLEASE H

Source
Escherichia coli, in many cases a recombinant strain carrying the RNase H gene on a pBR322 plasmid (219).

Description
Monomer, 17.5 kd.

Properties
Endoribonuclease, specific for the RNA component of a DNA–RNA hybrid (220):

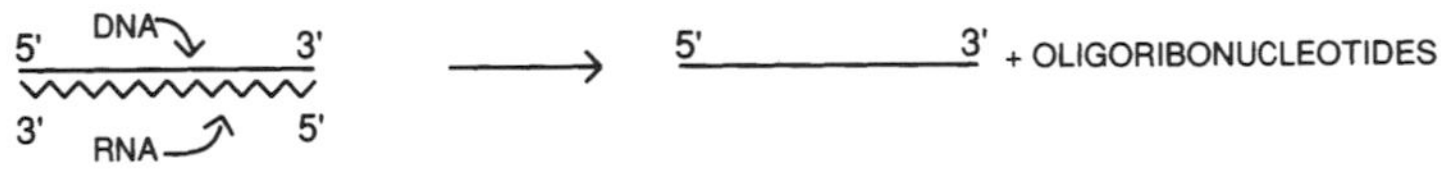

Applications
(i) Second strand cDNA synthesis (30, 32).

(ii) Removing poly(A) tails from mRNA, possibly to improve band resolution in gel electrophoresis (221, 222).

(iii) Detection of DNA–RNA hybrids (223, 224).

(iv) Studying RNA priming of DNA synthesis (225).

(v) Site-specific cleavage of RNA hybridized to a short oligonucleotide (226).

(vi) Analysing the products of *in vitro* polyadenylation (227).

Reaction conditions
Requires Mg^{2+}, 37°C. The pH optimum is 7.5–9.1 (220). RNase H is relatively insensitive to salt, being 50% active in 0.3 M NaCl.

10 × RNase H Buffer: 400 mM Tris-HCl (pH 7.5), 40 mM $MgCl_2$, 10 mM DTT.

Thermostable RNase H
A thermostable RNase H, functionally identical to the *E. coli* enzyme but with a temperature optimum of 65°C, is also available.

RIBONUCLEASE M1

Source
Cucumis melo.

Description
One of several endoribonucleases displaying partial sequence specificity and used in enzymatic sequencing of RNA. Ribonuclease M1 has relatively low specificity.

Properties
Endoribonuclease cutting at the arrow in the sequences N↑pA, N↑pU and N↑pG.

Applications
Enzymatic sequencing of RNA (57).

Reaction conditions
For RNA sequencing, use a reaction mixture containing 8 M urea and incubate at 50°C to reduce RNA secondary structure. Use the buffer recommended by the supplier.

RIBONUCLEASE ONE

Marketed by Promega Corporation

Source
Recombinant *Escherichia coli* overexpressing a cloned *E. coli* gene (228, 229)

Description
Periplasmic enzyme, 27 kd. The only commercially available enzyme that can cut phosphodiester bonds in RNA adjacent to all four nucleotides (228, 230). It has no activity towards ssDNA and dsDNA.

Properties
Single strand-specific ribonuclease, cleaving at any position and releasing cyclic NMPs, which are slowly hydrolysed to 3′-NMPs (228):

Applications
(i) Mapping single-stranded regions in RNA molecules.

(ii) Ribonuclease protection assays.

(iii) Detecting mismatches in DNA–RNA hybrids.

(iv) Removal of RNA from DNA preparations.

Reaction conditions
Use the conditions recommended by the supplier.

RIBONUCLEASE Phy1

Source
Physarum polycephalum.

Description
One of several endoribonucleases displaying partial sequence specificity and used in enzymatic sequencing of RNA. Ribonuclease Phy1 has relatively low specificity (231).

Properties
Endoribonuclease cutting at the arrow in the sequences Gp↑N, Ap↑N and Up↑N.

Applications
Enzymatic sequencing of RNA (57).

Reaction conditions
For RNA sequencing, use a reaction mixture containing 8 M urea and incubate at 50°C to reduce RNA secondary structure. Use the buffer recommended by the supplier.

RIBONUCLEASE PhyM

Source
Physarum polycephalum.

Description
One of several endoribonucleases displaying partial sequence specificity and used in enzymatic sequencing of RNA (232). Ribonuclease PhyM has relatively low specificity.

Properties
Endoribonuclease cutting primarily at the arrow in the sequences Up↑N and Ap↑N.

Applications
Enzymatic sequencing of RNA (57).

Reaction conditions
For RNA sequencing, use a reaction mixture containing 8 M urea and incubate at 50°C to reduce RNA secondary structure. Use the buffer recommended by the supplier.

RIBONUCLEASE T1

Source

Aspergillus oryzae.

Description

Single polynucleotide, 11 kd (233). One of several endoribonucleases displaying partial sequence specificity and used in enzymatic sequencing of RNA (234). Ribonuclease T1 has a relatively high degree of specificity for G residues. At salt concentrations below 100 mM, ribonuclease T1 is active with ssRNA, dsRNA and the RNA component of a DNA–RNA hybrid. At 300 mM NaCl and above it is specific for ssDNA.

Properties

Endoribonuclease cutting at the arrow in the sequence Gp↑N (235). Also cuts at inosine (Ip↑N) and xanthosine (Xp↑N) residues.

Applications

(i) Enzymatic sequencing of RNA (57).

(ii) Ribonuclease protection assay (236).

(iii) Transcript assay using the G-less cassette approach (237).

Reaction conditions

These depend on the application. For RNA sequencing, use a reaction mixture containing 8 M urea and incubate at 50°C to reduce RNA secondary structure. Note that RNase T1 is extremely difficult to inactivate, often requiring proteinase K treatment followed by phenol extraction.

10 × T1 Buffer: 200 mM trisodium citrate (pH 3.5), 10 mM $Na_2 \cdot EDTA$.

RIBONUCLEASE T2

Source
Aspergillus oryzae.

Description
One of several endoribonucleases displaying partial sequence specificity and used in enzymatic sequencing of RNA (234). Ribonuclease T2 has a relatively low degree of specificity.

Properties
Endoribonuclease cutting primarily at the arrow in the sequence Ap↑N but also with non-specific endoribonuclease activity (235).

Applications
(i) Enzymatic sequencing of RNA (57).

(ii) Hydrolysis of RNA to monoribonucleotides.

Reaction conditions
For RNA sequencing, use a reaction mixture containing 8 M urea and incubate at 50°C to reduce RNA secondary structure.

10 × T2 Buffer: 500 mM NaAc (pH 4.5), 20 mM Na_2·EDTA.

RIBONUCLEASE U2

Source
Ustilago sphaerogena.

Description
One of several endoribonucleases displaying partial sequence specificity and used in enzymatic sequencing of RNA (234). Ribonuclease U2 has a relatively high degree of specificity for A residues.

Properties
Endoribonuclease cutting primarily at the arrow in the sequence Ap↑N but with some activity also for Gp↑N (235).

Applications
Enzymatic sequencing of RNA (57).

Reaction conditions
For RNA sequencing, use a reaction mixture containing 8 M urea and incubate at 50°C to reduce RNA secondary structure.

10 × U2 Buffer: 200 mM trisodium citrate (pH 3.5), 10 mM Na_2·EDTA.

RNA POLYMERASE II

Source
Wheat germ.

Description
The eukaryotic RNA polymerase responsible for transcribing most protein-coding genes.

Properties
DNA-dependent RNA polymerase, able to recognize a variety of promoter sequences:

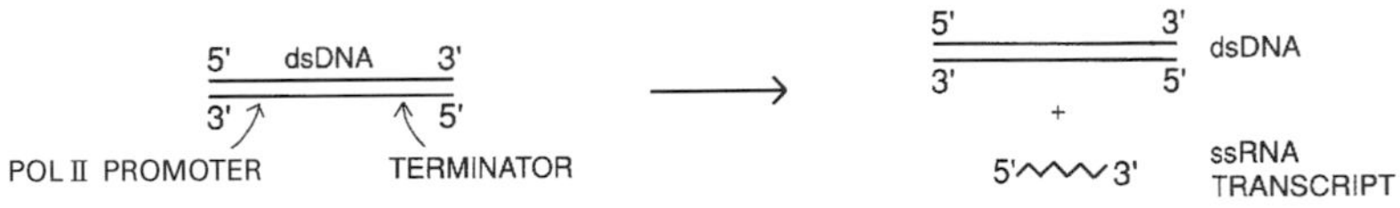

Applications
Eukaryotic transcription studies (238).

Reaction conditions
The enzyme has unusual requirements and works best at 25°C.

10 × Pol II Buffer: 100 mM Tris-HCl (pH 7.9), 10 mM $MnCl_2$, 200 mM $MgCl_2$, 500 mM $(NH_4)_2SO_4$, 1.25 mg ml^{-1} BSA.

RNAzyme RCH 1.0 AND 1.1

Source
Artificially synthesized.

Description
RCH 1.0 and 1.1 are versions of the hammerhead ribozymes (239). They differ by one nucleotide but this does not affect substrate specificity, only the kinetics of the reaction.

Properties
Sequence-specific endoribonuclease, cutting ssRNA at the arrow in the recognition sequence (239):

5′-ACGGUCUC↑ACGAGC-3′

Applications
None reported to date. Theoretically the enzyme has applications in:

(i) Direct restriction-type mapping of ssRNA.

(ii) Enzymatic RNA sequencing.

Reaction conditions
Use the supplier's recommendations when setting up reactions.

RNAzyme TET 1.0

Source
Tetrahymena (240).

Description
This is a modified version of the autocatalytic RNA contained within the intron of the *Tetrahymena* pre-LrRNA.

Properties
Sequence-specific endoribonuclease, cutting ssRNA 3′ of the sequence 5′-CUCU-3′ (241).

Applications (see ref. 242)
(i) Direct restriction-type mapping of ssRNA.

(ii) Enzymatic RNA sequencing.

(iii) Analysis of higher order structure in ssRNA.

Reaction conditions
GTP is required as a cofactor and reactions are carried out at 50°C. Urea is needed at 1.5 M to ensure specificity. Use the supplier's recommendations when setting up reactions.

S1 NUCLEASE

Source
Aspergillus oryzae.

Description
A Zn metalloprotein, glycosylated, 32 kd (243, 244). S1 nuclease is relatively thermostable.

Properties
Single strand-specific endonuclease, more active on ssDNA than ssRNA (184, 243):

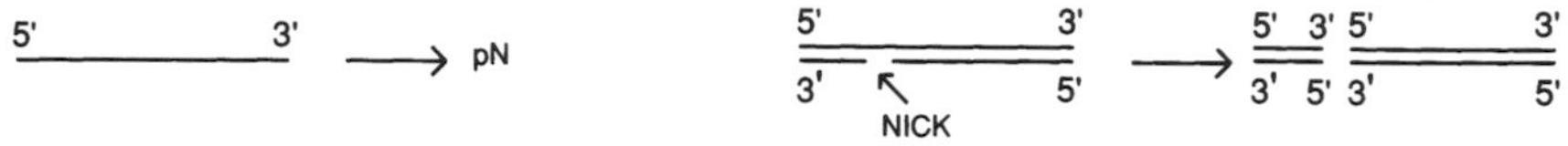

Applications
(i) Analysis of DNA–RNA hybrids (S1 nuclease mapping) to locate introns (147, 245) and transcript termini (150, 246, 247).

(ii) Generation of deletions for linker-scanning mutagenesis (55).

(iii) Removal of overhangs to produce blunt-ended molecules (135, 248).

(iv) Opening hairpins formed during cDNA synthesis (249).

Reaction conditions
S1 nuclease requires Zn^{2+} and low pH (4.0–4.5), being inactive at pH 6.0 (250, 251). Reactions can be stopped with EDTA (250, 251). High NaCl is used to prevent nicking of dsDNA (252). Incubate at 37°C.

5 × S1 Buffer: 250 mM NaAc (pH 4.5), 1000 mM NaCl, 5 mM $ZnSO_4$, 2.5% glycerol.

S7 NUCLEASE (MICROCOCCAL NUCLEASE)

Source
Staphylococcus aureus.

Description
Single polypeptide, 16.8 kd (253). Active on DNA and RNA. With DNA it displays a preference for ssDNA and AT-rich regions in dsDNA.

Properties
Endonuclease, active on DNA and RNA, both single- and double-stranded, with preference for A–T or A–U rich regions:

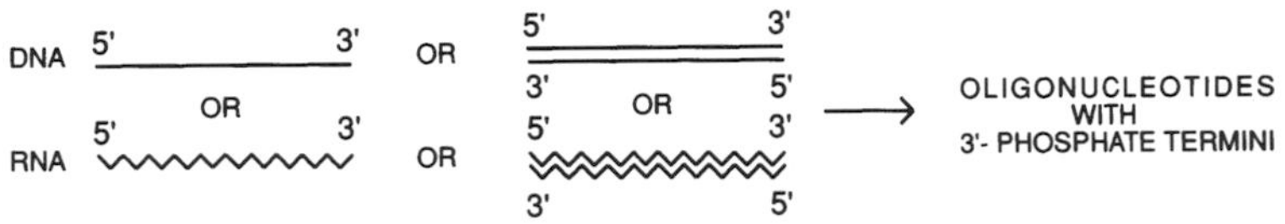

Applications
(i) Preparation of an mRNA-dependent protein synthesis system from rabbit reticulocyte lysates (254, 255).

(ii) Digestion of chromatin to prepare nucleosomes (256).

Reaction conditions
Requires Ca^{2+}, so reactions can be stopped with EGTA; 37°C. The pH optimum is high (pH 8.5–9.2), the exact figure depending on the Ca^{2+} concentration (257).

10 × S7 Buffer: 500 mM Na borate (pH 8.8), 50 mM NaCl, 25 mM $CaCl_2$.

SEQUENASE

Marketed by Amersham International

Source

Recombinant *E. coli* overexpressing a deleted version of the T7 gene 5 protein and native thioredoxin.

Description

Sequenase is a modified version of T7 DNA polymerase, lacking the 3′ to 5′ exonuclease activity of the normal enzyme. Sequenase 1.0 was prepared by chemical mutagenesis but has now been superseded by the engineered version 2.0 which is produced by a gene with a 28 codon deletion (258). Sequenase is functionally similar to the exo$^-$ version of the Klenow polymerase but has a faster rate of DNA synthesis (300 nucleotides per second). The enzyme has high processivity. A thermostable version (Thermosequenase) is also available.

Properties

5′ to 3′ DNA-dependent DNA polymerase, requiring a ssDNA template and a DNA or RNA primer with a 3′-OH terminus:

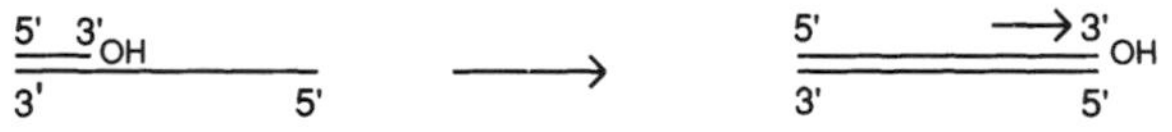

Applications

Chain termination DNA sequencing (259).

Reaction conditions

Requires Mg^{2+}, optimum pH 7.5. Reactions are usually carried out a room temperature.

5 × Sequenase buffer: 200 mM Tris-HCl (pH 7.5), 100 mM $MgCl_2$, 250 mM NaCl.

SHRIMP ALKALINE PHOSPHATASE

Source
Arctic shrimp, *Pandalus borealis*.

Description
Dimeric glycoprotein, two identical 69 kd subunits. Contains four atoms of zinc per molecule. Completely inactivated by heating at 65°C for 15 min.

Properties
5′ phosphatase, removing 5′ phosphate groups from ssDNA, dsDNA, ssRNA and dsRNA:

E.g.
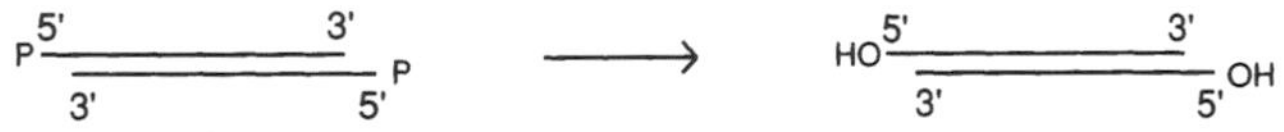

Applications
As for other alkaline phosphatases but the shrimp enzyme's heat lability makes it preferable for applications where residual phosphatase activity would interfere with subsequent manipulations.

(i) Dephosphorylation of dsDNA to prevent self-ligation (40, 41).

(ii) Dephosphorylation of DNA and RNA to allow 5′ end labelling with T4 polynucleotide kinase (42).

Reaction conditions
Alkaline phosphatases do not have particularly stringent buffer requirements.

10 × SAP Buffer: 200 mM Tris-HCl (pH 8.0), 100 mM $MgCl_2$.

SINGLE STRAND-BINDING PROTEIN

Source
Escherichia coli, including recombinant forms.

Description
Tetramer of four identical 18.9 kd subunits. Binds cooperatively to ssDNA but not to dsDNA.

Properties
Single strand DNA-binding protein.

Applications
(i) D-loop mutagenesis procedure (7, 8).

(ii) Visualizing ssDNA by electron microscopy.

(iii) DNA sequencing by the primer walking method.

(iv) Studying DNA helicase activity (129).

(v) Stimulating the transition of phage genomes from ssDNA to dsDNA (260).

(vi) Stimulating DNA polymerase activity (261).

(vii) Sequencing through regions of strong secondary structure (262).

Reaction conditions
Optimal attachment requires high salt (1 M NaCl), but this can subsequently be dialysed away.

SP6 RNA POLYMERASE

Source
Salmonella typhimurium LT2 infected with SP6 phage (263) or a recombinant *E. coli*.

Description
A single polypeptide, 96 kd. The SP6 promoter is highly efficient and recognized specifically by the SP6 polymerase (264–267). Vectors containing the SP6 promoter can direct synthesis of microgram amounts of RNA transcripts *in vitro* (8 mol RNA per mol DNA template, ref. 268). The SP6 polymerase does not terminate at nicks in the template.

Properties
DNA-dependent RNA polymerase, highly specific for the SP6 promoter:

Applications
Synthesis of RNA transcripts from sequences ligated downstream of the SP6 promoter, for:

(i) Studies of RNA processing (266, 268–270).

(ii) Production of antisense RNA (271).

(iii) Synthesis of mRNA for *in vitro* translation (272).

(iv) Obtaining highly labelled RNA probes (268, 273–276).

(v) RNA sequencing (277).

Reaction conditions
Requires Mg^{2+}; 37°C. Salt is not required but may be included in the reaction mix (up to 100 mM NaCl) to decrease non-specific transcript initiation. The polymerization will, however, be reduced by about 50%. The activity is greatly stimulated by spermidine (263).

10 × SP6 Buffer: 200 mM Tris-HCl (pH 7.9), 30 mM $MgCl_2$, 10 mM spermidine, 50 mM DTT.

STAPHYLOCOCCUS AUREUS RIBONUCLEASE

Source
Staphylococcus aureus.

Description
One of several endoribonucleases displaying partial sequence specificity and used in enzymatic sequencing of RNA. *S. aureus* ribonuclease has relatively low specificity.

Properties
Endoribonuclease cutting at the arrow in the sequences A↑pN and U↑pN at pH 7.5, and C↑pN, U↑pN at pH 3.5.

Applications
Enzymatic sequencing of RNA (57).

Reaction conditions
For RNA sequencing, use a reaction mixture containing 8 M urea and incubate at 50°C to reduce RNA secondary structure. Use the buffer recommended by the supplier since the conditions have to be adjusted correctly to achieve the desired sequence specificity.

SUPERSCRIPT REVERSE TRANSCRIPTASE

Marketed by Life Technologies

Source

Recombinant *E. coli* expressing a mutated form of the *pol* gene of Moloney murine leukaemia virus.

Description

Superscript and Superscript II are modified forms of the M-MuLV reverse transcriptase. The two enzymes have a 10^6–10^7-fold less exoribonuclease activity than the unmodified reverse transcriptase and so are more suitable for cDNA synthesis. Superscript is expressed from a deletion mutation of the *pol* gene (278) and has a lower polymerase activity than the unmodified version. Superscript II is expressed from a point-mutated *pol* gene and retains full polymerase activity.

Properties

(i) 5′ to 3′ DNA- or RNA-dependent DNA polymerase, requiring a ssDNA or ssRNA template and a DNA or RNA primer with a 3′-OH terminus:

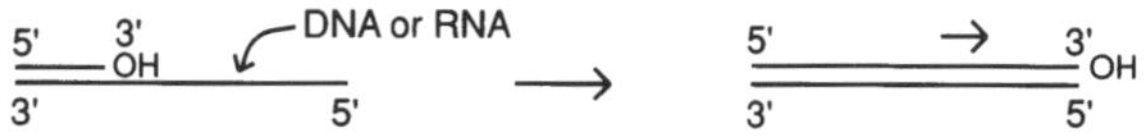

(ii) DNA unwinding activity.

Applications

(i) First strand cDNA synthesis (279).

(ii) Polymerase chain reaction with RNA substrates (RT-PCR) (173).

(iii) Chain termination RNA sequencing (34).

(iv) Modified chain termination DNA sequencing, especially to read GC-rich regions (35).

(v) Probing RNA secondary structure (36).

(vi) 3′ end labelling DNA fragments (37).

Reaction conditions

Requires Mg^{2+}. Superscript is incubated at 37°C. Actinomycin D (50 µg ml^{-1}) can be added to the reaction mix to inhibit the DNA-dependent DNA polymerase activity.

10 × Superscript RT Buffer: 500 mM Tris-HCl (pH 8.3), 400 mM KCl, 60 mM $MgCl_2$, 10 mM DTT, 1 mg ml^{-1} BSA.

T3 RNA POLYMERASE

Source
Recombinant *E. coli* containing plasmid pCM56, which carries the T3 polymerase gene downstream of the *lac* UV5 promoter (280).

Description
A single polypeptide, approximately 100 kd. Highly specific for the T3 promoter (281). Vectors carrying this promoter are available.

Properties
DNA-dependent RNA polymerase, highly specific for the T3 promoter:

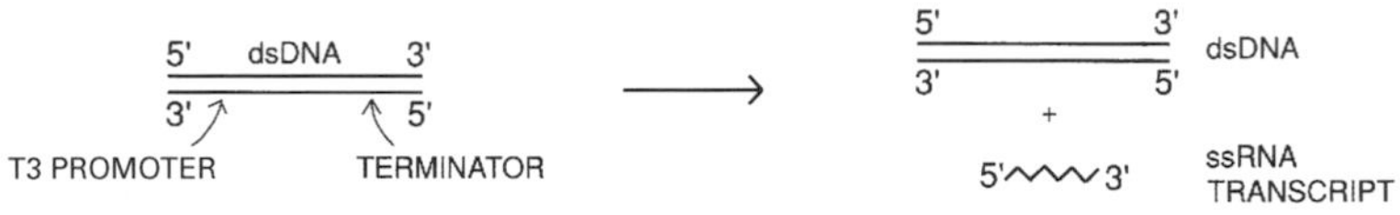

Applications
Synthesis of RNA transcripts from sequences ligated downstream of the T3 promoter, for (274, 282):

(i) Studies of RNA processing.

(ii) Production of antisense RNA.

(iii) Synthesis of mRNA for *in vitro* translation.

(iv) Obtaining highly labelled RNA probes.

(v) RNA sequencing.

Reaction conditions
Requires Mg^{2+}; 37°C. Salt is not required but may be included in the reaction mix (up to 100 mM NaCl) to decrease non-specific transcript initiation. The polymerization will, however, be reduced by about 50%. The activity is greatly stimulated by spermidine. The pH can be as low as 7.2.

5 × T3 Buffer: 200 mM Tris-HCl (pH 8.0), 250 mM NaCl, 40 mM $MgCl_2$, 10 mM spermidine, 150 mM DTT.

T4 DNA LIGASE

Source

Escherichia coli λ lysogens NM969 (283, 284) or 60 p*c*1857 p*PLc*28lig, or a recombinant *E. coli*.

Description

Monomer, 68 kd.

Properties

DNA ligase, catalysing the formation of a phosphodiester bond between adjacent 5′-P and 3′-OH termini (285).

(i) With cohesive-ended molecules:

(ii) With blunt-ended molecules:

(iii) With nicked molecules:

Applications

Joining DNA molecules (286–290).

Reaction conditions

Requires Mg^{2+} and ATP. For blunt-end ligations about 50 times more enzyme is needed than for cohesive ends. Blunt-end ligation is simulated by T4 RNA ligase (291) or PEG (292) and inhibited by more than 50 mM NaCl.

Ideal reactions conditions are reported as follows (293):

For blunt-end ligation (3:1 vector:insert ratio): 10 × buffer = 250 mM Tris-HCl (pH 7.5), 50 mM $MgCl_2$, 25% PEG8000, 5 mM DTT, 4 mM ATP. Use 1 unit enzyme, incubate for 4 h at 23°C in 20 μl total volume.

For cohesive-end ligation (1:2 vector:insert ratio): 10 × buffer = 250 mM Tris-HCl (pH 7.5), 100 mM $MgCl_2$, 100 mM DTT, 4 mM ATP. Use 0.01 units enzyme, incubate overnight at 8°C in 20 μl total volume.

For ligating linkers to blunt-ended DNA: 10 × buffer = 300 mM Tris-HCl (pH 7.5), 300 mM NaCl, 80 mM $MgCl_2$, 20 mM DTT, 2 mM Na_2·EDTA, 70 mM spermidine, 1 mg ml^{-1} BSA, 2.5 mM ATP. Use 2 units enzyme, 5 μg DNA, 500 ng 10–12mer linker or 12.5 μg 8mer linker, incubate overnight at 4°C in 20 μl total volume.

T4 DNA POLYMERASE

Source
T4-infected *E. coli* (294, 295) or a recombinant *E. coli* expressing the cloned T4 gene 43.

Description
A single polypeptide, 114 kd. Lacks the 5′ to 3′ exonuclease of the Kornberg polymerase and so is functionally similar to the Klenow fragment (296), though the T4 polymerase possesses a more active 3′ to 5′ exonuclease. It has low processivity.

Properties
(i) 5′ to 3′ DNA-dependent DNA polymerase, requiring a ssDNA template and a DNA or RNA primer with a 3′-OH terminus:

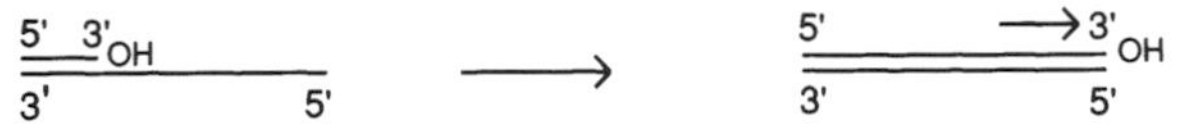

(ii) 3′ to 5′ exonuclease, degrading ssDNA (very active) or dsDNA from a 3′-OH terminus:

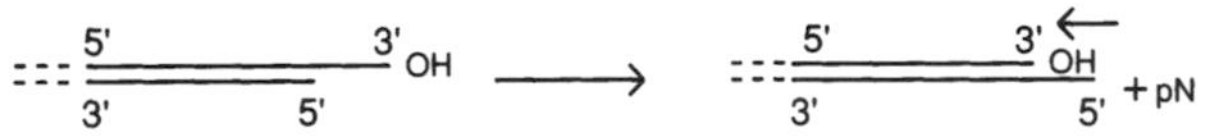

The active 3′ to 5′ exonuclease of T4 DNA polymerase enables the enzyme to carry out a replacement reaction with dsDNA molecules possessing blunt ends or 3′ overhangs:

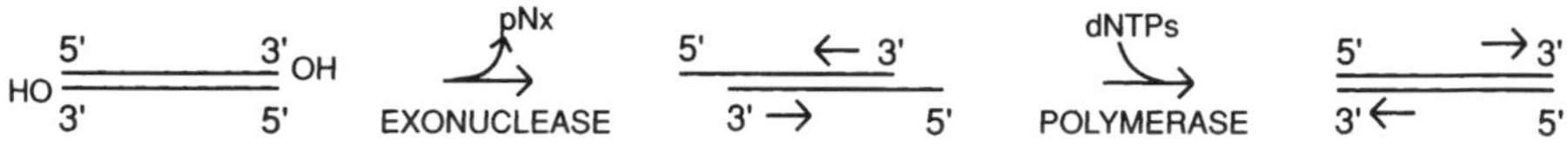

Applications
(i) Labelling DNA by the replacement reaction (297, 298).

(ii) End-filling and end-labelling dsDNA with a 5′ overhang (299), e.g. before cloning PCR products (300).

(iii) Gap-filling in site-directed mutagenesis with short mismatch oligonucleotides (9, 301).

(iv) Generation of overlapping subclones for DNA sequencing (302).

(v) Detection of thymine dimers (303).

Reaction conditions
pH optimum 8.0–9.0 with 50% maximal activity at pH 7.5. Requires Mg^{2+}, optimum 6 mM; 37°C.

10 × T4 Buffer: 330 mM Tris-acetate (pH 7.9), 660 mM KAc, 100 mM MgAc, 5 mM DTT, 1 mg ml^{-1} BSA.

T4 GENE 32 PROTEIN

Source

E. coli infected with a T4 triple mutant am*N*134 am*BL*292 am*E*219 (defective for genes 33, 35 and 58–61), or an overproducing recombinant *E. coli* (304).

Description

The gene 32 protein is a ssDNA-binding protein that stimulates the activity of T4 DNA polymerase 5- to 10-fold (305–309). The protein binds to ssRNA with a 10- to 1000-fold lower affinity.

Properties

Single strand DNA-binding protein.

Applications

(i) Stimulating the activity of T4 DNA polymerase in replicating a ssDNA template (307, 310).

(ii) Visualizing ssDNA by electron microscopy (10–13).

(iii) Site-directed mutagenesis protocols (9).

(iv) Improving the efficiency of restriction for large-scale DNA preparations (311).

(v) Recognition of damaged DNA (312).

(vi) Increasing the specificity of PCR.

Reaction conditions

For optimum attachment to most substrates use the following 10 × buffer: 500 mM Tris-HCl (pH 7.5), 100 mM $MgCl_2$, 1 mM Na_2·EDTA, 70 mM β-mercaptoethanol.

T4 POLYNUCLEOTIDE KINASE

Source

T4-infected *E. coli* (313, 314) or a recombinant *E. coli* (315).

Description

Tetramer of identical 33 kd subunits (295, 316). The phosphatase is active on ssDNA, dsDNA, ssRNA and dsRNA and also with 3′-dNMPs and 3′-NMPs.

Properties

(i) 5′ kinase, transferring the γ-phosphate of ATP to a 5′-OH terminus on a ssDNA, dsDNA, ssRNA or dsRNA molecule (295):

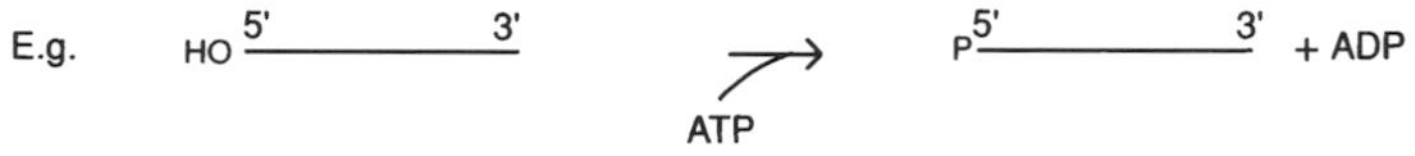

With excess ADP an exchange reaction occurs:

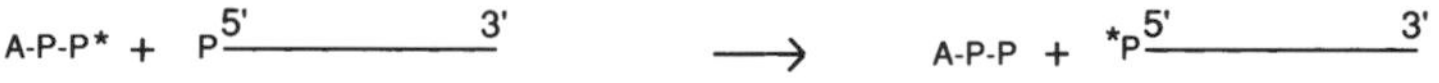

(ii) 3′ phosphatase, removing phosphate groups from the 3′ termini of ssDNA, dsDNA, ssRNA or dsRNA molecules:

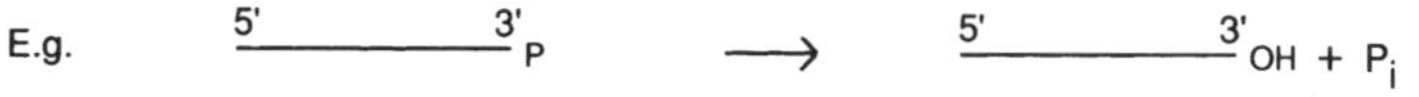

Applications

(i) Labelling the 5′ termini of DNA or RNA by the kinase or exchange reactions (313, 317, 318), especially prior to Maxam–Gilbert sequencing (319–321), enzymatic RNA sequencing (318, 322), restriction mapping (323, 324) and footprinting (66, 325).

(ii) Adding 5′ phosphates to synthetic oligonucleotides.

(iii) Removing 3′ phosphates (326).

Reaction conditions

Requires Mg^{2+}. The kinase and exchange reactions require different conditions, though both are carried out at 37°C. The kinase activity is maximal at pH 7.6 with at least 1 μM ATP and a 5:1 ratio of ATP to 5′-OH ends. The phosphatase has a pH optimum of 5.0–6.0 (327). Up to 10 mM spermidine can be added to the reaction mixture to stabilize the enzyme. For kinasing blunt-ended molecules the 10 × buffer should contain 50% glycerol.

10 × Kinase Buffer: 500 mM Tris-HCl (pH 7.6), 70 mM $MgCl_2$, 50 mM DTT.

10 × Exchange Buffer: 500 mM imidazole–HCl (pH 6.6), 100 mM $MgCl_2$, 50 mM DTT, 3 mM ADP.

T4 POLYNUCLEOTIDE KINASE (PHOSPHATASE-FREE)

Source
E. coli infected with a mutant T4 phage.

Description
Tetramer of identical 33 kd subunits (295, 316, 328). Lacks the 3′-phosphatase of the standard enzyme.

Properties
5′ kinase, transferring the γ-phosphate of ATP to a 5′-OH terminus on a ssDNA, dsDNA, ssRNA or dsRNA molecule (295):

With excess ADP an exchange reaction occurs:

Applications
Addition of a 5′-P group to 3′-CMP to produce labelled 5′-3′-CDP, used to 3′ end label RNA (329).

Reaction conditions
Requires Mg^{2+}. Up to 10 mM spermidine can be added to the reaction mixture to stabilize the enzyme.

10 × Kinase Buffer: 500 mM Tris-HCl (pH 7.6), 70 mM $MgCl_2$, 50 mM DTT.

T4 RNA LIGASE

Source
T4-infected *E. coli* or recombinant *E. coli* containing the plasmid pRF-E35 (330).

Description
Although called an RNA ligase this enzyme works on ssDNA as well as ssRNA. ATP is needed as a cofactor (331–333).

Properties
Ligase, catalysing the formation of a phosphodiester bond between 5′-P and 3′-OH groups on ssDNA and ssRNA molecules:

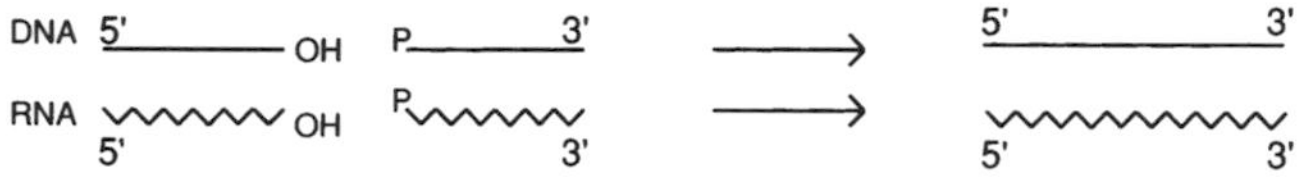

Applications
(i) Joining ssRNA or ssDNA molecules (334–336).

(ii) Increasing the efficiency of T4 DNA ligase in blunt-end ligation (291).

(iii) 3′ end labelling of RNA with 3′–5′-CDP (331, 333, 337, 338).

(iv) Stimulating misacylation of tRNAs (339).

(v) Modifying tRNA anticodons (340).

(vi) 5′ end tagging of mRNAs with oligos to map and sequence termini (16, 18).

(vii) Incorporating modified nucleotides into oligonucleotides (341).

Reaction conditions
Requires Mg^{2+} and ATP. Generally used at 37°C.

10 × RL Buffer: 500 mM Tris-HCl (pH 7.5), 100 mM $MgCl_2$, 100 mM DTT, 10 mM ATP, 100 μg ml^{-1} BSA.

T7 DNA POLYMERASE

Source

Escherichia coli infected with T7 phage or recombinant *E. coli* overexpressing the T7 gene 5 and native thioredoxin proteins.

Description

A dimer comprising the T7 gene 5 product (84 kd) and *E. coli* thioredoxin (12 kd) (342, 343). The T7 gene 5 product is the polymerase/exonuclease and the *E. coli* protein binds the complex tightly to the template, preventing early dissociation during complementary strand synthesis and giving high processivity. Functionally similar to the Klenow polymerase but has a faster rate of DNA synthesis (300 nucleotides per second).

Properties

(i) 5′ to 3′ DNA-dependent DNA polymerase, requiring a ssDNA template and a DNA or RNA primer with a 3′-OH terminus:

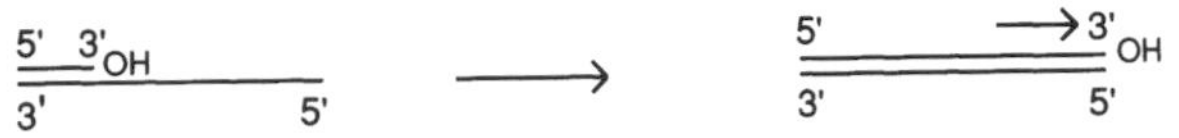

(ii) 3′ to 5′ exonuclease, degrading ssDNA or dsDNA from a 3′-OH terminus:

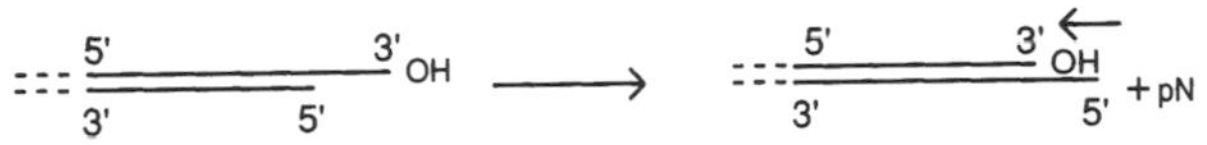

Applications

T7 DNA polymerase is used as an alternative to Klenow polymerase in those applications where the rapidity and template affinity of the T7 enzyme are an advantage.

(i) Chain termination DNA sequencing (259, 344–346).

(ii) Second strand synthesis in site-directed mutagenesis (143, 347, 348).

Reaction conditions

The supplier's recommendations should be followed since the conditions must be set up to minimize the exonuclease activity.

T7 ENDONUCLEASE

Source
E. coli infected with T7 phage, or a recombinant *E. coli* strain expressing T7 gene 3.

Description
The only ssDNA specific endonuclease that is active at neutral and alkaline pH (349). T7 endonuclease has a weak activity with dsDNA (1% of the ssDNA activity; refs 350, 351) with a preference for branched structures, probably because of short single-stranded regions within them (349).

Properties
Endodeoxyribonuclease, primarily specific for ssDNA:

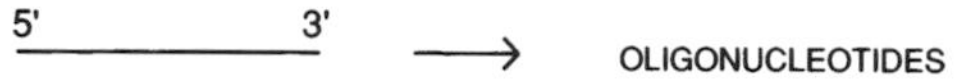

Applications
(i) Studying complex structures in dsDNA (351, 352).

(ii) Footprinting to locate proteins bound to DNA (353, 354).

(iii) Studying the unwinding of DNA caused by DNA-binding proteins (353).

(iv) Activation of the SOS response in *E. coli recA* (355).

Reaction conditions
Requires Mg^{2+}; 37°C.

10 × T7 Nuclease Buffer: 500 mM Tris-HCl (pH 8.0), 200 mM NaCl, 60 mM $MgCl_2$, 50 mM DTT, 1 mg ml^{-1} BSA.

T7 GENE 6 EXONUCLEASE

Source

E. coli HMS7 containing plasmids pJL23 and pGP6-1 (356).

Description

Removes mononucleotides in a stepwise fashion from 5′ termini in dsDNA. Considered a better choice than lambda exonuclease for some applications.

Properties

5′ to 3′ exodeoxyribonuclease, specific for dsDNA, active on 5′-P and 5′-OH termini, releasing dNMPs:

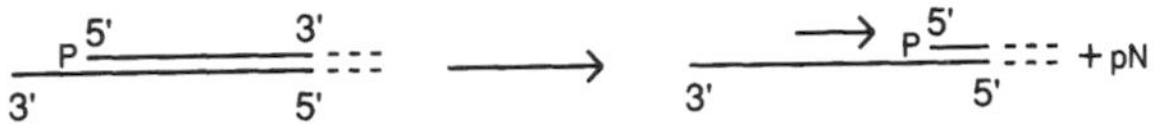

Applications

Generation of ssDNA templates for DNA sequencing (357).

Reaction conditions

Requires Mg^{2+}. Incubate at 37°C.

5 × T7 Exo Buffer: 200 mM Tris-HCl (pH 7.5), 100 mM $MgCl_2$, 250 mM NaCl.

T7 RNA POLYMERASE

Source
Recombinant *E. coli* containing plasmid pAR1219, which carries the T7 polymerase gene downstream of the *lac* UV5 promoter (358).

Description
A single polypeptide, 98 kd (359). Highly specific for the T7 promoter (360). Vectors carrying the T7 promoter can be used to obtain 30 µg of transcript per µg of DNA in 30 min.

Properties
DNA-dependent RNA polymerase, highly specific for the T7 promoter:

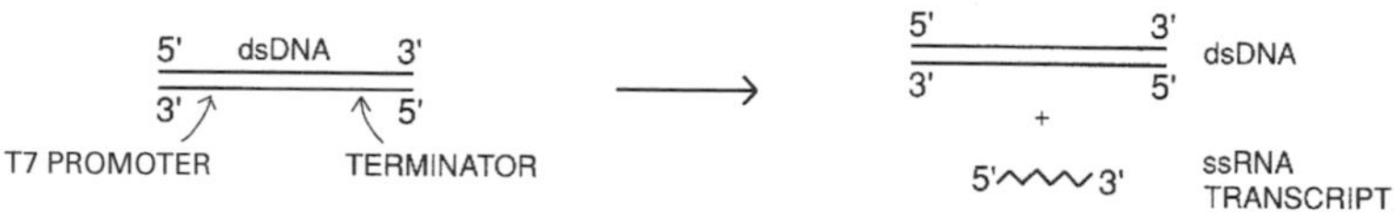

Applications
Synthesis of RNA transcripts from sequences ligated downstream of the T7 promoter, for (267, 268, 274–277, 361–365):

(i) Studies of RNA processing.

(ii) Production of antisense RNA.

(iii) Synthesis of mRNA for *in vitro* translation.

(iv) Obtaining highly labelled RNA probes.

(v) RNA sequencing.

Reaction conditions
Requires Mg^{2+}; 37°C. Salt is not required but may be included in the reaction mix (up to 100 mM NaCl) to decrease non-specific transcript initiation. The polymerization will, however, be reduced by about 50%. The activity is greatly stimulated by spermidine. The pH can be as low as 7.2.

10 × T7 Buffer: 200 mM Tris-HCl (pH 8.0), 125 mM NaCl, 40 mM $MgCl_2$, 10 mM spermidine, 25 mM DTT.

Taq DNA LIGASE

Source
Thermus aquaticus.

Description
An NAD-requiring DNA ligase, apparently functionally similar to the *E. coli* enzyme but active at high temperatures.

Properties
DNA ligase, catalyzing the formation of a phosphodiester bond between adjacent 5′-P and 3′-OH termini hybridized to a complementary target DNA:

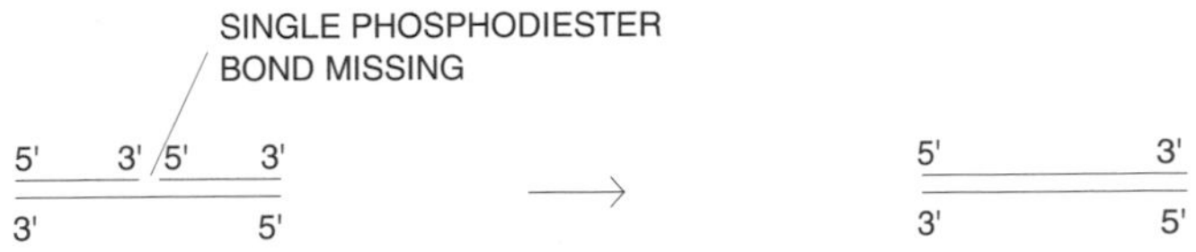

Applications
Because of the instability of base-paired structures at elevated temperatures, *Taq* DNA ligase cannot be used as a general purpose ligation enzyme. It has specialist applications:

(i) Ligase chain reaction (26–29, 31).

(ii) Repeat expansion detection (32).

(iii) High-fidelity gene synthesis (34).

(iv) *In vitro* mutagenesis procedures (35, 37).

(v) Padlock probe hybridization system (36).

Reaction conditions
Requires Mg^{2+} and NAD. Reaction conditions should be optimized for different substrates.

10 × Basic TaL Buffer: 200 mM Tris-HCl (pH 7.6), 250 mM KCl, 100 mM $MgCl_2$, 10 mM NAD, 100 mM DTT, 1.0% Triton X-100.

Taq DNA POLYMERASE

Source
Thermus aquaticus (366, 367) or from recombinant *E. coli* (368).

Description
The first thermostable DNA polymerase to be exploited commercially and the subject of seemingly endless legal proceedings. The enzyme has a single polypeptide chain and a molecular mass of approximately 95 kd. It is highly processive and has a 5′ to 3′ exonuclease activity but no activity in the 3′ to 5′ direction.

Properties
(i) 5′ to 3′ DNA-dependent DNA polymerase, requiring a ssDNA template and a DNA or RNA primer with a 3′-OH terminus:

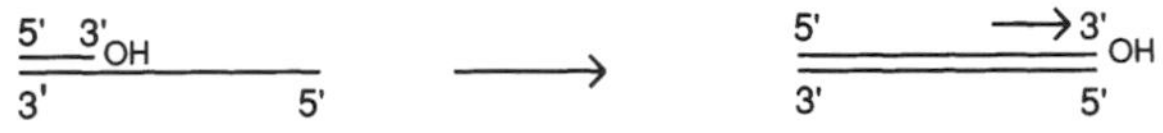

(ii) 5′ to 3′ exonuclease, degrading dsDNA from a 5′-P terminus:

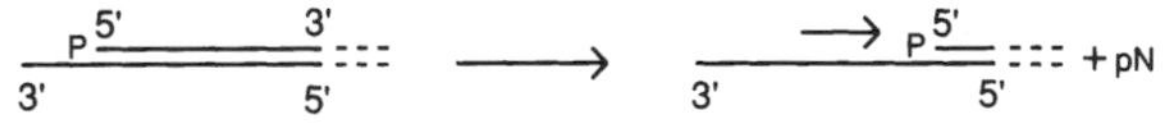

Applications
(i) DNA amplification by the polymerase chain reaction (369, 370).

(ii) Chain termination DNA sequencing at elevated temperatures to minimize problems with secondary structure (371).

(iii) Genomic footprinting (372).

Reaction conditions
The optimal Mg^{2+} concentration should be determined empirically for each experiment since it depends on the dNTP concentration and the nature of the template and primers. The optimum temperature is 75°C but a slightly lower temperature (e.g. 72°C) is generally used to minimize primer dissociation during the early stages of polymerization. At 72°C the optimum pH is 7.0–7.5, which corresponds to a pH of 8.5–9.0 for a Tris buffer prepared at 25°C. The provision of adequate buffering is critical to *Taq* polymerase activity and for this reason the supplier's buffer should be used, with the Mg^{2+} concentration adjusted as required.

TERMINAL DEOXYNUCLEOTIDYL TRANSFERASE

Source
Calf thymus (373).

Description
Dimer of 80 kd + 26 kd. The enzyme can polymerize ribonucleotides with limited efficiency. The 3′-OH acceptor can be an overhang, blunt or recessed end, though an overhang is most efficient (374, 375).

Properties
Template-independent DNA polymerase, adding deoxynucleotides to a 3′-OH terminus on ssDNA or dsDNA (374, 376).

(i) With ssDNA or dsDNA with a 3′ overhang of at least 2 nucleotides:

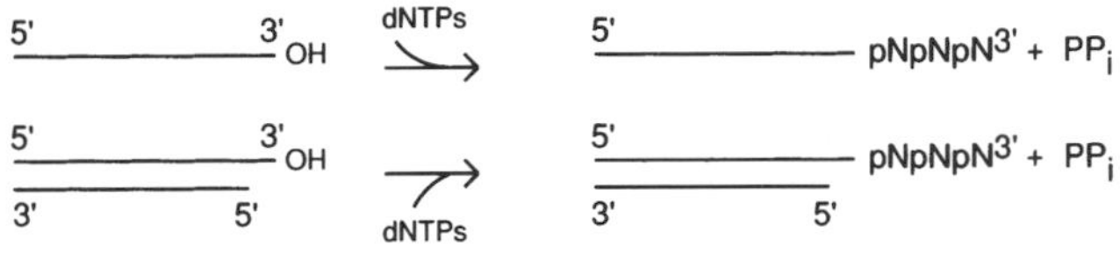

(ii) With dsDNA with a blunt or recessed 3′-OH terminus (377):

Applications
(i) Homopolymer tailing (32, 85, 378).

(ii) Labelling 3′ termini with labelled 3′-dNTP (379), 3′-NTP (380), cordycepin triphosphate (379) or a dideoxy-NTP (381).

(iii) Incorporating modified bases into oligonucleotides (382).

Reaction conditions
The choice of buffer is critical for the type of acceptor molecule being used, the type of nucleotides being polymerized and the purpose of the experiment. See refs 383, 384.

Tfl DNA POLYMERASE

Source

Thermus flavus (385).

Description

A thermostable enzyme, 82 kd, with greater temperature stability than *Taq* polymerase. There is no information on the presence or absence of exonuclease activities.

Properties

5′ to 3′ DNA-dependent DNA polymerase, requiring a ssDNA template and a DNA or RNA primer with a 3′-OH terminus:

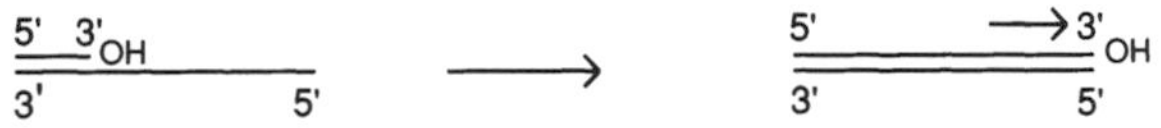

Applications

As permitted by the supplier.

Reaction conditions

The optimal Mg^{2+} concentration should be determined empirically for each experiment since it depends on the dNTP concentration and the nature of the template and primers. The provision of adequate buffering is critical and for this reason the supplier's buffer should be used, with the Mg^{2+} concentration adjusted as required.

THERMUS RNA POLYMERASE

Source
Thermophilic bacterium.

Description
Apparently functionally identical to the *E. coli* RNA polymerase recognizing similar promoter sequences, but with a temperature optimum greater than 65°C.

Properties
DNA-dependent RNA polymerase able to recognize a variety of promoter sequences:

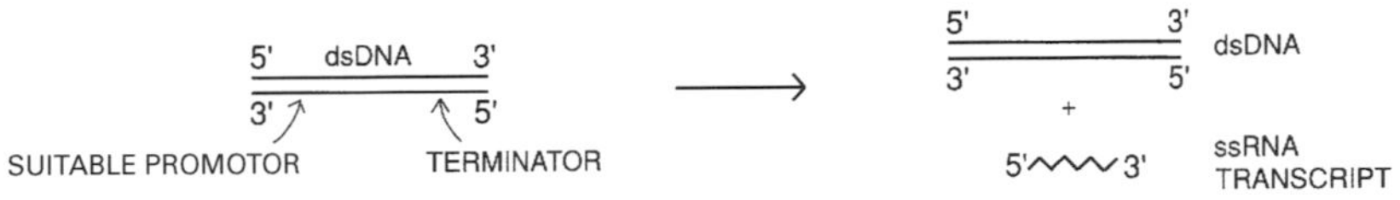

Applications
(i) *In vitro* transcription of genes with suitable promoters.

(ii) Synthesis of labelled RNA.

Reaction conditions
Use the conditions recommended by the supplier.

Tli DNA POLYMERASE

Source
Thermococcus litoralis (386).

Description
A thermostable enzyme, 85 kd, with greater temperature stability than *Taq* polymerase. It possesses a 3′ to 5′ exonuclease but no 5′ to 3′ exonuclease. It is prepared from the same bacterium as Vent DNA polymerase and is presumably the same enzyme.

Properties
(i) 5′ to 3′ DNA-dependent DNA polymerase, requiring a ssDNA template and a DNA or RNA primer with a 3′-OH terminus:

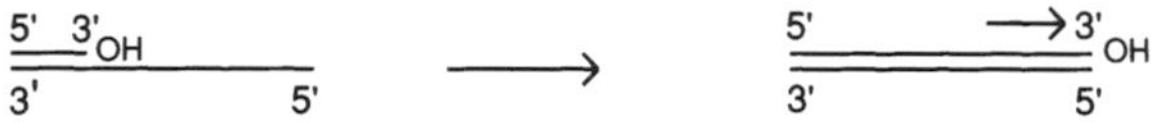

(ii) 3′ to 5′ exonuclease, degrading ssDNA or dsDNA from a 3′-OH terminus:

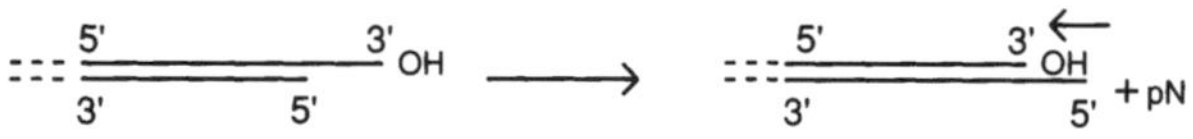

Applications
As permitted by the supplier.

Reaction conditions
The optimal Mg^{2+} concentration should be determined empirically for each experiment since it depends on the dNTP concentration and the nature of the template and primers. The provision of adequate buffering is critical and for this reason the supplier's buffer should be used, with the Mg^{2+} concentration adjusted as required.

Tth DNA POLYMERASE

Source

Thermus thermophilus HB8 (387).

Description

A thermostable enzyme, 94 kd. In the presence of Mg^{2+} it uses a DNA template, but with Mn^{2+} it efficiently reverse transcribes RNA (387). It lacks a 3′ to 5′ exonuclease and has no ribonuclease activity. There is no information on the presence or absence of a 5′ to 3′ exonuclease.

Properties

(i) In the presence of Mg^{2+}, 5′ to 3′ DNA-dependent DNA polymerase, requiring a ssDNA template and a DNA or RNA primer with a 3′-OH terminus:

(ii) In the presence of Mn^{2+}, 5′ to 3′ RNA-dependent DNA polymerase, requiring a ssRNA template and a DNA or RNA primer with a 3′-OH terminus:

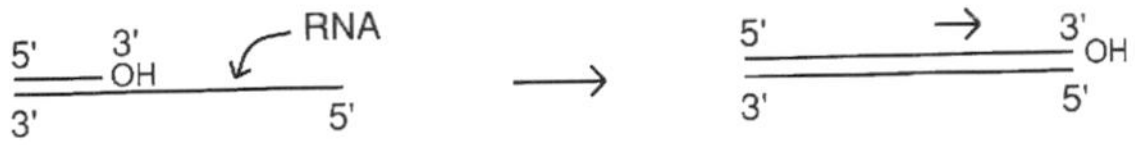

Applications

As permitted by the supplier.

Reaction conditions

The optimal Mg^{2+} and Mn^{2+} concentrations should be determined empirically for each experiment since they depend on the dNTP concentration, the nature of the primers and the desired specificity for a DNA or RNA template. The provision of adequate buffering is critical and for this reason the supplier's buffer should be used, with the Mg^{2+} and Mn^{2+} concentrations adjusted as required.

URACIL DNA-GLYCOSYLASE

Marketed by Life Technologies

Source

Recombinant *E. coli* overexpressing the *E. coli ung* gene (388, 389). There is also a heat-labile version isolated from a marine bacterium.

Description

Uracil DNA-glycosylase removes uracil residues from DNA leaving baseless sites that are sensitive to cleavage by heat or alkali treatment. It is active on ssDNA and dsDNA but not on RNA or short DNA oligonucleotides (less than 6 nucleotides in length).

Properties

N-glycosylase, specifically removing uracil bases from dsDNA or ssDNA (6):

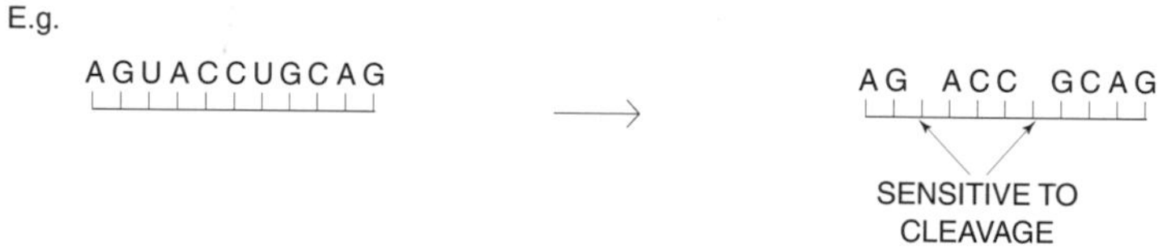

Applications

(i) Prevention of DNA carry-over in PCR reactions (390).

(ii) Increasing the efficiency of site-directed mutagenesis procedures.

(iii) Labelling oligonucleotides.

Reaction conditions

Uracil DNA-glycosylase is active over a wide pH range and does not require divalent cations. Salt at concentrations >150 mM reduces the enzyme activity. Incubate at 37°C.

10 × UDG Buffer: 200 mM Tris-HCl (pH 8.0), 10 mM Na_2·EDTA, 10 mM DTT, 1 mg ml^{-1} BSA.

VENT DNA POLYMERASE

Marketed by New England Biolabs

Source

Recombinant *E. coli* expressing a cloned gene from *Thermococcus litoralis* (386, 391).

Description

A thermostable enzyme with greater temperature stability than *Taq* polymerase. It is highly processive and possesses a 3′ to 5′ exonuclease but no 5′ to 3′ exonuclease.

Properties

(i) 5′ to 3′ DNA-dependent DNA polymerase, requiring a ssDNA template and a DNA or RNA primer with a 3′-OH terminus:

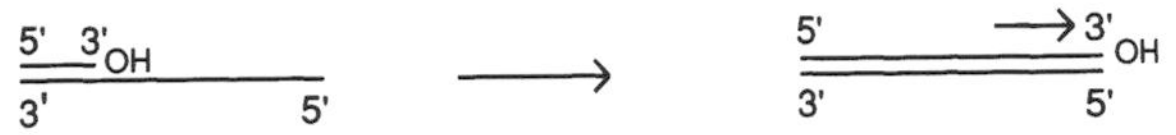

(ii) 3′ to 5′ exonuclease, degrading ssDNA or dsDNA from a 3′-OH terminus:

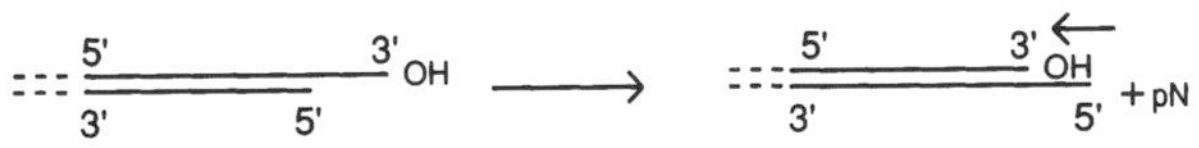

Applications

As permitted by the supplier.

Reaction conditions

The optimal Mg^{2+} concentration should be determined empirically for each experiment since it depends on the dNTP concentration and the nature of the template and primers. The provision of adequate buffering is critical and for this reason the supplier's buffer should be used, with the Mg^{2+} concentration adjusted as required.

VENT (exo$^-$) DNA POLYMERASE

Marketed by New England Biolabs

Source

Recombinant *E. coli* expressing a modified version of a cloned gene from *Thermococcus litoralis* (386, 391).

Description

A thermostable enzyme with greater temperature stability than *Taq* polymerase. Vent (exo$^-$) DNA polymerase is an engineered form of the Vent enzyme lacking the 3′ to 5′ exonuclease (62). It is highly processive.

Properties

5′ to 3′ DNA-dependent DNA polymerase, requiring a ssDNA template and a DNA or RNA primer with a 3′-OH terminus:

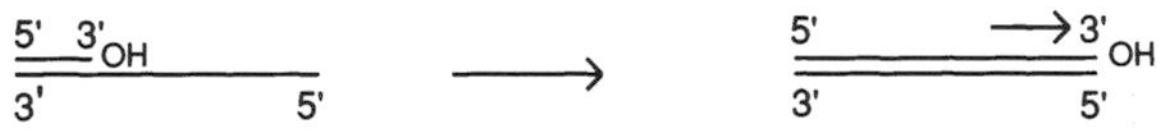

Applications

As permitted by the supplier.

Reaction conditions

The optimal Mg^{2+} concentration should be determined empirically for each experiment since it depends on the dNTP concentration and the nature of the template and primers. The provision of adequate buffering is critical and for this reason the supplier's buffer should be used, with the Mg^{2+} concentration adjusted as required.

3. REFERENCES

1. Landegren, U., Kaiser, R., Caskey, C.T. and Hood, L.E. (1988) *Science*, **242**, 229.

2. Wu, D.Y. and Wallace, R.B. (1989) *Genomics*, **4**, 560.

3. Barany, F. (1991) *Proc. Natl. Acad. Sci. USA*, **88**, 189.

4. Birkenmeyer, L. and Armstrong, A.S. (1992) *J. Clin. Microbiol.*, **30**, 3089.

5. Dille, B.J., Butzen, C.C. and Birkenmeyer, I.G. (1993) *J. Clin. Microbiol.*, **31**, 729.

6. Longo, M.C., Berninger, M.S. and Hartley, J.L. (1990) *Gene*, **93**, 125.

7. Shortle, D., Koshland, D., Weinstock, G.M. and Botstein, D. (1982) *Proc. Natl. Acad. Sci. USA*, **77**, 5375.

8. McEntee, K., Weinstock, G.M. and Lehman, I.R. (1980) *Proc. Natl. Acad. Sci. USA*, **77**, 857.

9. Craik, C.S., Largman, C., Fletcher, T., Roczniak, S., Barr, P.J., Fletterick, R. and Rutter, W.J. (1985) *Science*, **228**, 291.

10. Morris, C.F., Sinha, N.K. and Alberts, B.M. (1975) *Proc. Natl. Acad. Sci. USA*, **72**, 4800.

11. Delius, H., Mantell, N.J. and Alberts, B.M. (1972) *J. Mol. Biol.*, **67**, 341.

12. Brack, C., Bickle, T.A. and Yuan, R. (1975) *J. Mol. Biol.*, **96**, 693.

13. Wu, M. and Davidson, N. (1975) *Proc. Natl. Acad. Sci. USA*, **72**, 4506.

14. Wasserman, S.A. and Cozzarelli, N.R. (1985) *Proc. Natl. Acad. Sci. USA*, **82**, 1079.

15. Mandl, C.W., Heinz, F.X., Puchhammer-Stockl, E. and Kunz, C. (1991) *Biotechniques*, **10**, 484.

16. Fromont-Racine, M., Bertrand, E., Pictet, R. and Grange, T. (1993) *Nucl. Acids Res.*, **21**, 1683.

17. Maruyama, K. and Sugano, S. (1994) *Gene*, **138**, 171.

18. Volloch, V., Schweitzer, B. and Rits, S. (1994) *Nucl. Acids Res.*, **22**, 2507.

19. Xu, F.F. and Cohen, S.N. (1995) *Nature*, **374**, 180.

20. Kornberg, A. (1980) *DNA Replication*. Freeman, San Francisco.

21. Schalling, M., Hudson, T.J., Buetow, K.H. and Housman, D.E. (1993) *Nature Genetics*, **4**, 135.

22. Rouwendel, G.J.A., Wolpert, E.J.H., Zwiers, L.H. and Springer, J. (1993) *Biotechniques*, **15**, 68.

23. Michael, S.F. (1994) *Biotechniques*, **16**, 410.

24. Nilsson, M., Malmgren, H., Samiotaki, M., Kwiatkowski, M., Chowdary, B.P. and Landegren, U. (1994) *Science*, **265**, 2085.

25. Grandgenett, D.P., Gerard, G.F. and Green, M. (1973) *Proc. Natl. Acad. Sci. USA*, **70**, 230.

26. Breathnach, R., Mandel, J.L. and Chambon, P. (1977) *Nature*, **270**, 314.

27. Tilghman, S.M., Curtis, P.J., Tiemeier, D.C., Leder, P. and Weissman, C. (1978) *Proc. Natl. Acad. Sci. USA*, **75**, 1309.

28. Champoux, J.J., Gilboa, E. and Baltimore, D. (1984) *J. Virol.*, 49, 686.

29. Leis, J.P., Duyk, G., Johnson, S., Longiaru, M. and Skalka, A. (1983) *J. Virol.*, **45**, 727.

30. Gubler, U. and Hoffman, B.J. (1982) *Gene*, **25**, 263.

31. Verma, I.M. (1977) *Biochim. Biophys. Acta*, **473**, 1.

32. Okayama, H. and Berg, P. (1982) *Mol. Cell. Biol.*, **2**, 161.

33. Frohman, M.A., Dush, M.K. and Martin, G.R. (1988) *Proc. Natl. Acad. Sci. USA*, **85**, 8998.

34. Kaarinen, M., Griffiths, G.M., Hamlyn, P.M., Markham, A.F., Karjlainen, K., Pelkonen, J.L.T., Makela, O. and Milstein, C. (1983) *J. Immunol.*, **130**, 937.

35. Chen, E.Y. and Seeburg, P.H. (1985) *DNA*, **4**, 165.

36. Inoue, T. and Cech, T.R. (1985) *Proc. Natl. Acad. Sci. USA*, **82**, 648.

37. Oyama, F., Kikuchi, R., Omori, A. and Uchida, T. (1988) *Anal. Biochem.*, **172**, 444.

38. Fujinaga, K., Parsons, T., Beard, J.W., Beard, D. and Green, M. (1970) *Proc. Natl. Acad. Sci. USA*, **67**, 1432.

39. Efstratiadis, A., Vournakis, J.N., Donis-Keller, H., Chaconas, G., Dougall, D.K. and Kafatos, F.C. (1977) *Nucl. Acids Res.*, **4**, 4165.

40. Ullrich, A., Shine, J., Chirgwin, J., Pictet, R., Tischer, E., Rutter, W.J. and Goodman, H.M. (1977) *Science*, **196**, 1313.

41. Ish-Horowitz, D. and Burke, J.F. (1981) *Nucl. Acids Res.*, **9**, 2989.

42. Fosset, M., Chappelet-Tordo, D. and Lazdunski, M. (1974) *Biochemistry*, **13**, 1783.

43. McGadey, J. (1970) *Histochemie*, **23**, 180.

44. Blake, M.S., Johnston, K.H., Russell-Jones, G.J. and Gotsclich, E.C. (1984) *Anal. Biochem.*, **136**, 175.

45. Knoll, J.H.M. and Lichter, P. (1995) in *Current Protocols in Molecular Biology* (F.M. Ausubel, R. Brent, R.E. Kingston, D.D. Moore, J.G. Seidman, J.A. Smith and K. Struhl, eds). John Wiley, New York, Unit 14.7.

46. Lau, P.P. and Gray, H.B. (1979) *Nucl. Acids Res.*, **6**, 331.

47. Gray, H.B., Winston, T.P., Hodnett, J.L., Legerski, R.J., Nees, D.W., Wei, C.-F. and Robberson, D.L. (1981) in *Gene Amplification and Analysis: Structural Analysis of Nucleic Acids* (J.G. Chirikjian and T.S. Papas, eds). Elsevier, New York, Vol. 2, p. 169.

48. Gray, H.B., Ostrander, D.A., Hodnett, J.L., Legerski, R.J. and Robberson, D.L. (1975) *Nucl. Acids Res.*, **2**, 1459.

49. Legerski, R.J., Hodnett, J.L. and Gray, H.B. (1978) *Nucl. Acids Res.*, **5**, 1445.

50. Talmadge, K., Stahl, S. and Gilbert, W. (1980) *Proc. Natl. Acad. Sci. USA*, **77**, 3369.

51. Shishido, K. and Ando, T. (1982) in *Nucleases* (S. Linn and R. Roberts, eds). Cold Spring Harbor Laboratory Press, New York, p. 155.

52. Wei, C.-F., Alianell, G.A., Bencen, G.H. and Gray, H.B. (1983) *J. Biol. Chem.*, **258**, 13506.

53. Miele, E.A., Mills, D.R. and Kramer, F.R. (1983) *J. Mol. Biol.*, **171**, 282.

54. Sambrook, J., Fritsch, E.F. and Maniatis, T. (1989) *Molecular Cloning: A Laboratory Manual (2nd edn)*. Cold Spring Harbor Laboratory Press, New York.

55. Greene, J.M. (1988) in *Current Protocols in Molecular Biology* (F.M. Ausubel, R. Brent, R.E. Kingston, D.D. Moore, J.G. Seidman, J.A. Smith and K. Struhl, eds). John Wiley, New York, Unit 8.4.

56. Lockhard, R.E., Alzner-DeWeerd, B., Heckman, J.E., MacGee, J., Tabor, M.W. and RajBhandary, U.L. (1978) *Nucl. Acids Res.*, **5**, 37.

57. Beier, H. and Gross, H.J. (1991) in *Essential Molecular Biology: A Practical Approach* (T.A. Brown, ed). IRL Press at Oxford University Press, Oxford, Vol. 2, p. 221.

58. Ye, S.G. and Hong, G.F. (1987) *Scientia Sinica Ser. B*, **30**, 503.

59. Stenesh, J.A. and Roe, B.A. (1972) *Biochim. Biophys. Acta*, **272**, 156.

60. Chaconas, G. and van der Sande, J.H. (1980) *Meth. Enzymol.*, **65**, 75.

61. Jannasch, H.W., Wirsen, C.O., Molyneux, S.J. and Langworthy, T.A. (1992) *Appl. Environ. Microbiol.*, **58**, 3472.

62. Kong, H.M., Kucera, R.B. and Jack, W.E. (1993) *J. Biol. Chem.*, **268**, 1965.

63. Liao, T.-H. (1974) *J. Biol. Chem.*, **249**, 2354.

64. Matsuda, M. and Ogoshi, H. (1966) *J. Biochem.*, **59**, 230.

65. Rigby, P.W.J., Dieckmann, M., Rhodes, C. and Berg, P. (1977) *J. Mol. Biol.*, **113**, 237.

66. Galas, D.J. and Schmitz, A. (1978) *Nucl. Acids Res.*, **5**, 3157.

67. Jackson, P.D. and Felsenfeld, G. (1987) *Meth. Enzymol.*, **152**, 735.

68. Krieg, P.A. and Melton, D.A. (1987) *Meth. Enzymol.*, **155**, 397.

69. Weintraub, H. and Groudine, M. (1976) *Science*, **193**, 848.

70. Hong, G.-F. (1987) *Meth. Enzymol.*, **155**, 93.

71. Greenfield, L., Simpson, L. and Kaplan, D. (1975) *Biochim. Biophys. Acta*, **407**, 365.

72. Melgar, E. and Goldthwait, D.A. (1968) *J. Biol. Chem.*, **243**, 4409.

73. Klevan, L. and Wang, J.C. (1980) *Biochemistry*, **19**, 5229.

74. Perbal, B. (1988) *A Practical Guide to Molecular Cloning (2nd edn)*. Wiley, New York.

75. Gellert, M. (1981) *Annu. Rev. Biochem.*, **50**, 879.

76. Jacobsen, H., Klenow, H. and Overgaard-Hansen, K. (1974) *Eur. J. Biochem.*, **45**, 623.

77. Joyce, C.M. and Grindley, N.D.F. (1983) *Proc. Natl. Acad. Sci. USA*, **80**, 1830.

78. Brutlag, D., Atkinson, M.R., Setlow, P. and Kornberg, A. (1969) *Biochem. Biophys. Res. Commun.*, **37**, 982.

79. Klenow, H. and Henningsen, I. (1970) *Proc. Natl. Acad. Sci. USA*, **65**, 168.

80. Setlow, P., Brutlag, D. and Kornberg, A. (1972) *J. Biol. Chem.*, **247**, 224.

81. Setlow, P. and Kornberg, A. (1972) *J. Biol. Chem.*, **247**, 232.

82. Telford, J.L., Kressman, A., Koski, R.A., Grosschedl, R., Muller, F., Clarkson, S.G. and Birnstiel, M.L. (1979) *Proc. Natl. Acad. Sci. USA*, **76**, 2590.

83. Kramer, W., Drutsa, V., Jansen, H.-W., Kramer, B., Pflugfelder, M. and Fritz, H.-J. (1984) *Nucl. Acids Res.*, **12**, 9441.

84. Careter, P., Bedouelle, H. and Winter, G. (1985) *Nucl. Acids Res.*, **13**, 4431.

85. Anderson, S., Gait, M.J., Mayol, L. and Young, I.G. (1980) *Nucl. Acids Res.*, **8**, 1731.

86. Feinberg, A.P. and Vogelstein, B. (1983) *Anal. Biochem.*, **136**, 6.

87. Feinberg, A.P. and Vogelstein, B. (1984) *Anal. Biochem.*, **137**, 266.

88. Sanger, F., Nicklen, S. and Coulson, A.R. (1977) *Proc. Natl. Acad. Sci. USA*, **74**, 5463.

89. Houdebiene, L.-M. (1976) *Nucl. Acids Res.*, **3**, 615.

90. Goeddel, D.V., Shepherd, H.M., Yelverton, E., Leung, D. and Crea, R. (1980) *Nucl. Acids Res.*, **8**, 4057.

91. Rougeon, F., Kourilsky, P. and Mach, B. (1975) *Nucl. Acids Res.*, **2**, 2365.

92. Wickens, M.P., Buell, G.N. and Schimke, R.T. (1978) *J. Biol. Chem.*, **253**, 2483.

93. Wallace, R.B., Johnson, P.F., Tanaka, S., Schold, M., Itakura, K. and Abelson, J. (1980) *Science*, **209**, 1396.

94. Messing, J. and Vieira, J. (1982) *Gene*, **19**, 269.

95. Wallace, R.B., Johnson, M.J., Suggs, S.V., Miyoshi, K., Bhatt, R. and Itakura, K. (1981) *Gene*, **16**, 21.

96. Carter, P.J., Winter, G., Wilkinson, A.J. and Ferscht, A.R. (1984) *Cell*, **38**, 835.

97. Zoller, M.J. and Smith, M. (1982) *Nucl. Acids Res.*, **10**, 6487.

98. Meinkoth, J. and Wahl, G. (1984) *Anal. Biochem.*, **138**, 267.

99. Seiler-Tuyns, A., Eldridge, J.D. and Paterson, B.M. (1984) *Proc. Natl. Acad. Sci. USA*, **81**, 2980.

100. Studencki, A.B. and Wallace, R.B. (1984) *DNA*, **3**, 7.

101. Richardson, C.C., Schildkraut, C., Aposhian, H.V. and Kornberg, A. (1964) *J. Biol. Chem.*, **239**, 222.

102. Derbyshire, V., Freemont, P.S., Sanderson, M.R., Beese, L., Friedman, J.M., Joyce, C.M. and Steitz, T.A. (1988) *Science*, **240**, 199.

103. Walker, G.T. (1993) *PCR Meth. Appl.*, **3**, 1.

104. Maniatis, T., Jeffrey, A. and Kleid, D.G. (1975) *Proc. Natl. Acad. Sci. USA*, **72**, 1184.

105. Murray, N.E. and Kelley, W.S. (1979) *Mol. Gen. Genet.*, **175**, 77.

106. Lehman, I.R., Bessman, M.J., Simms, E.S. and Kornberg, A. (1958) *J. Biol. Chem.*, **233**, 163.

107. Jovin, T.M., Englund, P.T. and Bertsch, L.L. (1969) *J. Biol. Chem.*, **244**, 2996.

108. Liu, L.F. and Miller, K.G. (1981) *Proc. Natl. Acad. Sci. USA*, **78**, 3487.

109. Champoux, J.J. (1978) *Annu. Rev. Biochem.*, **47**, 449.

110. Wang, J.C. (1981) in *The Enzymes (3rd edn)* (P.D. Boyer, ed.). Academic Press, New York, Vol. 14, p. 331.

111. Been, M.D. and Champoux, J.J. (1981) *Proc. Natl. Acad. Sci. USA*, **78**, 2883.

112. Halligan, B.D., Davis, J.L., Edwards, K.A. and Liu, L.F. (1982) *J. Biol. Chem.*, **257**, 3995.

113. Prell, B. and Vosberg, H.P. (1980) *Eur. J. Biochem.*, **108**, 389.

114. Laskey, R.A., Mills, A.D. and Morris, N.R. (1977) *Cell*, **10**, 237.

115. Germond, J.-E., Vogt, V.M. and Hirt, B. (1974) *Eur. J. Biochem.*, **43**, 591.

116. Wang, J.C. (1979) *Proc. Natl. Acad. Sci. USA*, **76**, 200.

117. Peck, L.J. and Wang, J.C. (1981) *Nature*, **292**, 375.

118. Dynan, W.S. and Burgess, R.R. (1981) *J. Biol. Chem.*, **256**, 5866.

119. Merino, A., Madden, K.R., Lane, W.S., Champoux, J.J. and Reinberg, D. (1993) *Nature*, **365**, 227.

120. Luckow, B., Renkawitz, R. and Schutz, G. (1987) *Nucl. Acids Res.*, **15**, 417.

121. Panasenko, S.M., Cameron, J.R., Davis, R.W. and Lehman, I.R. (1977) *Science*, **196**, 188.

122. Panasenko, S.M., Alazard, R.J. and Lehman, I.R. (1978) *J. Biol. Chem.*, **253**, 4590.

123. Higgins, N.P. and Cozzarelli, N.R. (1979) *Meth. Enzymol.*, **68**, 50.

124. Sgaramella, V. and Khorana, H.G. (1972) *J. Mol. Biol.*, **72**, 493.

125. Sano, H. and Feix, G. (1974) *Biochemistry*, **13**, 5110.

126. Zimmerman, S.B. and Pheiffer, B.H. (1983) *Proc. Natl. Acad. Sci. USA*, **80**, 5852.

127. Travers, A.A. and Burgess, R.R. (1969) *Nature*, **222**, 537.

128. Calva, E., Rosenvold, E.C., Szybalski, W. and Burgess, R.R. (1980) *J. Biol. Chem.*, **255**, 11011.

129. Rosamond, J., Telander, K.M. and Linn, S. (1979) *J. Biol. Chem.*, **254**, 8646.

130. Goldmarek, P.J. and Linn, S. (1972) *J. Biol. Chem.*, **247**, 1849.

131. Rao, R.N. and Rogers, S.G. (1978) *Gene*, **3**, 247.

132. Rogers, S.G. and Weiss, B. (1980) *Gene*, **11**, 187.

133. Weiss, B. (1976) *J. Biol. Chem.*, **251**, 1896.

134. Rogers, S.G. and Weiss, B. (1980) *Meth. Enzymol.*, **65**, 201.

135. Roberts, T.M., Kacich, R. and Ptashne, M. (1979) *Proc. Natl. Acad. Sci. USA*, **76**, 760.

136. Guo, L.-H. and Wu, R. (1982) *Nucl. Acids Res.*, **10**, 2065.

137. Yang, C., Guo, L. and Wu, R. (1983) in *Frontiers in Biochemical and Biophysical Studies of Proteins and Membranes* (T. Lie, S. Sakakibara, A. Schechter, K. Yagi, H. Yajima and K. Tasanobu, eds). Elsevier, New York, p. 5.

138. Ozkaynak, E. and Putney, S.D. (1987) *Biotechniques*, **5**, 770.

139. Henikoff, S. (1987) *Meth. Enzymol.*, **155**, 156.

140. Smith, A.J.H. (1979) *Nucl. Acids Res.*, **6**, 831.

141. Donelson, J.E. and Wu, R. (1972) *J. Biol. Chem.*, **247**, 4661.

142. Guo, L.-H., Yang, R.C.A. and Wu, R. (1983) *Nucl. Acids Res.*, **11**, 5521.

143. Vandeyar, M.A., Weiner, M.P., Hutton, C.J. and Batt, C.A. (1988) *Gene*, **65**, 129.

144. Shalloway, D., Kleinberger, T. and Livingston, D.M. (1980) *Cell*, **20**, 411.

145. von der Ahe, D., Janich, S., Scheidereit, C., Renkawitz, R., Schutz, G. and Beato, M. (1985) *Nature*, **313**, 706.

146. Wu, C. (1985) *Nature*, **317**, 84.

147. Chase, J.W. and Richardson, C.C. (1974) *J. Biol. Chem.*, **249**, 4545.

148. Chase, J.W. and Richardson, C.C. (1974) *J. Biol. Chem.*, **249**, 4553.

149. Chase, J.W and Vales, L.D. (1981) in *Gene Amplification and Analaysis: Structural Analysis of Nucleic Acids* (J.G. Chirikjian and T.S. Papas, eds). Elsevier, New York, Vol. 2, p. 148.

150. Berk, A.J. and Sharp, P.A. (1978) *Proc. Natl. Acad. Sci. USA*, **75**, 1274.

151. Berk, A.J. and Sharp, P.A. (1977) *Cell*, **12**, 721.

152. Goff, S.P. and Berg, P. (1978) *Proc. Natl. Acad. Sci. USA*, **75**, 1763.

153. Hoffman, L.M. and Jendrisak, J. (1990) *Gene*, **88**, 97.

154. Kobori, H., Sullivan, C.W. and Shizuya, H. (1984) *Proc. Natl. Acad. Sci. USA*, **81**, 6691.

155. Gauthier, A., Turmel, M. and Lemieux, C. (1991) *Curr. Genet.*, **19**, 43.

156. Marshall, P. and Lemieux, C. (1991) *Gene*, **104**, 241.

157. Marshall, P. and Lemieux, C. (1992) *Nucl. Acids Res.*, **20**, 6401.

158. Little, J.W. (1981) in *Gene Amplification and Analaysis: Structural Analysis of Nucleic Acids* (J.G. Chirikjian and T.S. Papas, eds). Elsevier, New York, Vol. 2, p. 136.

159. Radding, C.M. (1966) *J. Mol. Biol.*, **18**, 235.

160. Little, J.W., Lehman, I.R. and Kaiser, A.D. (1967) *J. Biol. Chem.*, **242**, 672.

161. Jackson, D.A., Symons, R.H. and Berg, P. (1972) *Proc. Natl. Acad. Sci. USA*, **69**, 2904.

162. Higuchi, R.G. and Ochman, H. (1989) *Nucl. Acids Res.*, **17**, 5865.

163. Chow, S., Daub, E. and Murialdo, H. (1987) *Gene*, **60**, 277.

164. Rackwitz, H.R., Zehetner, G., Murialdo, H., Delius, H., Chai, J.H., Poustka, A., Frischauf, A. and Lehrach, H. (1985) *Gene*, **40**, 259.

165. Wang, Y. and Wu, R. (1993) *Nucl. Acids Res.*, **21**, 2143.

166. Shortle, D., Grisafi, P., Benkovic, S.J. and Botstein, D. (1982) *Proc. Natl. Acad. Sci. USA*, **79**, 1588.

167. Roth, M.J., Tanese, N. and Goff, S.P. (1985) *J. Biol. Chem.*, **260**, 9326.

168. Moelling, K., (1974) *Virology*, **62**, 46.

169. Krug, M.S. and Berger, S.L. (1987) *Meth. Enzymol.*, **152**, 316.

170. Kotewicz, M.L., D'Alessio, J.M., Driftmier, K.M., Blodgett, K.P. and Gerard, G.F. (1985) *Gene*, **35**, 249.

171. Tanese, N., Roth, M. and Goff, S.P. (1985) *Proc. Natl. Acad. Sci. USA*, **82**, 4944.

172. Gerard, G.F. (1986) *Focus*, **7(1)**, 1.

173. Frohman, M.A., Dush, M.K. and Martin, G.R. (1988) *Proc. Natl. Acad. Sci. USA*, **85**, 8998.

174. Ahlquist, P., Dasgupta, R., Shih, D.S., Zimmern, D. and Kaesberg, P. (1979) *Nature*, **281**, 277.

175. Financsek, I., Mizumoto, K. and Muramatsu, M. (1982) *Gene*, **18**, 115.

176. Handa, H., Mizumoto, K., Oda, K., Okamoto, T. and Fukasawa, T. (1985) *Gene*, **33**, 159.

177. Itoh, N., Mizumoto, K. and Kaziro, Y. (1984) *J. Biol. Chem.*, **259**, 13923.

178. Itoh, N., Yamada, H., Kaziro, Y. and Mizumoto, K. (1987) *J. Biol. Chem.*, **262**, 1989.

179. Laskowski, M. (1980) *Meth. Enzymol.*, **65**, 263.

180. Sung, S. and Laskowski, M. (1962) *J. Biol. Chem.*, **237**, 506.

181. Johnson, P.H. and Laskowski, M. (1968) *J. Biol. Chem.*, **243**, 3421.

182. Johnson, P.H. and Laskowski, M. (1970) *J. Biol. Chem.*, **245**, 891.

183. Mikulski, A.A. and Laskowski, M. (1970) *J. Biol. Chem.*, **245**, 5026.

184. Kroeker, W.D. and Kowalski, D. (1978) *Biochemistry*, **17**, 3236.

185. Green, M.R. and Roeder, R.G. (1980) *Cell*, **22**, 231.

186. Ghangas, G.S. and Wu, R. (1975) *J. Biol. Chem.*, **250**, 4601.

187. Hammond, A.W. and D'Alessio, J. (1986) *Focus*, **8(4)**, 4.

188. Wesink, P.C., Finnegan, D.J., Donelson, J.E. and Hogness, D.S. (1974) *Cell*, **3**, 315.

189. Gubler, U. (1987) *Meth. Enzymol.*, **152**, 330.

190. Fraser, M.J., Tjeerde, R. and Matsumoto, K. (1976) *Can. J. Biochem.*, **54**, 971.

191. Fraser, M.J. (1980) *Meth. Enzymol.*, **65**, 255.

192. Krupp, G. and Gross, H.J. (1979) *Nucl. Acids Res.*, **6**, 3481.

193. Brederode, F.T., Koper-Zwarthoff, E.C. and Bol, J.F. (1980) *Nucl. Acids Res.*, **8**, 2213.

194. Linn, S. and Lehman, I.R. (1965) *J. Biol. Chem.*, **240**, 1287.

195. Furuichi, Y. and Miura, K. (1975) *Nature*, **253**, 374.

196. Kihara, K., Nomiyama, H., Yukuhiro, M. and Mukai, J.-I. (1976) *Anal. Biochem.*, **75**, 672.

197. Lundberg, K.S., Shoemaker, D.D., Adams, M.W.W., Short, J.M., Sorge, J.A. and Mathur, E.J. (1991) *Gene*, **108**, 1.

198. Flaman, J.-M., Frebourg, T., Moreau, V., Charbonnier, F., Martin, C., Ishioka, C., Friend, S.H. and Iggo, R. (1994) *Nucl. Acids Res.*, **22**, 3259.

199. Cline, J., Braman, J.C. and Hogrefe, H.H. (1996) *Nucl. Acids Res.*, **24**, 3546.

200. Bjork, W. (1963) *J. Biol. Chem.*, **238**, 2487.

201. Philipps, G.R. (1975) *Z. Physiol. Chem.*, **356**, 1085.

202. Xu, M.Q., Southworth, M.W., Mersha, F.B., Hornstra, L.J. and Perler, F.B. (1993) *Cell*, **75**, 1371.

203. Sippel, A.E. (1973) *Eur. J. Biochem.*, **37**, 31.

204. Beltz, W.R. and Ashton, S.H. (1982) *Fed. Proc.*, **41**, 1450, Abstract 6896.

205. Gething, M.-J., Bye, J., Skehel, J. and Waterfield, M. (1980) *Nature*, **287**, 301.

206. Clark, A.J. (1973) *Annu. Rev. Genet.*, **7**, 67.

207. Radding, C.M. (1978) *Annu. Rev. Biochem.*, **47**, 847.

208. Soltis, D.A. and Lehman, I.R. (1983) *J. Biol. Chem.*, **258**, 6073.

209. Taidi-Laskowski, B., Tyan, D., Honigberg, S.M., Radding, C.R. and Grumet, F.C. (1988) *Nucl. Acids Res.*, **16**, 8157.

210. Honigberg, S.M., Rao, B.J. and Radding, C.M. (1986) *Proc. Natl. Acad. Sci. USA*, **83**, 9586.

211. Rigas, B., Welcher, A.A., Ward, D.C. and Weissman, S.M. (1986) *Proc. Natl. Acad. Sci. USA*, **83**, 9591.

212. Ferrin, L.J. and Camerini-Otero, R.D. (1991) *Science*, **254**, 1494.

213. Koob, M., Burkiewicz, A., Kur, J. and Szybalski, W. (1992) *Nucl. Acids Res.*, **20**, 5831.

214. Hirs, C.H.W., Moore, S. and Stein, W.H. (1956) *J. Biol. Chem.*, **219**, 623.

215. Davidson, J.N. (1972) *The Biochemistry of the Nucleic Acids (7th edn)*. Academic Press, New York.

216. Myers, R.M., Larin, Z. and Maniatis, T. (1985) *Science*, **230**, 1242.

217. Winter, E., Yamamoto, F., Almoguera, C. and Perucho, M. (1985) *Proc. Natl. Acad. Sci. USA*, **82**, 7575.

218. Levy, C.C. and Karpetsky, T.P. (1980) *J. Biol. Chem.*, **255**, 2153.

219. Kanaya, S. and Crouch, R.J. (1983) *J. Biol. Chem.*, **258**, 1276.

220. Berkower, I., Leis, J. and Hurwitz, J. (1973) *J. Biol. Chem.*, **248**, 5914.

221. Vournakis, J.N., Efstratiadis, A. and Kafatos, F.C. (1975) *Proc. Natl. Acad. Sci. USA*, **72**, 2959.

222. Stavrianopoulos, J.G., Gambino-Giuffrida, A. and Chargaff, E. (1976) *Proc. Natl. Acad. Sci. USA*, **73**, 1081.

223. Keller, W. and Crouch, R. (1972) *Proc. Natl. Acad. Sci. USA*, **69**, 3360.

224. Grossman, L.I., Watson, R. and Vinograd, J. (1973) *Proc. Natl. Acad. Sci. USA*, **70**, 3339.

225. Hillenbrand, G. and Staudenbauer, W.L. (1982) *Nucl. Acids Res.*, **10**, 833.

226. Donis-Keller, H. (1979) *Nucl. Acids Res.*, **7**, 179.

227. Goodwin, E.C. and Rottman, F.M. (1992) *Nucl. Acids Res.*, **20**, 916.

228. Shen, V. and Schlessinger, D. (1982) in *The Enzymes (3rd edn)* (P.D. Boyer, ed.). Academic Press, New York, Vol. 15, p. 501.

229. Meador, J. and Kennell, D. (1990) *Gene*, **95**, 1.

230. Meador, J., Cannon, B., Cannistraro, V.J. and Kennell, D. (1990) *Eur. J. Biochem.*, **187**, 549.

231. Pilly, D., Niemeyer, A., Schmidt, M, and Bargetzi, J.P. (1978) *J. Biol. Chem.*, **253**, 437.

232. Donis-Keller, H. (1980) *Nucl. Acids Res.*, **8**, 3133.

233. Takahashi, K. (1971) *J. Biochem.*, **70**, 945.

234. Donis-Keller, H., Maxam, A. and Gilbert, W. (1977) *Nucl. Acids Res.*, **4**, 2527.

235. Uchida, T. and Egami, R. (1971) in *The Enzymes (3rd edn)* (P.D. Boyer, ed.). Academic Press, New York, Vol. 4, p. 205.

236. Gilman, M. (1993) in *Current Protocols in Molecular Biology* (F.M. Ausubel, R. Brent, R.E. Kingston, D.D. Moore, J.G. Seidman, J.A. Smith and K. Struhl, eds). John Wiley, New York, Unit 4.7.

237. Sawadogo, M. and Roeder, R.G. (1985) *Proc. Natl. Acad. Sci. USA*, **82**, 4394.

238. Lewis, M.K. and Burgess, R.R. (1980) *J. Biol. Chem.*, **255**, 4928.

239. Perreault, J.P., Wu, T.F., Cousineau, B., Ogilvie, K.K. and Cedergren, R. (1990) *Nature*, **344**, 565.

240. Zaug, A.J., Been, M.D. and Cech, T.R. (1986) *Nature*, **324**, 429.

241. Zaug, A.J., Grosshans, C.A. and Cech, T.R. (1988) *Biochemsitry*, **27**, 8924.

242. Cech, T.R. (1989) *USB Comments*, **16(2)**, 13.

243. Vogt, V.M. (1973) *Eur. J. Biochem.*, **33**, 192.

244. Oleson, A.E. and Sasakuma, M. (1980) *Arch. Biochem. Biophys.*, **204**, 361.

245. Gannon, F., Jeltsch, J.M. and Perrin, F. (1980) *Nucleic Acids. Res.*, **8**, 4405.

246. Berk, A.J. and Sharp, P.A. (1978) *Cell*, **14**, 695.

247. Grosschedl, R. and Birnstiel, M.L. (1980) *Proc. Natl. Acad. Sci. USA*, **77**, 1432.

248. Weaver, R.F. and Weissman, C. (1979) *Nucl. Acids Res.*, **7**, 1175.

249. Maniatis, T., Gee, S.G., Efstratiadis, A. and Kafatos, F.C. (1976) *Cell*, **8**, 163.

250. Zechel, K. and Weber, K. (1977) *Eur. J. Biochem.*, **77**, 133.

251. Kedzierski, W., Laskowski, M and Mandel, M. (1973) *J. Biol. Chem.*, **248**, 1277.

252. Vogt, V.M. (1980) *Meth. Enzymol.*, **65**, 263.

253. Taniuchi, H., Anfinsen, C.B. and Sodja, A. (1967) *J. Biol. Chem.*, **242**, 4752.

254. Pelham, H.R.B. and Jackson, R.J. (1976) *Eur. J. Biochem.*, 67, 247.

255. Jackson, R.J. and Hunt, T. (1983) *Meth. Enzymol.*, **96**, 50.

256. Noll, M. (1974) *Nature*, **251**, 249.

257. Heins, J.N., Suriano, J.R., Taniuchi, H. and Anfinsen, C.B. (1967) *J. Biol. Chem.*, **242**, 1016.

258. Tabor, S. and Richardson, C.C. (1989) *J. Biol. Chem.*, **264**, 6447.

259. Tabor, S. and Richardson, C.C. (1987) *Proc. Natl. Acad. Sci. USA*, **84**, 4767.

260. Christiansen, C. and Baldwin, R.L. (1977) *J. Mol. Biol.*, **115**, 441.

261. Kowalczykowski, S.C., Bear, D.G. and von Hippel, P.H. (1981) in *The Enzymes (3rd edn)* (P.D. Boyer, ed.). Academic Press, New York, Vol. 14, p. 373.

262. Anon (1994) *Molecular Biology Reagents Catalogue*. United States Biochemical Corp., Cleveland, p. 266.

263. Butler, E.T. and Chamberlain, M.J. (1982) *J. Biol. Chem.*, **257**, 5772.

264. Melton, D.A., Krieg, P.A., Rebagliati, M.R., Maniatis, T., Zinn, K. and Green, M.R. (1984) *Nucl. Acids Res.*, **12**, 7035.

265. Green, M.R., Maniatis, T. and Melton, D.A. (1983) *Cell*, **32**, 681.

266. Kassavetis, G.A., Butler, E.T., Roulland, D. and Chamberlain, M.J. (1982) *J. Biol. Chem.*, **257**, 5779.

267. Kotani, H., Ishizaki, Y., Hiraoka, N. and Obayashi, A. (1987) *Nucl. Acids Res.*, **15**, 2653,

268. Johnson, M.T. and Johnson, B. (1984) *Biotechniques*, **2**, 156.

269. Krainer, A.F., Maniatis, T., Ruskin, B. and Green, M.R. (1984) *Cell*, **36**, 993.

270. Krieg, P.A. and Melton, D.A. (1984) *Nature*, **308**, 203.

271. Melton, D.A. (1985) *Proc. Natl. Acad. Sci. USA*, **82**, 144.

272. Krieg, P.A. and Melton, D.A. (1984) *Nucl. Acids. Res.*, **12**, 7057.

273. Zinn, K., DiMaio, D. and Maniatis, T. (1983) *Cell*, **34**, 865.

274. Church, G.M. and Gilberts, W. (1984) *Proc. Natl. Acad. Sci. USA*, **81**, 1991.

275. Cox, K.H., DeLeon, D.V., Angerer, L.M. and Angerer, R.C. (1984) *Dev. Biol.*, **101**, 485.

276. Looney, J.E., Han, J.H. and Harding, J.D. (1984) *Gene*, **27**, 67.

277. Parvin, J.D., Smith, F.I. and Palese, P. (1986) *DNA*, **5**, 167.

278. Kotewicz, M.L., Sampson, C.M., D'Alessio, J.M. and Gerard, G.F. (1988) *Nucl. Acids Res.*, **16**, 265.

279. Gerard, G.F., D'Alessio, J.M. and Kotewicz, M.L. (1989) *Focus*, **11**, 66.

280. Morris, C.E., Klement, J.F. and McAllister, W.T. (1986) *Gene*, **41**, 193.

281. Brown, J.E., Klement, J.F. and McAllister, W.T. (1986) *Nucl. Acids Res.*, **14**, 3521.

282. Peebles, C.L. (1980) *Gene Anal. Tech.*, **3**, 59.

283. Wilson, G.G. and Murray, N.E. (1979) *J. Mol. Biol.*, **132**, 471.

284. Murray, N.E., Bruce, S.A. and Murray, K. (1979) *J. Mol. Biol.*, **132**, 493.

285. Nilsson, S.V. and Magnusson, G. (1982) *Nucl. Acids Res.*, **10**, 1425.

286. Ferretti, L. and Sgaramella, V. (1981) *Nucl. Acids Res.*, **9**, 85.

287. Pohl, F.M., Thomae, R. and Karst, A. (1982) *Eur. J. Biochem.*, **123**, 141.

288. Dugaiczyk, A., Boyer, H.W. and Goodman, H.M. (1975) *J. Mol. Biol.*, **96**, 171.

289. Graf, H. (1979) *Biochim. Biophys. Acta*, **564**, 225.

290. Sgaramella, V. (1972) *Proc. Natl. Acad. Sci. USA*, **69**, 3389.

291. Sugino, A., Goodman, H.M., Heyneker, H.L., Shine, J., Boyer, H.W. and Cozzarelli, N.Z. (1977) *J. Biol. Chem.*, **252**, 3987.

292. Pheiffer, B.H. and Zimmerman, S.B. (1983) *Nucl. Acids Res.*, **11**, 7853.

293. Cobianchi, F. and Wilson, S.H. (1987) *Meth. Enzymol.*, **152**, 94.

294. Challberg, M.D. and Englund, P.T. (1979) *J. Biol. Chem.*, **254**, 7820.

295. Panet, A., van de Sande, J.H., Louwen, P.C., Khorana, H.G., Raae, A.J., Lillehaug, J.R. and Kleppe, K. (1973) *Biochemistry*, **12**, 5045,

296. Englund, P.T. (1971) *J. Biol. Chem.*, **246**, 3269.

297. O'Farrell, P.H., Kutter, E. and Nakanishi, M., (1980) *Mol. Gen. Genet.*, **179**, 421.

298. Morris, C.F., Hama-Inaba, H., Mace, D., Sinha, N.K. and Alberts, B. (1979) *J. Biol. Chem.*, **254**, 6787.

299. Burd, J.F. and Wells, R.D. (1974) *J. Biol. Chem.*, **249**, 7094.

300. Wang, K., Koop, B.F. and Hood, I. (1994) *Biotechniques*, **17**, 236.

301. Nossal, N.G. (1974) *J. Biol. Chem.*, **249**, 5668.

302. Dale, R.M.K., McClure, B.A. and Houchins, J.P. (1985) *Plasmid*, **12**, 31.

303. Doetsch, P.W., Chan, G.L. and Haseltine, W.A. (1985) *Nucl. Acids Res.*, **13**, 3285.

304. Shamoo, Y., Adari, H., Konigsberg, W.H., Williams, K.R. and Chase, J.W. (1986) *Proc. Natl. Acad. Sci. USA*, **83**, 8844.

305. Bittner, M., Burke, R.L. and Alberts, B.M. (1979) *J. Biol. Chem.*, **254**, 9565.

306. Alberts, B. and Sternglanz, R. (1977) *Nature*, **269**, 655.

307. Huberman, J.A., Kornberg, A. and Alberts, B.M. (1971) *J. Mol. Biol.*, **62**, 39.

308. Chase, J.W. and Williams, K.R. (1986) *Annu. Rev. Biochem.*, **55**, 103.

309. Williams, K.R., LoPresti, M.B. and Setoguchi, M. (1981) *J. Biol. Chem.*, **256**, 1754.

310. Topal, M.D. and Sinha, N.K. (1983) *J. Biol. Chem.*, **258**, 12274.

311. Dombroski, D.F. and Morgan, A.R. (1985) *J. Biol. Chem.*, **260**, 415.

312. Toulme, J.J., Behmoaras, T., Guigues, M. and Helene, C. (1983) *EMBO J.*, **2**, 505.

313. Richardson, C.C. (1965) *Proc. Natl. Acad. Sci. USA*, **54**, 158.

314. Depew, R.E., Snopek, T.J. and Cozzarelli, N.R. (1975) *Virology*, **64**, 144.

315. Midgley, C.A. and Murray, N.E. (1985) *EMBO J.*, **4**, 2695.

316. Lillehaug, J.R. (1977) *Eur. J. Biochem.*, **73**, 499.

317. Berkner, K.L. and Folk, W.R. (1977) *J. Biol. Chem.*, **252**, 3176.

318. Simoncsits, A., Brownlee, G.G., Brown, R.S., Rubin, J.R. and Guilley, H. (1977) *Nature*, **269**, 833.

319. Maxam, A.M. and Gilbert, W. (1980) *Meth. Enzymol.*, **65**, 499.

320. Reddy, M.V., Gupta, R.C. and Randerath, K. (1981) *Anal. Biochem.*, **117**, 271.

321. Randerath, K., Reddy, M.V. and Gupta, R.C. (1981) *Proc. Natl. Acad. Sci. USA*, **78**, 6126.

322. Gross, H.J., Krupp, G., Domdey, H., Raba, M., Jank, P., Lossow, C., Alberty, H., Ramm, K. and Sanger, H.L. (1981) *Eur. J. Biochem.*, **21**, 249.

323. Smith, H.O. and Birnstiel, M.L. (1976) *Nucl. Acids Res.*, **3**, 2387.

324. Maat, J. and Smith, A.J.H. (1978) *Nucl. Acids Res.*, **3**, 2387.

325. Johnsrud, L., (1978) *Proc. Natl. Acad. Sci. USA*, **75**, 5314.

326. Royer-Pokora, B., Gordon, L.K. and Hasletine, W.A. (1981) *Nucl. Acids Res.*, **9**, 4595.

327. Cameron, V. and Uhlenbeck, O.C. (1977) *Biochemistry*, **16**, 5120.

328. Soltis, D.A. and Uhlenbeck, O.C. (1982) *J. Biol. Chem.*, **257**, 11340.

329. Ohtsuka, E., Tanaka, S, Tanaka, T., Miyake, T., Markham, A.F., Nakagawa, E., Wakabayashi, T., Taniyama, Y., Nishikawa, S., Fukumoto, R., Uemura, H., Doi, T., Tokunaga, T. and Ikehara, M. (1981) *Proc. Natl. Acad. Sci. USA*, **78**, 5493.

330. Rand, K.N. and Gait, M.J. (1984) *EMBO J.*, **3**, 397.

331. Sugino, A., Snopek, T.J. and Cozzarelli, N.R. (1977) *J. Biol. Chem.*, **252**, 1732.

332. Englund, T.E., Gumport, R.I. and Uhlenbeck, O.C. (1977) *Proc. Natl. Acad. Sci. USA*, **74**, 4839.

333. Walker, G.L., Uhlenbeck, O.C., Bedows, E. and Gumport, R.I. (1975) *Proc. Natl. Acad. Sci. USA*, **72**, 122.

334. Tessier, D.C., Brousseau, R. and Vernet, T. (1986) *Anal. Biochem.*, **158**, 171.

335. Romaniuk, P.J. and Uhlenbeck, O.C. (1983) *Meth. Enzymol.*, **100**, 52.

336. Middleton, T., Herlihy, W.C., Schimmel, P.R. and Munro, H.N. (1985) *Anal. Biochem.*, **144**, 110.

337. Bruce, A.G. and Uhlenbeck, O.C. (1978) *Nucl. Acids Res.*, **5**, 3665.

338. Englund, T.E. and Uhlenbeck, O.C. (1978) *Nature*, **275**, 560.

339. Noren, C.J., Anthony-Cahill, S.J., Griffith, M.C. and Schultz, P.G. (1989) *Science*, **244**, 182.

340. Brude, A.G. and Uhlenbeck, O.C. (1982) *Biochemistry*, **21**, 855.

341. Uhlenbeck, O.L. and Gumport, R.I. (1982) in *The Enzymes (3rd edn)* (P.D. Boyer, ed.). Academic Press, New York, Vol. 15, p. 31.

342. Modrich, P. and Richardson, C.C. (1975) *J. Biol. Chem.*, **250**, 5515.

343. Mark, D.F. and Richardson, C.C. (1976) *Proc. Natl. Acad. Sci. USA*, **73**, 780.

344. Tabor, S. (1987) *PhD Dissertation*. Harvard University, Boston.

345. Huber, H.E., Tabor, S. and Richardson, C.C. (1987) *J. Biol. Chem.*, **262**, 16224.

346. Tabor, S., Huber, H.E. and Richardson, C.C. (1987) *J. Biol. Chem.*, **262**, 16212.

347. Lechner, R.L., Engler, M.J. and Richardson, C.C. (1983) *J. Biol. Chem.*, **258**, 11174.

348. Bebenek, K. and Kunkel, T.A. (1989) *Nucl. Acids Res.*, **17**, 5408.

349. Pham, T.T. and Coleman, J.E. (1985) *Biochemistry*, 24, 5672.

350. Center, M.S., Studier, F.W. and Richardson, C.C. (1970) *Proc. Natl. Acad. Sci. USA*, **65**, 242.

351. de Massy, B., Weisberg, R.A. and Studier, F.W. (1987) *J. Mol. Biol.*, **193**, 359.

352. Panayotatos, N. and Wells, R.D. (1981) *Nature*, **289**, 466.

353. Straney, D.C. and Crothers, D.M. (1987) *J. Mol. Biol.*, **193**, 279.

354. Osterman, H.L. and Coleman, J.E. (1981) *Biochemistry*, **20**, 4884.

355. Panayotatos, N. and Fontaine, A. (1985) *J. Biol. Chem.*, **260**, 3173.

356. Engler, M.J. and Richardson, C.C. (1983) *J. Biol. Chem.*, **258**, 11197.

357. Straus, N.A. and Zagursky, R.J. (1991) *Biotechniques*, **10**, 376.

358. Davanloo, P., Rosenberg, A.H., Dunn, J.J. and Studier, F.W. (1984) *Proc. Natl. Acad. Sci. USA*, **81**, 2035.

359. Stahl, S.J. and Zinn, K. (1981) *J. Mol. Biol.*, **148**, 481.

360. Dunn, J.J. and Studier, F.W. (1983) *J. Mol. Biol.*, **166**, 477.

361. Tabor, S. and Richardson, C.C. (1985) *Proc. Natl. Acad. Sci. USA*, **82**, 1074.

362. Southern, E.M. (1975) *J. Mol. Biol.*, **98**, 503.

363. Smith, G.E. and Summers, M.D. (1980) *Anal. Biochem.*, **109**, 123.

364. Schenborn, E.T. and Meirendorf, R.C. (1985) *Nucl. Acids Res.*, 13, 6223.

365. Langer, P.R., Waldrop, A.A. and Ward, D.C. (1981) *Proc. Natl. Acad. Sci. USA*, **78**, 6633.

366. Chien, A., Edgar, D.B. and Trela, J.M. (1976) *J. Bacteriol.*, **127**, 1550.

367. Saiki, R.K. and Gelfand, D.H. (1989) *Amplifications*, **1**, 4.

368. Lawyer, F.C., Stoffel, S, Saiki, R.K., Myambo, K., Drummond, R. and Gelfand, D.H. (1989) *J. Biol. Chem.*, **264**, 6427.

369. Saiki, R.K., Scharf, S., Fallona, F., Mullis, K.B., Horn, G.T., Erlich, H.A. and Arnheim, N. (1985) *Science*, **230**, 1350.

370. Saiki, R.K., Gelfand, D.H., Stoffel, S., Scharf, S.J., Higuchi, R., Horn, G.T., Mullis, K.B. and Erlich, H.A. (1988) *Science*, **239**, 487.

371. Innis, M.A., Myambo, K.B., Gelfand, D.H. and Brow, M.A.D. (1988) *Proc. Natl. Acad. Sci. USA*, **85**, 9436.

372. Saluz, H.P. and Jost, J.P. (1989) *Nature*, **338**, 277.

373. Chang, L.M.S. and Bollum, F.J. (1971) *J. Biol. Chem.*, **246**, 909.

374. Deng, G. and Wu, R. (1981) *Nucl. Acids Res.*, **9**, 4173.

375. Michelson, A.M. and Orkin, S.H. (1982) *J. Biol. Chem.*, **257**, 14773.

376. Chang, L.M.S. and Bollum, F.J. (1986) *Crit. Rev. Biochem.*, **21**, 27.

377. Roychoudhury, R., Jay, E. and Wu, R. (1976) *Nucl. Acids Res.*, **3**, 863.

378. Heidecker, G. and Messing, J. (1983) *Nucl. Acids Res.*, **11**, 4891.

379. Tu, C.-P.D. and Cohen, S.N. (1980) *Gene*, **10**, 177.

380. Wu, R., Jay, E. and Roychoudhury, R. (1976) *Methods Cancer Res.*, **12**, 87.

381. Yousaf, S.I., Carroll, A.R. and Clarke, B.E. (1984) *Gene*, **27**, 309.

382. Ratliff, R.L. (1982) in *The Enzymes (3rd edn)* (P.D. Boyer, ed.). Academic Press, New York, Vol. 15, p. 105.

383. Deng, G. and Wu, R. (1983) *Meth. Enzymol.*, **100**, 96.

384. Eschenfeldt, W.H., Puskas, R.S. and Berger, S.L. (1987) *Meth. Enzymol.*, **152**, 337.

385. Kaledin, A.S., Slyuisarenko, A.G. and Gorodetskii, S.I. (1981) *Biochemistry (USSR)*, **46**, 1247.

386. Mattila, P., Korpela, J., Tenkanen, T. and Pitkanen, K. (1991) *Nucl. Acids Res.*, **19**, 4967.

387. Rüttiman, C., Cotoras, M., Zaldivar, J. and Vicuma, R. (1985) *Eur. J. Biochem.*, **149**, 41.

388. Duncan, B.K. and Chamgers, J.A. (1984) *Gene*, **28**, 211.

389. Varshney, U., Hutcheon, T. and van de Sande, J.H. (1988) *J. Biol. Chem.*, **263**, 7776.

390. Longo, M.C., Berninger, M.S. and Hartley, J.L. (1990) *Gene*, **93**, 125.

391. Perler, F.B., Comb, D.G., Jack, W.E., Moran, L.S., Qiang, B.Q., Kucera, R.B., Benner, J., Statko, B.E., Nwankwo, D.O., Hempstead, S.K., Carlow, C.K.S. and Jannasch, H. (1992) *Proc. Natl. Acad. Sci. USA*, **89**, 5577.

CHAPTER 4
GENOMES

This chapter provides information on the genomes of selected organisms. The data are not comprehensive (it is obviously impossible to include all species), but the examples that are chosen are representative of different groups.

1. GENOMES

1.1 Genome sizes

Table 1 shows the sizes of the nuclear genomes of a range of organisms. Most of these data have been obtained by reassociation kinetics and related analyses, and as such should be treated with caution. Sequencing projects are beginning to show that sizes estimated by these indirect methods rarely give more than an approximation of the true genome size.

The practical importance of nuclear genome size lies with the influence it has on the number of clones needed for a 'complete' genomic library. The relevant formula is:

$$N = \frac{\ln(1-P)}{\ln(1-a/b)}$$

where N = the number of clones required, P = probability that a given sequence will be present, a = average size of the DNA fragments inserted in the vector and b = total size of the genome. In *Figure 1* the relationship between N and b is plotted for a = 17 kb (fragments suitable for a phage λ replacement vector), a = 35 kb (cosmid vector) and a = 100 kb (P1 vector), at probabilities of $P = 0.95$ (95% probability) and $P = 0.99$ (99% probability).

Mitochondrial and chloroplast genome sizes have been determined for a wide range of organisms, in some cases by complete DNA sequencing, in others by direct examination of the molecules by electron microscopy or gel electrophoresis. Data for representative species are given in *Table 2*.

Table 1. Haploid genome sizes for representative organisms[a]

Group	Species	Estimated genome size (kb)
Microorganisms		
Bacteria		
	Mycoplasma genitalium	580[b]
	Mycoplasma pneumoniae	816[b]
	Methanoccocus jannaschii	1740[b,c]
	Haemophilus influenzae	1830[b]
	Synechocystis sp.	3573[b]
	Escherichia coli	4639[b]
	Bacillus megaterium	30 000

Genomes

Figure 1. Graph for the calculation of the approximate number of clones required for a gene library. Each line represents a different combination of *a* (the average size in kb of the fragments inserted into the vector) and *P* (the probability that a given sequence will be present). The values used are $a = 17$ kb (fragments suitable for a phage λ replacement vector), $a = 35$ kb (cosmid vector) and $a = 100$ kb (P1 vector), at probabilities of $P = 0.95$ (95% probability) and $P = 0.99$ (99% probability).

Worked example. Imagine you wish to make an *E. coli* gene library where $a = 35$ and $P = 0.95$. The *E. coli* genome is 4.6×10^3 kb, so find the position corresponding to 4.6 on the x-axis. A vertical drawn from this point intersects the line for $a = 35$, $P = 0.95$ at a position corresponding to 4.0 on the y-axis. The value for *n* in this example is 3 (the exponent from the genome size), so the number of clones required is $4.0 \times 10^2 = 400$.

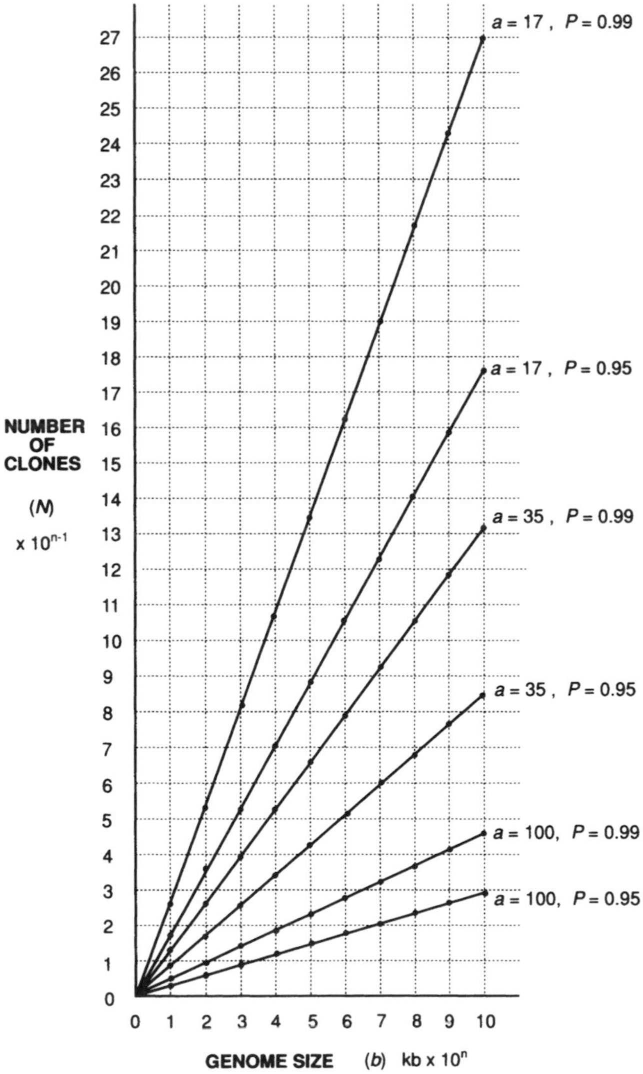

Table 1. Genome sizes (continued)

Group	Species	Estimated genome size (kb)
Fungi		9400–175 000
	Hansenula holstii	10 300
	Saccharomyces cerevisiae	12 069[b]
	Saccharomyces carlsbergensis	15 000
	Torulopsis holmii	21 600
	Aspergillus nidulans	25 400
	Neurospora crassa	27 000
	Phycomyces blakesleeanus	30 100
	Achyla bisexualis	41 400
	Dictyostelium discoideum	54 000
Algae		37 500–190 000 000
Protozoa		37 500–330 000 000
	Tetrahymena pyriformis	190 000
Animals		
Porifera		56 500
Coelenterata		280 000–685 000
	Cassiopeia sp.	310 000
Nematoda		75 000–620 000
	Caenorhabditis elegans	100 000
Annelida		660 000–6 750 000
	Nereis sp.	1 400 000
Arthropoda		
Crustacea		660 000–21 250 000
	Plagusia depressa	1 400 000
	Limulus polyphemus	2 650 000
Insecta		47 000–12 000 000
Diptera		125 000–835 000
	Drosophila melanogaster	140 000
	Chironomus tetans	197 500
	Culex pipens	820 000
	Musca domestica	840 000
Lepidoptera		470 000–565 000
	Bombyx mori	490 000
Orthoptera		3 100 000–11 900 000
	Locusta migratoria	5 000 000
Mollusca		375 000–5 100 000
	Fissurella bandadensis	470 000
	Tectorius muricatus	630 000
	Aplysia californica	1 700 000
Echinodermata		470 000–4 150 000
	Strongylocentrotus purpuratus	845 000
Protochordata		140 000–565 000
	Asidea atra	150 000

Table 1. Genome sizes (continued)

Group	Species	Estimated genome size (kb)
Agnatha		1 320 000–2 650 000
Pisces		
Chondrichthyes		2 650 000–6 950 000
	Carcharias obscurus	2 650 000
Osteichthyes		375 000–140 000 000
	Fugu rupribes	400 000
	Rutilus rutilus	4 500 000
	Protopterus aethiopicus	140 000 000
Amphibia		
Anura		950 000–10 150 000
	Xenopus laevis	3 100 000
	Bufo bufo	6 600 000
Urodela		14 200 000–78 500 000
	Ambystoma mexicanum	35 700 000
	Notophthalamus viridescens	42 500 000
	Amphiuma sp.	78 500 000
Reptilia		1 600 000–5 100 000
	Python reticulatus	1 600 000
	Natrix natrix	2 350 000
	Caiman crocodylus	2 450 000
	Terrapene carolina	3 850 000
Aves		1 125 000–1 975 000
	Gallus domesticus	1 125 000
Mammalia		
Marsupalia		2 800 000–4 420 000
Placentalia		2 350 000–5 550 000
	Homo sapiens	3 000 000
	Microtus agrestis	3 000 000
	Mus musculus	3 300 000
	Peromyscus eremicus	4 420 000
	Dipodomys ordii monoensis	5 200 000
Plants		
Bryophyta		600 000–4 050 000
Pteridophyta		950 000–300 000 000
Gymnospermae		4 900 000–47 000 000
Angiospermae		95 000–120 000 000
	Arabidopsis thaliana	100 000
	Oryza sativa	565 000
	Lycopersicon esculentum	700 000
	Daucus carota	950 000
	Brassica napus	1 500 000
	Medicago sativa	1 600 000
	Solanum tuberosum	2 000 000

Table 1. Genome sizes (continued)

Group	Species	Estimated genome size (kb)
	Nicotiana tabacum	3 500 000
	Pisum sativum	4 800 000
	Zea mays	5 000 000
	Secale cereale	8 275 000
	Triticum aestivum	17 000 000
	Allium cepa	17 000 000
	Tulipa polychroma	23 000 000
	Lilium davidii	40 000 000
	Fritillaria assyriaca	120 000 000

[a]Data taken from refs 1 and 2 and papers cited therein.
[b]Completely sequenced genomes.
[c]The *M. jannaschii* genome is tripartite consisting of a large circular chromosome of 1665 kb, a large circular extrachromosomal element of 58.4 kb and a small circular extrachromosomal element of 16.6 kb (3).

Table 2. Sizes of organelle genomes of representative organisms[a]

Group	Species	Estimated genome size (kb)
Mitochondrial genomes		
Fungi		18.0–175
Myxomycota	*Physarum polycephalum*	69
Oomycota	*Phytophora infestans*	36.2
	Saprolegnia ferax	46.5
	Achlya sp.	50.0
	Pythium ultimum	57
Zygomycota	*Phycomyces blakesleeanus*	25.6
Ascomycota	*Cephalosporium acremonium*	27
	Aspergillus nidulans	33.3
	Claviceps purpurea	45
	Penicillium chrysogenum	49.2
	Neurospora crassa	60
	Podospora anserina	94
	Brettanomyces custersii	108
	Cochliobolus heterostrophus	115
Basidiomycota	*Coprimus cinereus*	43.3
	Schiziphyllum commune	50.3–52.2
	Ustilago cynodontis	76.5
	Coprinus stercorarius	91
	Suillus grisellus	121
	Agaricus bitorquis	148–174
Yeasts	*Torulopsis glabrata*	18.9
	Schizosaccharomyces pombe	19
	Saccharomyces exiguus	23.7
	Kloechera africana	27.1
	Kluveromyces lactis	37
	Hansenula petersonii	42
	Yarrowia lipolytica	44–48

Table 2. Organelle genomes (continued)

Group	Species	Estimated genome size (kb)
	Hansenula mrakii	55
	Saccharomyces cerevisiae	68–78
Other lower eukaryotes		
	Hydra sp.	15
	Chlamydomonas reinhardtii	15.8
	Trypanosoma brucei (maxicircle)	22
	Tetrahymena sp.	40–50
	Paramecium sp.	40–50
Animals		14.5–23.0
	Ascaris suum	14.5
	Paracentrotus lividus	15.697
	Mus musculus	16.295
	Bos taurus	16.338
	Homo sapiens	16.569
	Cnemidophorus exsanguis	17.4
	Xenopus laevis	18.4
	Drosophila melanogaster	19.0
Plants		150–2500
	Brassica oleracea	160
	Brassica hirta	208
	Brassica campestris	218
	Brassica nigra	231
	Raphanus sativa	242
	Chenopodium album	270
	Lycopersicon esculentum	270
	Zea mays	570
	Cucumis melo	2500
Chloroplast genomes		
Chlorophyta	*Codium fragile*	85
	Chlorella ellipsoidea	174
	Chlamydomonas reinhardtii	195
	Chlamydomonas smithii	199
	Chlamydomonas eugametos	243
	Chlamydomonas moewusii	292
	Acetabularia mediterranea	~2000
Other algae	*Dictyota dichotoma*	123
	Vaucheria sessilus	125
	Cyanophora paradoxa	127
	Euglena gracilis	130–152
	Plylaiella littorali	140
	Sphacelaria sp.	150
	Olisthodiscus luteus	154
Bryophyta	*Marchantia polymorpha*	121
	Sphaerocarpus donnellii	125

Table 2. Organelle genomes (continued)

Group	Species	Estimated genome size (kb)
Pteridophyta	*Osmunda cinnamomea*	144
	Asplenium nidus	150
	Pteris vittata	150
	Adiantum capillus-veneris	153
Gymnospermae	*Ginkgo biloba*	158
Angiospermae		120–217
	Pisum sativum	120
	Brassica oleracea	150
	Nicotiana tabacum	156
	Lycopersicon esculentum	158
	Spirodela sp.	180

[a]Data taken from various sources, including refs 4–6 and papers cited therein.

1.2 Physical data

Molecular biologists need to be aware of the GC contents and melting temperatures of the genomes that they work with (*Table 3*) since these are relevant to techniques that require separation of the two strands of genomic DNA molecules (e.g. hybridization analysis, polymerase chain reaction). Buoyant densities (*Table 4*) must be taken into account when purifying DNA by density gradient centrifugation. Sequence complexity (the relative amounts of single copy and repetitive DNA in a genome) varies greatly even among a single group of organisms (*Table 5*) and influences the feasibility of large-scale sequencing projects and the ease with which desired clones can be isolated from a genomic library.

Table 3. GC contents and melting temperatures for representative organisms[a]

Organism	GC content (%)	T_m (°C)
Bacteriophages		
λ	48.6	89.0
T1	48.0	–
T2	34.6	83.0
T3	49.6	90.0
T4	34.5	84.0
T5	39.0	–
T6	34.5	83.0
T7	47.4	89.5
Viruses		
Adenovirus	56.0–57.0	88.8–94.2
Herpesvirus	–	97.0
Polyomavirus	–	89.2
Prokaryotes		
Agrobacterium tumefaciens	58.0–59.7	94.2–94.8
Anacystis nidulans	54.3	–
Bacillus amyloliquefaciens	44.0	87.1–87.6
Bacillus brevis	–	87.5

Genomes

Table 3. GC contents and melting temperatures (continued)

Organism	GC content (%)	T_m (°C)
Bacillus cereus	–	83.0
Bacillus licheniformis	46.9	88.6
Bacillus megaterium	–	85.0
Bacillus subtilis	42.6	86.3–87.1
Bacillus thuringiensis	–	83.5
Clostridium butyricum	37.4	82.1
Clostridium perfringens	–	80.5
Desulphovibrio desulphuricans	56.3–57.4	90.5
Diplococcus pneumoniae	–	85.5
Erwinia amylovora	–	91.0
Erwinia herbicola	–	92.1–92.2
Escherichia coli	51.0	90.5–91.9
Haemophilus aegyptius	–	86.0
Haemophilus influenzae	38.0	85.6
Klebsiella pneumoniae	–	92.5–92.9
Lactobacillus brevis	43.2–43.9	–
Lactobacillus casei	43.4–49.4	–
Lactobacillus lactis	48.2–48.3	–
Listeria monocytogenes	–	85.3
Methanococcus jannaschii	31.4, 28.2, 28.8[b]	
Micrococcus luteus	65.4–74.3	95.1–98.9
Mycobacterium tuberculosis	65.0	–
Mycoplasma gallisepticum	–	84.0
Mycoplasma genitalium	32.0	–
Neisseria gonorrheae	–	89.5
Neisseria meningitidis	51.3	91.0
Nocardia sp.	68.6	96.6
Pasteurella pestis	–	88.5
Proteus mirabilis	–	85.3
Pseudomonas aeruginosa	–	97.5–97.7
Pseudomonas denitrificans	–	92.9
Pseudomonas fluorescens	59.5	94.5–95.3
Pseudomonas putida	–	93.4
Rhizobium japonicum	–	95.0
Rhodospirillum rubrum	–	94.5
Salmonella typhimurium	–	91.8
Serratia marcescens	–	93.5–94.9
Shigella dysenteriae	–	90.5
Staphylococcus aureus	32.4–37.7	82.9–83.5
Streptomyces albus	72.3	100.5
Streptomyces griseolus	72.4	–
Streptomyces griseus	72.1	–
Vibrio cholerae	–	88.5–88.7
Wolbachia persica	–	81.6
Xanthomonas campestris	–	97.3
Eukaryotic microorganisms		
Chlorella ellipsoidea	58.5	–
Crithidia fasciculata	–	97.4

Table 3. GC contents and melting temperatures (continued)

Organism	GC content (%)	T_m (°C)
Dictyostelium discoideum	23.0	79.5
Euglena gracilis	49.9	90.0
Saccharomyces cerevisiae	39.0	85.2
Schizosaccharomyces pombe	44.8	87.0
Tetrahymena pyriformis	–	81.2
Trichomonas gallinae	–	80.6
Invertebrates		
Drosophila melanogaster	39.1	86.5
Paracentrotus lividus	35.4	–
Strongylocentrotus pupuratus	36.6	–
Vertebrates		
Bos taurus	43.2	87.0
Cancer gracilis	–	84.0
Canis familiaris	44.1	–
Felis domesticus	42.8	–
Gallus domesticus	45.0	87.5
Homo sapiens	40.3	86.5
Iguana iguana	43.9	–
Mus musculus	40.3	86.5
Rattus sp.	41.8	87.5
Salmo salar	43.5	87.5
Xenopus laevis	40.9	–
Plants		
Nicotiana tabacum	–	85.5
Triticum vulgare	45.8	88.5

[a]Data taken from various sources, including refs 7 and 8 and papers cited therein.
[b]The GC contents are given for each component of the tripartite *Methanococcus jannaschii* genome. See *Table 1*, footnote c.

Table 4. Buoyant densities for representative organisms[a]

Organism	Buoyant density (g cm^{-3})	Organism	Buoyant density (g cm^{-3})
Bacteriophages		Herpesvirus	1.727
λ	1.710	Polyomavirus	1.709
T1	1.705	**Prokaryotes**	
T2	1.700	*Agrobacterium tumefaciens*	1.718–1.719
T3	1.712	*Anacystis nidulans*	1.714
T4	1.698	*Bacillus amyloliquefaciens*	1.707–1.708
T5	1.702	*Bacillus brevis*	1.704
T6	1.707	*Bacillus cereus*	1.696
T7	1.710	*Bacillus licheniformis*	1.705
Viruses		*Bacillus megaterium*	1.697
Adenovirus	1.708–1.720	*Bacillus subtilis*	1.705–1.706

Genomes

Table 4. Buoyant densities (continued)

Organism	Buoyant density (g cm^{-3})
Bacillus thuringiensis	1.695
Clostridium perfringens	1.691
Desulphovibrio desulphuricans	1.716–1.720
Diplococcus pneumoniae	1.701
Erwinia amylovora	1.713
Erwinia herbicola	1.714–1.715
Escherichia coli	1.710
Haemophilus aegyptius	1.698
Haemophilus influenzae	1.696
Klebsiella pneumoniae	1.712–1.715
Lactobacillus brevis	1.701–1.706
Lactobacillus casei	1.705–1.706
Lactobacillus lactis	1.710
Listeria monocytogenes	1.697
Micrococcus luteus	1.731
Mycobacterium tuberculosis	1.724
Mycoplasma gallisepticum	1.694
Neisseria gonorrheae	1.710
Neisseria meningitidis	1.710
Pasteurella pestis	1.706
Proteus mirabilis	1.700
Pseudomonas aeruginosa	1.726–1.727
Pseudomonas denitrificans	1.716
Pseudomonas fluorescens	1.721–1.724
Pseudomonas putida	1.719
Rhizobium japonicum	1.722
Rhodospirillum rubrum	1.726
Salmonella typhimurium	1.712
Serratia marcescens	1.717–1.718
Shigella dysenteriae	1.710
Staphylococcus aureus	1.693
Streptomyces albus	1.730
Streptomyces griseolus	1.729
Streptomyces griseus	1.730
Vibrio cholerae	1.706
Wolbachia persica	1.690
Xanthomonas campestris	1.727
Eukaryotic microorganisms	
Chlorella ellipsoidea	1.716
Crithidia fasciculata	1.717
Dictyostelium discoideum	1.682
Euglena gracilis	1.706
Saccharomyces cerevisiae	1.699
Tetrahymena pyriformis	1.690
Trichomonas gallinae	1.693
Invertebrates	
Drosophila melanogaster	1.702
Paracentrotus lividus	1.697
Strongylocentrotus pupuratus	1.699
Vertebrates	
Bos taurus	1.700
Cancer gracilis	1.700
Canis familiaris	1.700
Felis domesticus	1.700
Gallus domesticus	1.700
Homo sapiens	1.699
Iguana iguana	1.702
Mus musculus	1.701
Rattus sp.	1.701
Salmo salar	1.703
Xenopus laevis	1.699
Plants	
Nicotiana tabacum	1.700
Triticum vulgare	1.702

[a]Data taken from various sources, including refs 7 and 8 and papers cited therein.

Table 5. Sequence complexity of the genomes of representative organisms[a]

Group	Species	Approximate single copy DNA content (% total genome)
Fungi	*Phycomyces blakesleeanus*	65
	Dictyostelium discoideum	70
	Achyla bisexualis	82
	Torulopsis holmii	84
	Saccharomyces carlsbergensis	89
	Hansenula holstii	92
	Aspergillus nidulans	97
Protozoa	*Tetrahymena pyriformis*	90
Coelenterata	*Aurelia aurita*	70
Mollusca	*Aplysia californica*	55
Arthropoda		
Crustacea	*Limulus polyphemus*	70
Insecta	*Drosophila melanogaster*	60
	Musca domestica	90
	Chironomus tetans	95
Echinodermata	*Strongylocentrotus purpuratus*	75
Protochordata	*Ciona intestinalis*	70
Pisces	*Rutilus rutilus*	54
Amphibia	*Bufo bufo*	20
	Xenopus laevis	75
Reptilia	*Natrix natrix*	47
	Terrapene carolina	54
	Caiman crocodylus	66
	Python reticulatus	71
Aves	*Gallus domesticus*	80
Mammalia	*Homo sapiens*	64
	Mus musculus	70
Planta	*Lolium multiflorum*	36

[a]Data taken from ref. 1 and papers cited therein.

1.3 Genome projects

Rapid progress is being made with the sequence analysis of 'small' genomes. At the time of writing, complete sequences have been published for seven prokaryotes and one eukaryote (*Table 6*). Genome sequencing projects are also underway for a substantial number of other prokaryotes with antipated completion dates between 1997 and 1999 (*Table 7*). Updated information on these various projects can be found on the web pages for TIGR (The Institute for Genome Research, Gaithersburg, Maryland; ref. 9).

The Human Genome Project has set 2005 as the target date for sequence completion. The most important electronic databases relating to the project are listed in *Table 8*.

All nucleic acid and protein sequence information is collected and made available by the three major sequence databases – Genbank, EMBL and DDBJ – and protein data are also available from PIR and SWISS-PROT. These resources are described in *Table 9*.

Table 10 gives a list of eukaryotes other than humans for which genome projects are underway. In some cases, such as with *Caenorhabditis elegans* and *Plasmodium falciparum*, the immediate objective is the complete genome sequence and this is likely to be achieved in the near future. With other organisms there is, as yet, no clear initiative towards obtaining a complete sequence, with current research being directed towards genetic and physical mapping.

Table 6. Complete genome sequences[a,b]

Organism	Genome size (bp)	Number of genes		References
		Protein	RNA	
Prokaryotes				
Escherichia coli	4 638 858	4285	106	9–18
Haemophilus influenzae	1 830 135	1717	74	19–21
Helicobacter pylori	1 667 867	1590	44	22
Mycoplasma genitalium	580 073	468	36	23, 24
Mycoplasma pneumoniae	816 394	678	33	25, 26
Methanococcus jannaschii[b]	1 739 933	1738	43	3, 27
Synechocystis sp.	3 573 470	3170	49	28–30
Eukaryote				
Saccharomyces cerevisiae	12 069 313	6156	262	31–35

[a]Status at November 1997.
[b]Taken from ref. 9 and the references cited in the right hand column of this table.
[c]Data refer to all three components of the tripartite genome. See *Table 1*, footnote c.

Table 7. Prokaryote genome sequencing projects that are nearing completion[a,b]

Organism	Expected completion date	References
Actinobacillus actinomycetemcomitans		9
Archaeoglobus fulgidus	1997	36
Aquifex aeolicus	1997	9
Bacillus subtilis	1997	37, 38
Borrelia burgdorferi	1997	39
Caulobacter crecentus		9
Chlamydia trachomatis		40
Clostridium acetobutylicum		41
Deinococcus radiodurans	1998	42
Enterococcus faecalis	1999	43
Halobacterium salinarium		9
Legionella pneumophila		9
Methanobacterium thermoautotrophicum	1997	44
Mycobacterium avium	2000	9
Mycobacterium leprae		33, 45, 46
Mycobacterium tuberculosis	1998	33, 45
Neisseria gonorrhoeae		47
Neisseria meningitidis		48
Porphyromonas gingivalis		9
Pseudomonas aeruginosa		49
Pyrobaculum aerophilum		9
Pyrococcus furiosus		9
Pyrococcus shinkaj		9

Table 7. Prokaryote projects (continued)

Organism	Expected completion date	References
Rickettsia prowazekii		9
Salmonella typhimurium		9
Shewanella putrefaciens		9
Streptococcus pneumoniae	1998	9
Streptococcus pyogenes		50
Sulfolobus solfataricus		51
Thermotoga maritima	1998	9
Treponema denticula	1997	52
Thermoplasma acidophilum		9
Ureaplasma urealyticum	1997	9
Vibrio cholerae	1998	53

[a]Status at November 1997.
[b]Taken from ref. 9.

Table 8. Major databases of relevance to the Human Genome Project[a]

Database and description	References
GDB (Genome Database) The major human mapping database. Housed at Johns Hopkins University, School of Medicine, Baltimore, MD. Permits interaction with the OMIM database (see below).	54, 55
dbEST Collection of partial sequences from cDNA clones (ESTs).	56
OMIM (Online Mendelian Inheritance in Man) Electronic catalogue of Victor McKusick's *Mendelian Inheritance in Man*, listing all known inherited human disorders. Housed at the Johns Hopkins University, and interactive with GDB (see above).	54
MGD (Mouse Genome Database) The major mouse database. Based at the Jackson Laboratories, Bar Harbor, ME.	57, 58
Généthon Human genetic maps based on linked $(CA)_n$ repeat markers, based in the Généthon Laboratory near Paris.	59
CHLC (Co-operative Human Linkage Center) Collaborative human genetic map database containing genotypes, marker data and linkage server, co-ordinated by Jeff Murray (University of Iowa).	60
CEPH YAC-based physical maps of the human genome, based at the CEPH Laboratory in Paris.	61
Whitehead Institute YAC-based physical maps of the human genome, based at the Whitehead Institute for Biomedical Reseach, Cambridge, MA.	62

[a]Reproduced, with permission of the publishers, from Strachan, T. and Read, A.R. (1996) *Human Molecular Genetics*. BIOS Scientific Publishers, Oxford.

Table 9. Major nucleic acid and protein sequence databases[a]

Database and description	References
Genbank DNA and protein sequences. One of many databases distributed by the US National Center for Biotechnology Information (NCBI).	63, 64
EMBL DNA sequences. Formerly at Heidelberg, but now distributed by the European Bioinformatics Institute (EBI) at Cambridge, UK, together with various other molecular biology databases.	65, 66
DDBJ DNA Database of Japan (Mishima).	67, 68
PIR Protein sequences. A single database distributed as a collaboration by: the US National Biomedical Research Foundation, Washington; the Martinsried Institute for Protein Sequences, Germany; and the Japan International Protein Information Database, Tokyo.	63, 69
SWISS–PROT Protein sequences. Maintained collaboratively by the University of Geneva and the EMBL data library.	63, 65, 70

[a]Reproduced, with permission of the publishers, from Strachan, T. and Read, A.R. (1996) *Human Molecular Genetics*. BIOS Scientific Publishers, Oxford.

Table 10. Eukaryotic genome projects

Organism	References	Organism	References
Microbial eukaryotes		Cow	86
Candida albicans	71	Dog	87
Dictyostelium discoideum	72	Mouse	58, 59, 88, 89
Plasmodium falciparum	32, 73–75	Pig	90
Schizosaccharomyces pombe	32	Rat	91
Invertebrates		Sheep	92
Bombyx mori	76	**Plants**	
Caenorhabditis elegans	77, 78	*Arabidopsis thaliana*	93–98
Drosophila melanogaster	32, 79, 80	Beans	99
Mosquito	81	Cotton	100
Fish		Forest trees	101
Fugu fugu	82	Maize	102, 103
Mekada fish	83	Rice	104
Zebrafish	84	Soybean	105
Vertebrates			
Chicken	85		

2. THE GENETIC CODE

The 'universal' genetic code is given in *Table 11* with variations described in *Table 12*. Three-letter and one-letter abbreviations for the amino acids are listed in *Table 13*.

Table 11. The genetic code

UUU	phenylalanine	UCU	serine	UAU	tyrosine	UGU	cysteine
UUC	phenylalanine	UCC	serine	UAC	tyrosine	UGC	cysteine
UUA	leucine	UCA	serine	UAA	stop (ochre)	UGA	stop (umber)
UUG	leucine	UCG	serine	UAG	stop (amber)	UGG	tryptophan
CUU	leucine	CCU	proline	CAU	histidine	CGU	arginine
CUC	leucine	CCC	proline	CAC	histidine	CGC	arginine
CUA	leucine	CCA	proline	CAA	glutamine	CGA	arginine
CUG	leucine	CCG	proline	CAG	glutamine	CGG	arginine
AUU	isoleucine	ACU	threonine	AAU	asparagine	AGU	serine
AUC	isoleucine	ACC	threonine	AAC	asparagine	AGC	serine
AUA	isoleucine	ACA	threonine	AAA	lysine	AGA	arginine
AUG	methionine	ACG	threonine	AAG	lysine	AGG	arginine
GUU	valine	GCU	alanine	GAU	aspartic acid	GGU	glycine
GUC	valine	GCC	alanine	GAC	aspartic acid	GGC	glycine
GUA	valine	GCA	alanine	GAA	glutamic acid	GGA	glycine
GUG	valine	GCG	alanine	GAG	glutamic acid	GGG	glycine

Table 12. Variations on the genetic code

Organism[a]	Genes	Codon[b]	Universal meaning	Actual meaning	References
Prokaryotes					
Various	Selenoproteins[c]	UGA	Stop	SeCys	106
Mycoplasma sp.	All genes	UGA	Stop	Trp	107
Organellar genomes					
Mammals	All mitochondrial	UGA	Stop	Trp	108
		AGA, AGG	Arg	Stop	
		AUA	Ile	Met	
Drosophila	All mitochondrial	UGA	Stop	Trp	109
		AGA	Arg	Ser	
		AUA	Ile	Met	
Saccharomyces cerevisiae	All mitochondrial	UGA	Stop	Trp	110
		CUN	Leu	Thr	
		AUA	Ile	Met	
Fungi	All mitochondrial (?)	UGA	Stop	Trp	111
Maize[d]	All mitochondrial (?)	CGG	Arg	Trp	112
Eukaryotic nuclear genomes					
Protozoa	All nuclear	UAA, UAG	Stop	Gln	113–116
Candida cylindracea	All nuclear	CUG	Leu	Ser	117
Various mammals	Selenoproteins[c]	UGA	Stop	SeCys	106

[a]Where a single species is given it is possible that related organisms also display the same code modifications.
[b]N = any nucleotide.
[c]The following are known to be selenoproteins: formate dehydrogenase (*Escherichia coli*, *Enterobacter aerogenes*, *Clostridium thermoaceticum*, *Clostridium thermoautotrophicum*, *Methanococcus vannielii*), NiFeSe hydrogenase (*Desulphomicrobium baculatum*, *Methanococcus voltae*), glycine reductase (*Clostridium sticklandii*, *Clostridium purinolyticum*), cellular glutathione peroxidase (human, cow, rat, mouse), plasma glutathione peroxidase (human), phospholipid hydroperoxide glutathione peroxidase (pig, rat), selenoprotein P (human, cow, rat), selenoprotein W (rat), type 1 deiodinase (human, rat, mouse, dog), type 2 deiodinase (*Rana catesbiana*), type 3 deiodinase (human, rat, *R. catesbiana*). See ref. 98.
[d]In maize and other plants the CGG codon is probably converted into UGG (the correct codon for tryptophan) by RNA editing (118, 119).

Table 13. Amino acid abbreviations

Amino acid	Three-letter abbreviation	One-letter abbreviation
Alanine	Ala	A
Arginine	Arg	R
Asparagine	Asn	N
Aspartic acid	Asp	D
Cysteine	Cys	C
Glutamic acid	Glu	E
Glutamine	Gln	Q
Glycine	Gly	G
Histidine	His	H
Isoleucine	Ile	I
Leucine	Leu	L
Lysine	Lys	K
Methionine	Met	M
Phenylalanine	Phe	F
Proline	Pro	P
Selenocysteine	SeCys	–
Serine	Ser	S
Threonine	Thr	T
Tryptophan	Trp	W
Tyrosine	Tyr	Y
Valine	Val	V

3. CODON USAGE

(Data kindly provided by Yasukazu Nakamura, Laboratory of Gene Structure 2, Kazusa DNA Research Institute, Japan)

The Codon Usage Database (120, 121) contains continually updated information on codon usage for all organisms for which genes appear in Genbank. The codon usage for a single organism can be extracted using the online search facility, and lists of all species contained in the database can also be accessed. The information available for each species includes the GC content of the coding regions and the GC frequencies at each nucleotide position in the codon. The following pages give the codon usages for the best-studied genomes.

Codon usage for PLASMIDS — ColE1

Frequencies per thousand codons — Based on 26 genes containing 3483 codons

UUU	19.8	UCU	11.8	UAU	13.8	UGU	9.2
UUC	12.6	UCC	10.6	UAC	10.9	UGC	8.6
UUA	13.2	UCA	10.9	UAA	3.7	UGA	3.2
UUG	12.3	UCG	6.6	UAG	1.1	UGG	11.8
CUU	21.2	CCU	8.3	CAU	10.3	CGU	16.9
CUC	13.8	CCC	5.5	CAC	6.6	CGC	15.8
CUA	8.0	CCA	6.3	CAA	9.5	CGA	10.0
CUG	27.8	CCG	11.2	CAG	36.5	CGG	15.8
AUU	16.9	ACU	10.6	AAU	21.5	AGU	11.2
AUC	14.9	ACC	11.5	AAC	17.8	AGC	17.5
AUA	12.3	ACA	20.4	AAA	42.5	AGA	13.8
AUG	19.8	ACG	11.8	AAG	26.4	AGG	13.2
GUU	16.9	GCU	33.6	GAU	27.6	GGU	16.1
GUC	10.6	GCC	18.7	GAC	15.2	GGC	13.2
GUA	10.0	GCA	29.3	GAA	45.4	GGA	18.7
GUG	19.5	GCG	16.1	GAG	28.4	GGG	14.6

GC content of genes = 48.9%
GC frequency for first nucleotide = 55.8%, for second nucleotide = 43.3%, for third nucleotide = 47.7%

Codon usage for PLASMIDS — F plasmid

Frequencies per thousand codons — Based on 31 genes containing 7184 codons

UUU	27.0	UCU	12.4	UAU	18.5	UGU	6.5
UUC	17.1	UCC	14.3	UAC	15.2	UGC	3.5
UUA	9.2	UCA	12.0	UAA	2.2	UGA	1.5
UUG	7.1	UCG	6.4	UAG	0.6	UGG	14.1
CUU	14.1	CCU	7.7	CAU	11.1	CGU	17.4
CUC	10.2	CCC	7.0	CAC	6.4	CGC	12.2
CUA	2.6	CCA	4.9	CAA	6.4	CGA	4.2
CUG	51.2	CCG	20.0	CAG	40.4	CGG	11.8
AUU	27.8	ACU	9.9	AAU	24.5	AGU	16.7
AUC	22.4	ACC	16.3	AAC	21.2	AGC	14.6
AUA	9.5	ACA	12.0	AAA	33.7	AGA	9.0
AUG	32.7	ACG	15.7	AAG	16.0	AGG	5.7
GUU	19.5	GCU	16.3	GAU	28.4	GGU	17.8
GUC	13.5	GCC	32.0	GAC	22.6	GGC	18.0
GUA	8.9	GCA	21.4	GAA	32.2	GGA	10.0
GUG	23.9	GCG	17.5	GAG	21.6	GGG	13.5

GC content of genes = 49.7%
GC frequency for first nucleotide = 54.6%, for second nucleotide = 40.2%, for third nucleotide = 54.5%

Codon usage for PLASMIDS

Ti plasmid

Frequencies per thousand codons

Based on 60 genes containing 18 185 codons

UUU	17.0	UCU	9.7	UAU	14.8	UGU	4.1
UUC	18.7	UCC	12.6	UAC	9.8	UGC	7.1
UUA	6.8	UCA	11.6	UAA	0.9	UGA	1.2
UUG	18.1	UCG	15.1	UAG	1.2	UGG	10.3
CUU	20.1	CCU	9.8	CAU	12.6	CGU	13.8
CUC	21.1	CCC	11.7	CAC	10.4	CGC	21.6
CUA	8.7	CCA	12.2	CAA	20.1	CGA	11.5
CUG	21.6	CCG	17.6	CAG	21.0	CGG	15.5
AUU	19.7	ACU	10.4	AAU	16.8	AGU	10.1
AUC	24.8	ACC	18.0	AAC	16.7	AGC	17.1
AUA	8.2	ACA	10.3	AAA	16.1	AGA	7.4
AUG	21.2	ACG	17.2	AAG	18.5	AGG	8.6
GUU	19.3	GCU	21.3	GAU	31.0	GGU	16.7
GUC	20.6	GCC	26.3	GAC	26.4	GGC	27.2
GUA	8.1	GCA	21.7	GAA	32.4	GGA	14.9
GUG	15.8	GCG	27.9	GAG	27.9	GGG	12.9

GC content of genes = 53.8%
GC frequency for first nucleotide = 60.0%, for second nucleotide = 45.4%, for third nucleotide = 56.1%

Codon usage for BACTERIOPHAGES

λ

Frequencies per thousand codons

Based on 78 genes containing 15 729 codons

UUU	19.5	UCU	7.7	UAU	17.5	UGU	3.9
UUC	14.9	UCC	11.2	UAC	12.2	UGC	9.2
UUA	8.5	UCA	13.4	UAA	1.8	UGA	2.6
UUG	5.9	UCG	8.8	UAG	0.5	UGG	16.0
CUU	14.1	CCU	7.9	CAU	10.8	CGU	16.7
CUC	9.2	CCC	5.0	CAC	7.0	CGC	16.6
CUA	3.4	CCA	9.5	CAA	10.4	CGA	6.9
CUG	34.9	CCG	16.9	CAG	32.4	CGG	9.5
AUU	23.0	ACU	9.5	AAU	18.4	AGU	11.6
AUC	20.2	ACC	20.1	AAC	20.5	AGC	17.2
AUA	8.6	ACA	14.2	AAA	37.1	AGA	9.4
AUG	28.2	ACG	18.2	AAG	20.2	AGG	4.7
GUU	20.0	GCU	18.2	GAU	31.1	GGU	19.1
GUC	11.3	GCC	27.7	GAC	24.5	GGC	21.1
GUA	10.0	GCA	30.3	GAA	36.8	GGA	13.0
GUG	24.1	GCG	25.0	GAG	27.3	GGG	14.6

GC content of genes = 51.2%
GC frequency for first nucleotide = 56.6%, for second nucleotide = 43.6%, for third nucleotide = 53.5%

Codon usage for BACTERIOPHAGES — φX174

Frequencies per thousand codons — Based on 22 genes containing 4592 codons

UUU	30.1	UCU	28.1	UAU	23.7	UGU	5.9
UUC	15.2	UCC	8.1	UAC	8.9	UGC	7.2
UUA	15.2	UCA	9.8	UAA	1.5	UGA	3.3
UUG	21.8	UCG	10.2	UAG	0.0	UGG	13.5
CUU	28.7	CCU	19.2	CAU	13.1	CGU	23.7
CUC	9.4	CCC	4.8	CAC	4.6	CGC	18.9
CUA	5.4	CCA	5.0	CAA	19.6	CGA	4.6
CUG	17.2	CCG	11.5	CAG	26.1	CGG	4.8
AUU	32.0	ACU	28.3	AAU	29.0	AGU	7.4
AUC	10.2	ACC	11.5	AAC	18.1	AGC	4.8
AUA	3.7	ACA	9.1	AAA	36.8	AGA	6.1
AUG	32.2	ACG	16.1	AAG	24.2	AGG	1.1
GUU	34.6	GCU	49.7	GAU	34.0	GGU	31.8
GUC	7.8	GCC	15.0	GAC	24.8	GGC	19.2
GUA	7.4	GCA	8.7	GAA	16.8	GGA	9.6
GUG	8.7	GCG	11.3	GAG	28.1	GGG	2.6

GC content of genes = 45.2%
GC frequency for first nucleotide = 52.7%, for second nucleotide = 41.1%, for third nucleotide = 41.2%

Codon usage for BACTERIOPHAGES — T4

Frequencies per thousand codons — Based on 235 genes containing 43 253 codons

UUU	33.8	UCU	24.7	UAU	33.5	UGU	8.4
UUC	11.4	UCC	3.9	UAC	9.9	UGC	4.4
UUA	26.9	UCA	16.9	UAA	3.2	UGA	1.6
UUG	10.8	UCG	4.0	UAG	0.5	UGG	14.5
CUU	20.2	CCU	14.2	CAU	13.6	CGU	19.2
CUC	4.2	CCC	1.1	CAC	4.6	CGC	5.5
CUA	7.1	CCA	13.5	CAA	22.4	CGA	6.2
CUG	6.0	CCG	4.9	CAG	10.7	CGG	1.2
AUU	52.3	ACU	25.9	AAU	41.3	AGU	9.8
AUC	11.0	ACC	6.1	AAC	14.1	AGC	4.9
AUA	12.8	ACA	16.8	AAA	67.8	AGA	9.5
AUG	28.4	ACG	4.4	AAG	18.5	AGG	1.8
GUU	31.0	GCU	29.6	GAU	48.0	GGU	26.7
GUC	5.4	GCC	5.2	GAC	14.2	GGC	7.4
GUA	18.4	GCA	19.9	GAA	60.1	GGA	18.2
GUG	5.4	GCG	6.1	GAG	12.0	GGG	3.7

GC content of genes = 35.1%
GC frequency for first nucleotide = 46.6%, for second nucleotide = 34.0%, for third nucleotide = 24.6%

Codon usage for BACTERIOPHAGES — T7

Frequencies per thousand codons — Based on 88 genes containing 18 425 codons

UUU	10.7	UCU	17.9	UAU	11.9	UGU	6.7
UUC	25.1	UCC	11.5	UAC	20.5	UGC	7.1
UUA	11.7	UCA	8.5	UAA	2.8	UGA	1.5
UUG	9.4	UCG	4.2	UAG	0.4	UGG	16.9
CUU	15.8	CCU	14.8	CAU	6.7	CGU	22.6
CUC	11.1	CCC	1.7	CAC	15.0	CGC	15.5
CUA	9.7	CCA	9.1	CAA	16.3	CGA	7.3
CUG	22.9	CCG	8.6	CAG	18.3	CGG	2.9
AUU	23.6	ACU	19.2	AAU	12.3	AGU	8.4
AUC	24.6	ACC	18.1	AAC	32.9	AGC	7.7
AUA	4.8	ACA	9.3	AAA	24.3	AGA	5.0
AUG	31.2	ACG	7.4	AAG	44.6	AGG	3.4
GUU	19.1	GCU	43.3	GAU	24.0	GGU	37.1
GUC	11.9	GCC	13.0	GAC	37.8	GGC	14.4
GUA	17.7	GCA	18.3	GAA	29.9	GGA	11.3
GUG	14.9	GCG	14.8	GAG	42.1	GGG	8.5

GC content of genes = 49.0%
GC frequency for first nucleotide = 55.7%, for second nucleotide = 39.6%, for third nucleotide = 51.8%

Codon usage for VIRUSES — Cauliflower mosaic virus

Frequencies per thousand codons — Based on 52 genes containing 19 198 codons

UUU	15.3	UCU	13.0	UAU	12.0	UGU	7.0
UUC	21.5	UCC	10.5	UAC	17.6	UGC	8.3
UUA	19.9	UCA	15.9	UAA	0.8	UGA	1.4
UUG	7.4	UCG	7.8	UAG	0.5	UGG	8.1
CUU	19.2	CCU	12.1	CAU	12.1	CGU	2.7
CUC	17.6	CCC	10.1	CAC	8.8	CGC	0.8
CUA	15.2	CCA	19.7	CAA	34.8	CGA	4.6
CUG	5.1	CCG	4.6	CAG	11.4	CGG	0.8
AUU	25.1	ACU	16.0	AAU	33.1	AGU	10.5
AUC	31.9	ACC	16.7	AAC	25.7	AGC	15.1
AUA	21.1	ACA	17.8	AAA	55.2	AGA	19.6
AUG	21.6	ACG	6.9	AAG	51.0	AGG	9.0
GUU	16.5	GCU	14.7	GAU	27.5	GGU	11.0
GUC	10.5	GCC	13.5	GAC	24.1	GGC	8.3
GUA	11.3	GCA	18.8	GAA	54.5	GGA	27.3
GUG	6.6	GCG	5.3	GAG	24.1	GGG	3.3

GC content of genes = 40.4%
GC frequency for first nucleotide = 45.7%, for second nucleotide = 34.1%, for third nucleotide = 41.4%

Codon usage for VIRUSES

Hepatitis B virus

Frequencies per thousand codons — Based on 251 genes 77 876 containing codons

UUU	26.6	UCU	28.6	UAU	16.9	UGU	23.1
UUC	24.9	UCC	24.8	UAC	9.6	UGC	13.4
UUA	14.1	UCA	20.3	UAA	1.4	UGA	0.7
UUG	22.1	UCG	10.2	UAG	1.1	UGG	29.3
CUU	22.9	CCU	32.2	CAU	14.7	CGU	8.5
CUC	25.4	CCC	22.4	CAC	14.7	CGC	10.9
CUA	18.1	CCA	23.6	CAA	22.1	CGA	4.8
CUG	22.3	CCG	10.9	CAG	14.6	CGG	8.6
AUU	16.9	ACU	20.2	AAU	17.6	AGU	9.5
AUC	17.5	ACC	22.7	AAC	16.0	AGC	6.8
AUA	6.4	ACA	13.4	AAA	16.3	AGA	15.6
AUG	17.8	ACG	5.3	AAG	9.4	AGG	14.2
GUU	17.5	GCU	18.4	GAU	9.1	GGU	9.7
GUC	11.9	GCC	17.5	GAC	18.7	GGC	13.5
GUA	8.5	GCA	14.0	GAA	13.3	GGA	23.2
GUG	16.1	GCG	6.4	GAG	13.4	GGG	19.0

GC content of genes = 50.0%
GC frequency for first nucleotide = 50.7%, for second nucleotide = 50.2%, for third nucleotide = 49.2%

Codon usage for VIRUSES

Human cytomegalovirus

Frequencies per thousand codons — Based on 105 genes containing 41 747 codons

UUU	18.9	UCU	12.5	UAU	10.6	UGU	10.0
UUC	20.3	UCC	18.0	UAC	27.3	UGC	13.0
UUA	5.6	UCA	7.7	UAA	0.9	UGA	1.3
UUG	13.0	UCG	17.2	UAG	0.3	UGG	10.7
CUU	8.9	CCU	8.4	CAU	9.2	CGU	14.3
CUC	26.6	CCC	19.5	CAC	22.6	CGC	25.2
CUA	12.8	CCA	8.5	CAA	16.0	CGA	10.1
CUG	39.6	CCG	18.8	CAG	23.6	CGG	9.0
AUU	8.5	ACU	11.6	AAU	11.2	AGU	7.5
AUC	25.2	ACC	29.3	AAC	26.6	AGC	20.9
AUA	6.3	ACA	10.2	AAA	16.2	AGA	5.8
AUG	21.8	ACG	24.6	AAG	15.7	AGG	3.4
GUU	9.4	GCU	11.4	GAU	13.7	GGU	12.3
GUC	22.0	GCC	34.5	GAC	35.2	GGC	23.2
GUA	12.4	GCA	8.8	GAA	24.5	GGA	9.0
GUG	29.1	GCG	18.4	GAG	24.3	GGG	6.3

GC content of genes = 55.8%
GC frequency for first nucleotide = 56.8%, for second nucleotide = 44.1%, for third nucleotide = 66.5%

Codon usage for VIRUSES — Human herpesvirus 1

Frequencies per thousand codons — Based on 239 genes containing 123 706 codons

UUU	17.6	UCU	4.3	UAU	5.2	UGU	6.3
UUC	16.9	UCC	19.0	UAC	21.6	UGC	11.8
UUA	1.7	UCA	2.0	UAA	0.5	UGA	0.9
UUG	8.9	UCG	15.9	UAG	0.5	UGG	11.6
CUU	6.6	CCU	5.3	CAU	4.5	CGU	5.3
CUC	21.9	CCC	46.5	CAC	20.9	CGC	39.9
CUA	4.7	CCA	7.5	CAA	5.2	CGA	6.6
CUG	51.0	CCG	28.8	CAG	24.5	CGG	25.2
AUU	5.9	ACU	2.6	AAU	2.6	AGU	2.8
AUC	20.8	ACC	31.0	AAC	21.5	AGC	16.9
AUA	3.5	ACA	3.8	AAA	5.3	AGA	1.7
AUG	18.0	ACG	22.3	AAG	14.7	AGG	4.8
GUU	6.5	GCU	5.8	GAU	9.2	GGU	5.9
GUC	25.3	GCC	67.3	GAC	43.8	GGC	32.4
GUA	3.8	GCA	6.3	GAA	9.5	GGA	7.5
GUG	33.3	GCG	43.5	GAG	38.9	GGG	33.3

GC content of genes = 67.8%
GC frequency for first nucleotide = 67.7%, for second nucleotide = 52.5%, for third nucleotide = 83.3%

Codon usage for VIRUSES — HIV-1

Frequencies per thousand codons — Based on 1102 genes containing 261 325 codons

UUU	18.2	UCU	6.8	UAU	15.7	UGU	12.1
UUC	9.9	UCC	5.0	UAC	14.3	UGC	7.8
UUA	21.9	UCA	9.9	UAA	0.9	UGA	1.8
UUG	10.8	UCG	1.2	UAG	1.5	UGG	32.3
CUU	9.7	CCU	17.7	CAU	20.4	CGU	1.0
CUC	5.3	CCC	5.6	CAC	13.5	CGC	2.3
CUA	17.5	CCA	28.7	CAA	26.5	CGA	7.2
CUG	19.9	CCG	4.2	CAG	20.5	CGG	1.5
AUU	14.8	ACU	13.0	AAU	26.8	AGU	15.5
AUC	11.2	ACC	9.8	AAC	16.6	AGC	17.5
AUA	26.7	ACA	26.8	AAA	31.0	AGA	37.8
AUG	24.0	ACG	2.0	AAG	26.5	AGG	16.1
GUU	8.7	GCU	17.5	GAU	22.7	GGU	8.2
GUC	7.8	GCC	12.0	GAC	18.5	GGC	10.7
GUA	26.7	GCA	31.5	GAA	42.6	GGA	33.2
GUG	15.5	GCG	2.6	GAG	32.5	GGG	21.5

GC content of genes = 44.5%
GC frequency for first nucleotide = 51.4%, for second nucleotide = 42.1%, for third nucleotide = 40.0%

Codon usage for VIRUSES — Influenza A virus

Frequencies per thousand codons — Based on 709 genes containing 297 274 codons

UUU	16.4	UCU	15.0	UAU	15.7	UGU	8.7
UUC	20.0	UCC	9.9	UAC	12.8	UGC	11.4
UUA	7.3	UCA	17.7	UAA	1.2	UGA	0.8
UUG	12.4	UCG	4.0	UAG	0.4	UGG	16.2
CUU	14.7	CCU	11.5	CAU	10.4	CGU	2.4
CUC	13.0	CCC	7.2	CAC	6.5	CGC	3.4
CUA	12.6	CCA	13.4	CAA	21.9	CGA	6.3
CUG	15.9	CCG	4.1	CAG	18.1	CGG	5.8
AUU	22.3	ACU	18.4	AAU	34.2	AGU	14.6
AUC	20.3	ACC	13.6	AAC	23.4	AGC	15.1
AUA	23.7	ACA	25.9	AAA	32.4	AGA	33.6
AUG	36.0	ACG	4.7	AAG	19.5	AGG	19.0
GUU	13.4	GCU	16.3	GAU	25.3	GGU	11.6
GUC	10.9	GCC	12.1	GAC	21.1	GGC	8.8
GUA	12.4	GCA	26.7	GAA	39.9	GGA	33.9
GUG	19.5	GCG	5.0	GAG	28.8	GGG	20.2

GC content of genes = 44.3%
GC frequency for first nucleotide = 47.3%, for second nucleotide = 41.8%, for third nucleotide = 43.9%

Codon usage for VIRUSES — Tobacco mosaic virus

Frequencies per thousand codons — Based on 34 genes containing 19 534 codons

UUU	28.3	UCU	20.6	UAU	17.1	UGU	11.2
UUC	19.5	UCC	9.5	UAC	19.3	UGC	7.1
UUA	18.5	UCA	15.5	UAA	0.9	UGA	0.4
UUG	23.1	UCG	13.1	UAG	0.7	UGG	9.4
CUU	18.5	CCU	10.4	CAU	12.5	CGU	5.5
CUC	9.2	CCC	5.7	CAC	8.0	CGC	3.8
CUA	8.5	CCA	8.4	CAA	17.0	CGA	5.2
CUG	14.1	CCG	9.7	CAG	16.1	CGG	1.9
AUU	22.1	ACU	19.2	AAU	29.3	AGU	14.8
AUC	14.5	ACC	11.8	AAC	16.6	AGC	9.8
AUA	12.4	ACA	17.4	AAA	30.7	AGA	21.7
AUG	27.9	ACG	11.4	AAG	33.1	AGG	15.1
GUU	31.3	GCU	19.3	GAU	35.6	GGU	16.5
GUC	16.8	GCC	13.5	GAC	24.3	GGC	6.1
GUA	12.7	GCA	21.2	GAA	28.3	GGA	15.4
GUG	28.3	GCG	15.0	GAG	32.4	GGG	6.6

GC content of genes = 43.5%
GC frequency for first nucleotide = 47.8%, for second nucleotide = 37.2%, for third nucleotide = 45.4%

Codon usage for VIRUSES

Vaccinia virus

Frequencies per thousand codons

Based on 406 genes containing 103 167 codons

UUU	33.2	UCU	24.8	UAU	35.9	UGU	15.9
UUC	14.2	UCC	10.0	UAC	16.0	UGC	4.9
UUA	25.8	UCA	15.6	UAA	2.3	UGA	0.9
UUG	18.6	UCG	8.7	UAG	0.8	UGG	7.6
CUU	11.6	CCU	11.1	CAU	14.5	CGU	6.3
CUC	6.4	CCC	4.3	CAC	6.2	CGC	2.5
CUA	18.5	CCA	13.1	CAA	15.3	CGA	5.6
CUG	7.4	CCG	6.4	CAG	6.3	CGG	1.8
AUU	37.2	ACU	22.2	AAU	47.4	AGU	15.2
AUC	18.4	ACC	8.7	AAC	20.2	AGC	6.0
AUA	37.3	ACA	21.9	AAA	48.8	AGA	22.4
AUG	28.1	ACG	8.9	AAG	22.2	AGG	4.1
GUU	20.1	GCU	13.1	GAU	49.2	GGU	12.6
GUC	8.7	GCC	6.9	GAC	16.3	GGC	3.9
GUA	23.4	GCA	11.6	GAA	38.4	GGA	20.6
GUG	9.1	GCG	7.6	GAG	14.5	GGG	2.9

GC content of genes = 34.4%
GC frequency for first nucleotide = 39.6%, for second nucleotide = 32.8%, for third nucleotide = 30.8%

Codon usage for PROKARYOTES

Anabaena

Frequencies per thousand codons

Based on 104 genes containing 32 771 codons

UUU	23.4	UCU	15.7	UAU	15.8	UGU	6.3
UUC	13.5	UCC	9.6	UAC	15.4	UGC	5.0
UUA	32.3	UCA	8.5	UAA	1.7	UGA	0.6
UUG	24.3	UCG	3.9	UAG	0.9	UGG	12.5
CUU	7.3	CCU	14.7	CAU	9.3	CGU	16.6
CUC	9.9	CCC	11.3	CAC	9.7	CGC	14.5
CUA	14.3	CCA	13.9	CAA	37.5	CGA	5.5
CUG	13.8	CCG	3.4	CAG	14.7	CGG	6.9
AUU	35.6	ACU	18.0	AAU	24.2	AGU	13.0
AUC	22.4	ACC	17.1	AAC	19.0	AGC	9.6
AUA	7.5	ACA	17.9	AAA	36.6	AGA	8.0
AUG	20.0	ACG	4.3	AAG	15.2	AGG	2.6
GUU	21.4	GCU	34.4	GAU	32.7	GGU	33.5
GUC	11.2	GCC	16.8	GAC	16.8	GGC	13.9
GUA	21.3	GCA	23.1	GAA	49.6	GGA	12.3
GUG	13.2	GCG	12.4	GAG	15.0	GGG	8.6

GC content of genes = 44.0%
GC frequency for first nucleotide = 54.0%, for second nucleotide = 39.4%, for third nucleotide = 38.7%

Codon usage for PROKARYOTES *Bacillus subtilis*

Frequencies per thousand codons Based on 1721 genes containing 549 454 codons

UUU	27.8	UCU	13.7	UAU	22.2	UGU	3.7
UUC	14.4	UCC	8.0	UAC	12.0	UGC	4.2
UUA	19.0	UCA	14.6	UAA	1.9	UGA	0.8
UUG	14.9	UCG	6.1	UAG	0.4	UGG	9.8
CUU	23.1	CCU	11.1	CAU	14.8	CGU	8.6
CUC	10.5	CCC	2.9	CAC	7.8	CGC	9.1
CUA	4.7	CCA	7.2	CAA	20.1	CGA	3.9
CUG	21.8	CCG	16.1	CAG	18.8	CGG	6.4
AUU	36.6	ACU	9.0	AAU	22.2	AGU	6.5
AUC	26.9	ACC	8.1	AAC	18.3	AGC	13.9
AUA	8.8	ACA	23.0	AAA	49.3	AGA	10.9
AUG	26.7	ACG	14.6	AAG	20.4	AGG	3.8
GUU	20.0	GCU	19.5	GAU	33.4	GGU	13.5
GUC	16.8	GCC	15.1	GAC	19.3	GGC	23.5
GUA	14.0	GCA	22.1	GAA	49.5	GGA	22.2
GUG	17.3	GCG	20.4	GAG	23.2	GGG	10.8

GC content of genes = 44.4%

GC frequency for first nucleotide = 52.7%, for second nucleotide = 36.3%, for third nucleotide = 44.2%

Codon usage for PROKARYOTES *Escherichia coli*

Frequencies per thousand codons Based on 9097 genes containing 2 867 975 codons

UUU	21.7	UCU	9.3	UAU	16.4	UGU	5.2
UUC	16.7	UCC	8.9	UAC	12.5	UGC	6.4
UUA	13.4	UCA	7.6	UAA	2.0	UGA	1.0
UUG	13.0	UCG	8.6	UAG	0.3	UGG	14.3
CUU	11.1	CCU	7.1	CAU	12.6	CGU	21.0
CUC	10.6	CCC	5.2	CAC	9.9	CGC	21.4
CUA	3.9	CCA	8.4	CAA	14.6	CGA	3.6
CUG	51.6	CCG	22.6	CAG	29.0	CGG	5.4
AUU	29.6	ACU	9.5	AAU	18.2	AGU	8.9
AUC	25.0	ACC	23.1	AAC	22.0	AGC	15.6
AUA	5.0	ACA	7.7	AAA	34.7	AGA	2.6
AUG	27.2	ACG	13.8	AAG	11.4	AGG	1.5
GUU	19.0	GCU	16.3	GAU	32.2	GGU	25.4
GUC	14.8	GCC	25.1	GAC	19.7	GGC	29.1
GUA	11.2	GCA	20.5	GAA	40.0	GGA	8.3
GUG	25.6	GCG	32.4	GAG	18.5	GGG	10.9

GC content of genes = 51.5%

GC frequency for first nucleotide = 58.7%, for second nucleotide = 40.7%, for third nucleotide – 55.2%

Codon usage for PROKARYOTES — *Haemophilus influenzae*

Frequencies per thousand codons — Based on 1916 genes containing 607 699 codons

UUU	32.1	UCU	16.9	UAU	25.5	UGU	6.5
UUC	12.3	UCC	4.5	UAC	7.2	UGC	3.3
UUA	48.8	UCA	12.9	UAA	2.4	UGA	0.3
UUG	17.4	UCG	4.0	UAG	0.4	UGG	10.9
CUU	19.5	CCU	12.4	CAU	13.2	CGU	22.9
CUC	5.1	CCC	2.9	CAC	7.0	CGC	9.7
CUA	6.8	CCA	16.6	CAA	38.2	CGA	5.2
CUG	4.3	CCG	4.8	CAG	7.4	CGG	1.2
AUU	48.6	ACU	17.1	AAU	38.7	AGU	12.7
AUC	14.0	ACC	11.7	AAC	13.2	AGC	9.0
AUA	6.3	ACA	16.8	AAA	57.8	AGA	4.3
AUG	22.8	ACG	8.8	AAG	7.3	AGG	0.6
GUU	20.8	GCU	21.2	GAU	42.2	GGU	29.2
GUC	6.4	GCC	10.7	GAC	7.8	GGC	19.8
GUA	19.0	GCA	32.6	GAA	53.0	GGA	11.1
GUG	20.2	GCG	16.2	GAG	10.5	GGG	7.2

GC content of genes = 38.6%
GC frequency for first nucleotide = 50.5%, for second nucleotide = 36.4%, for third nucleotide = 28.9%

Codon usage for PROKARYOTES — *Methanococcus jannaschii*

Frequencies per thousand codons — Based on 1737 genes containing 500 513 codons

UUU	33.3	UCU	10.5	UAU	33.5	UGU	8.7
UUC	8.8	UCC	2.8	UAC	9.9	UGC	4.0
UUA	52.2	UCA	14.4	UAA	2.7	UGA	0.4
UUG	19.0	UCG	0.8	UAG	0.3	UGG	7.3
CUU	9.3	CCU	8.5	CAU	10.2	CGU	0.3
CUC	3.2	CCC	1.3	CAC	4.1	CGC	0.1
CUA	8.5	CCA	22.3	CAA	9.2	CGA	0.4
CUG	2.2	CCG	1.4	CAG	5.3	CGG	0.1
AUU	48.3	ACU	14.6	AAU	37.3	AGU	11.0
AUC	10.6	ACC	4.2	AAC	15.5	AGC	5.2
AUA	45.3	ACA	19.8	AAA	72.8	AGA	27.6
AUG	22.0	ACG	1.7	AAG	30.9	AGG	9.9
GUU	43.1	GCU	24.4	GAU	45.5	GGU	13.0
GUC	4.7	GCC	5.4	GAC	9.7	GGC	4.2
GUA	14.9	GCA	22.4	GAA	51.9	GGA	35.7
GUG	5.8	GCG	2.3	GAG	34.7	GGG	10.3

GC content of genes = 31.9%
GC frequency for first nucleotide = 41.5%, for second nucleotide = 29.5%, for third nucleotide = 24.8%

Codon usage for PROKARYOTES — *Mycobacterium leprae*

Frequencies per thousand codons — Based on 692 genes containing 180 669 codons

UUU	9.2	UCU	5.6	UAU	8.1	UGU	4.4
UUC	20.8	UCC	11.6	UAC	13.8	UGC	7.7
UUA	5.4	UCA	6.8	UAA	0.9	UGA	1.9
UUG	23.5	UCG	18.3	UAG	1.0	UGG	13.4
CUU	9.3	CCU	7.1	CAU	8.5	CGU	13.2
CUC	14.8	CCC	12.4	CAC	14.4	CGC	21.5
CUA	8.9	CCA	9.2	CAA	11.3	CGA	9.7
CUG	37.4	CCG	23.2	CAG	22.3	CGG	19.7
AUU	12.3	ACU	10.8	AAU	8.3	AGU	6.8
AUC	30.8	ACC	27.7	AAC	18.1	AGC	13.5
AUA	5.1	ACA	8.2	AAA	9.2	AGA	2.7
AUG	20.1	ACG	14.5	AAG	17.6	AGG	4.0
GUU	14.9	GCU	21.4	GAU	21.8	GGU	23.9
GUC	29.1	GCC	38.6	GAC	36.0	GGC	33.2
GUA	9.6	GCA	17.5	GAA	21.2	GGA	11.7
GUG	37.8	GCG	34.8	GAG	29.2	GGG	14.5

GC content of genes = 59.4%
GC frequency for first nucleotide = 63.8%, for second nucleotide = 46.9%, for third nucleotide = 67.5%

Codon usage for PROKARYOTES — *Mycoplasma genitalium*

Frequencies per thousand codons — Based on 475 genes containing 174 589 codons

UUU	52.3	UCU	12.4	UAU	23.9	UGU	6.5
UUC	8.3	UCC	4.1	UAC	8.3	UGC	1.6
UUA	49.7	UCA	16.5	UAA	2.0	UGA	6.3
UUG	14.2	UCG	1.1	UAG	0.7	UGG	3.5
CUU	19.8	CCU	14.8	CAU	10.3	CGU	6.9
CUC	5.0	CCC	3.8	CAC	5.5	CGC	3.0
CUA	12.7	CCA	11.1	CAA	38.4	CGA	1.3
CUG	4.4	CCG	1.0	CAG	9.0	CGG	1.0
AUU	51.0	ACU	25.4	AAU	45.7	AGU	26.1
AUC	18.0	ACC	10.6	AAC	29.3	AGC	6.7
AUA	12.5	ACA	16.6	AAA	69.9	AGA	14.1
AUG	15.1	ACG	1.7	AAG	24.4	AGG	4.7
GUU	37.6	GCU	27.4	GAU	42.3	GGU	22.9
GUC	3.5	GCC	4.1	GAC	6.9	GGC	5.0
GUA	13.1	GCA	21.4	GAA	45.2	GGA	11.5
GUG	7.2	GCG	2.7	GAG	11.2	GGG	7.0

GC content of genes = 31.7%
GC frequency for first nucleotide = 41.7%, for second nucleotide = 30.3%, for third nucleotide = 23.3%

Codon usage for PROKARYOTES — *Pseudomonas aeruginosa*

Frequencies per thousand codons — Based on 480 genes containing 148 069 codons

UUU	4.4	UCU	2.4	UAU	6.9	UGU	1.4
UUC	31.0	UCC	13.8	UAC	19.7	UGC	7.6
UUA	1.5	UCA	1.9	UAA	0.6	UGA	2.3
UUG	10.9	UCG	14.2	UAG	0.3	UGG	13.4
CUU	5.4	CCU	4.1	CAU	6.1	CGU	8.6
CUC	22.1	CCC	12.0	CAC	13.4	CGC	39.8
CUA	2.6	CCA	3.3	CAA	8.1	CGA	3.4
CUG	67.0	CCG	29.4	CAG	36.4	CGG	12.4
AUU	6.2	ACU	4.0	AAU	6.0	AGU	4.2
AUC	37.0	ACC	34.8	AAC	26.0	AGC	24.1
AUA	1.9	ACA	2.5	AAA	7.9	AGA	1.2
AUG	21.8	ACG	7.0	AAG	30.7	AGG	2.4
GUU	6.3	GCU	9.6	GAU	13.0	GGU	11.4
GUC	27.6	GCC	58.5	GAC	40.1	GGC	54.6
GUA	5.3	GCA	8.1	GAA	25.0	GGA	5.3
GUG	31.7	GCG	34.7	GAG	35.1	GGG	9.5

GC content of genes = 63.7%
GC frequency for first nucleotide = 65.0%, for second nucleotide = 44.2%, for third nucleotide = 81.9%

Codon usage for PROKARYOTES — *Rhizobium meliloti*

Frequencies per thousand codons — Based on 237 genes containing 83 304 codons

UUU	7.5	UCU	3.6	UAU	11.2	UGU	2.0
UUC	30.2	UCC	15.2	UAC	10.4	UGC	7.9
UUA	1.1	UCA	3.5	UAA	0.6	UGA	1.8
UUG	8.6	UCG	21.4	UAG	0.4	UGG	12.2
CUU	15.4	CCU	5.0	CAU	9.4	CGU	9.1
CUC	35.9	CCC	11.0	CAC	11.1	CGC	33.5
CUA	3.3	CCA	4.8	CAA	6.3	CGA	5.4
CUG	32.5	CCG	25.4	CAG	26.4	CGG	15.5
AUU	9.5	ACU	4.0	AAU	10.1	AGU	2.4
AUC	40.3	ACC	21.7	AAC	19.5	AGC	15.3
AUA	4.0	ACA	5.0	AAA	8.0	AGA	3.2
AUG	25.3	ACG	20.5	AAG	28.9	AGG	5.7
GUU	11.0	GCU	13.6	GAU	19.2	GGU	11.7
GUC	38.1	GCC	49.6	GAC	34.7	GGC	47.4
GUA	4.8	GCA	16.8	GAA	27.2	GGA	10.3
GUG	20.8	GCG	39.7	GAG	33.3	GGG	10.6

GC content of genes = 61.4%
GC frequency for first nucleotide = 63.9%, for second nucleotide = 45.5%, for third nucleotide = 74.9%

Codon usage for PROKARYOTES *Salmonella typhimurium*

Frequencies per thousand codons — Based on 476 genes containing 157 245 codons

UUU	20.9	UCU	8.2	UAU	16.0	UGU	4.3
UUC	15.9	UCC	11.3	UAC	12.6	UGC	6.4
UUA	12.9	UCA	6.7	UAA	1.9	UGA	0.8
UUG	11.6	UCG	9.3	UAG	0.2	UGG	12.1
CUU	10.6	CCU	7.3	CAU	11.5	CGU	20.0
CUC	10.3	CCC	6.5	CAC	10.0	CGC	23.0
CUA	4.3	CCA	5.7	CAA	12.3	CGA	3.7
CUG	52.8	CCG	24.2	CAG	31.7	CGG	6.4
AUU	27.7	ACU	7.6	AAU	17.3	AGU	7.0
AUC	25.1	ACC	24.8	AAC	22.0	AGC	16.9
AUA	5.3	ACA	6.4	AAA	35.2	AGA	2.8
AUG	26.2	ACG	16.8	AAG	12.2	AGG	1.8
GUU	15.5	GCU	14.4	GAU	32.1	GGU	18.6
GUC	18.0	GCC	28.1	GAC	22.4	GGC	35.6
GUA	11.9	GCA	13.2	GAA	38.0	GGA	7.8
GUG	24.8	GCG	39.4	GAG	21.8	GGG	11.4

GC content of genes = 53.1%
GC frequency for first nucleotide = 59.4%, for second nucleotide = 40.9%, for third nucleotide = 59.2%

Codon usage for PROKARYOTES *Staphylococcus aureus*

Frequencies per thousand codons — Based on 338 genes containing 114 179 codons

UUU	32.6	UCU	14.7	UAU	33.7	UGU	4.0
UUC	11.1	UCC	2.0	UAC	9.0	UGC	1.3
UUA	47.5	UCA	20.3	UAA	2.0	UGA	0.4
UUG	13.0	UCG	4.0	UAG	0.5	UGG	8.7
CUU	11.3	CCU	12.4	CAU	16.7	CGU	11.8
CUC	1.9	CCC	1.1	CAC	4.7	CGC	2.7
CUA	8.2	CCA	15.7	CAA	31.2	CGA	4.4
CUG	2.6	CCG	4.1	CAG	5.1	CGG	0.7
AUU	47.2	ACU	18.1	AAU	48.4	AGU	18.0
AUC	12.5	ACC	2.7	AAC	17.8	AGC	6.7
AUA	20.6	ACA	29.3	AAA	70.7	AGA	13.5
AUG	22.7	ACG	8.2	AAG	16.5	AGG	2.3
GUU	25.7	GCU	18.9	GAU	45.6	GGU	30.5
GUC	6.4	GCC	4.4	GAC	13.4	GGC	8.9
GUA	22.4	GCA	25.1	GAA	55.2	GGA	16.3
GUG	9.3	GCG	7.2	GAG	11.5	GGG	4.5

GC content of genes = 33.0%
GC frequency for first nucleotide = 44.0%, for second nucleotide = 32.3%, for third nucleotide = 22.8%

Codon usage for PROKARYOTES *Streptomyces coelicolor*

Frequencies per thousand codons — Based on 145 genes containing 48 978 codons

UUU	0.3	UCU	0.5	UAU	1.0	UGU	0.5
UUC	27.6	UCC	21.1	UAC	20.5	UGC	5.5
UUA	0.1	UCA	0.7	UAA	0.3	UGA	2.3
UUG	2.2	UCG	15.1	UAG	0.3	UGG	13.0
CUU	1.6	CCU	1.1	CAU	1.4	CGU	5.2
CUC	36.5	CCC	22.6	CAC	20.2	CGC	39.1
CUA	0.4	CCA	1.0	CAA	1.2	CGA	2.3
CUG	58.4	CCG	32.1	CAG	30.1	CGG	27.0
AUU	0.8	ACU	1.4	AAU	0.6	AGU	1.1
AUC	32.0	ACC	39.3	AAC	19.1	AGC	11.6
AUA	0.8	ACA	1.4	AAA	0.9	AGA	0.5
AUG	16.6	ACG	18.6	AAG	29.6	AGG	3.1
GUU	1.5	GCU	2.4	GAU	2.4	GGU	9.9
GUC	47.7	GCC	74.8	GAC	58.6	GGC	60.7
GUA	2.8	GCA	5.2	GAA	7.7	GGA	6.9
GUG	32.9	GCG	46.8	GAG	56.4	GGG	14.9

GC content of genes = 71.1%
GC frequency for first nucleotide = 71.2%, for second nucleotide = 48.8%, for third nucleotide = 93.4%

Codon usage for MICROBIAL EUKARYOTES *Aspergillus nidulans*

Frequencies per thousand codons — Based on 201 genes containing 119 005 codons

UUU	13.0	UCU	15.4	UAU	10.5	UGU	4.6
UUC	25.1	UCC	16.8	UAC	19.3	UGC	8.0
UUA	5.0	UCA	11.4	UAA	0.6	UGA	0.5
UUG	12.9	UCG	13.7	UAG	0.6	UGG	13.3
CUU	18.6	CCU	16.0	CAU	10.6	CGU	12.1
CUC	22.6	CCC	15.5	CAC	12.8	CGC	16.7
CUA	8.1	CCA	12.4	CAA	14.9	CGA	9.7
CUG	20.6	CCG	13.0	CAG	25.5	CGG	9.9
AUU	18.5	ACU	15.2	AAU	13.9	AGU	9.0
AUC	27.4	ACC	19.7	AAC	25.9	AGC	15.0
AUA	5.3	ACA	12.7	AAA	15.0	AGA	5.7
AUG	22.3	ACG	12.3	AAG	33.1	AGG	5.3
GUU	18.5	GCU	23.8	GAU	26.0	GGU	20.3
GUC	24.7	GCC	28.3	GAC	29.2	GGC	24.4
GUA	5.7	GCA	15.7	GAA	24.6	GGA	14.9
GUG	14.8	GCG	17.3	GAG	35.5	GGG	10.3

GC content of genes = 53.5%
GC frequency for first nucleotide = 57.3%, for second nucleotide = 43.9%, for third nucleotide = 59.2%

Codon usage for MICROBIAL EUKARYOTES — *Candida albicans*

Frequencies per thousand codons — Based on 154 genes containing 82 955 codons

UUU	25.9	UCU	23.9	UAU	24.2	UGU	10.4
UUC	16.7	UCC	10.7	UAC	11.9	UGC	1.3
UUA	34.6	UCA	22.8	UAA	1.2	UGA	0.2
UUG	36.9	UCG	5.1	UAG	0.5	UGG	11.0
CUU	9.0	CCU	13.2	CAU	13.9	CGU	6.4
CUC	2.1	CCC	3.2	CAC	6.1	CGC	0.5
CUA	2.9	CCA	28.3	CAA	38.0	CGA	3.1
CUG	2.3	CCG	2.2	CAG	6.2	CGG	0.6
AUU	40.6	ACU	33.1	AAU	36.9	AGU	14.4
AUC	15.3	ACC	14.1	AAC	19.0	AGC	3.5
AUA	9.1	ACA	14.3	AAA	50.2	AGA	24.1
AUG	18.6	ACG	2.6	AAG	19.3	AGG	2.1
GUU	36.0	GCU	34.6	GAU	44.7	GGU	39.9
GUC	11.6	GCC	13.6	GAC	14.4	GGC	4.4
GUA	6.7	GCA	13.6	GAA	51.5	GGA	12.1
GUG	8.0	GCG	1.5	GAG	8.1	GGG	6.7

GC content of genes = 36.8%
GC frequency for first nucleotide = 44.6%, for second nucleotide = 37.8%, for third nucleotide = 28.0%

Codon usage for MICROBIAL EUKARYOTES — *Chlamydomonas reinhardtii*

Frequencies per thousand codons — Based on 170 genes containing 57 759 codons

UUU	4.9	UCU	4.8	UAU	2.8	UGU	2.0
UUC	29.2	UCC	15.9	UAC	24.3	UGC	14.7
UUA	1.5	UCA	2.3	UAA	2.0	UGA	0.4
UUG	2.7	UCG	17.6	UAG	0.4	UGG	12.3
CUU	4.4	CCU	6.6	CAU	2.5	CGU	6.7
CUC	11.3	CCC	29.5	CAC	15.7	CGC	39.5
CUA	2.4	CCA	3.5	CAA	3.8	CGA	1.4
CUG	64.3	CCG	13.6	CAG	33.2	CGG	7.5
AUU	11.3	ACU	6.1	AAU	2.7	AGU	2.0
AUC	31.0	ACC	33.9	AAC	32.1	AGC	18.1
AUA	1.1	ACA	3.4	AAA	2.2	AGA	0.7
AUG	25.7	ACG	13.0	AAG	59.8	AGG	2.0
GUU	5.8	GCU	19.0	GAU	8.2	GGU	11.3
GUC	17.3	GCC	54.5	GAC	44.1	GGC	58.2
GUA	2.1	GCA	6.5	GAA	2.6	GGA	2.9
GUG	47.5	GCG	27.8	GAG	58.4	GGG	4.9

GC content of genes = 64.0%
GC frequency for first nucleotide = 61.7%, for second nucleotide = 44.2%, for third nucleotide = 86.0%

Codon usage for MICROBIAL EUKARYOTES *Dictyostelium discoideum*

Frequencies per thousand codons Based on 198 genes containing 105 658 codons

UUU	20.4	UCU	18.5	UAU	23.5	UGU	14.0
UUC	17.9	UCC	5.9	UAC	7.6	UGC	2.3
UUA	45.2	UCA	43.9	UAA	1.7	UGA	0.0
UUG	12.9	UCG	2.1	UAG	0.1	UGG	9.1
CUU	12.3	CCU	4.4	CAU	15.5	CGU	13.2
CUC	8.1	CCC	0.8	CAC	4.3	CGC	0.2
CUA	3.0	CCA	40.3	CAA	46.8	CGA	0.5
CUG	0.5	CCG	0.4	CAG	1.2	CGG	0.2
AUU	42.6	ACU	25.3	AAU	62.7	AGU	18.1
AUC	17.6	ACC	13.2	AAC	13.0	AGC	2.8
AUA	9.5	ACA	23.0	AAA	54.7	AGA	20.6
AUG	19.2	ACG	0.7	AAG	15.2	AGG	1.0
GUU	32.9	GCU	21.6	GAU	48.5	GGU	49.6
GUC	8.1	GCC	9.7	GAC	5.7	GGC	2.6
GUA	13.4	GCA	19.7	GAA	51.4	GGA	7.4
GUG	2.1	GCG	0.4	GAG	10.5	GGG	0.5

GC content of genes = 33.5%
GC frequency for first nucleotide = 43.6%, for second nucleotide = 37.2%, for third nucleotide = 19.6%

Codon usage for MICROBIAL EUKARYOTES *Neurospora crassa*

Frequencies per thousand codons Based on 232 genes containing 107850 codons

UUU	9.9	UCU	11.1	UAU	7.8	UGU	2.7
UUC	27.4	UCC	20.6	UAC	21.0	UGC	9.3
UUA	2.4	UCA	5.4	UAA	1.2	UGA	0.7
UUG	12.9	UCG	13.1	UAG	0.4	UGG	13.2
CUU	16.6	CCU	12.7	CAU	7.3	CGU	11.2
CUC	33.3	CCC	23.1	CAC	16.1	CGC	19.6
CUA	4.2	CCA	7.4	CAA	11.2	CGA	4.6
CUG	15.9	CCG	10.9	CAG	27.8	CGG	6.2
AUU	16.7	ACU	11.6	AAU	9.0	AGU	5.5
AUC	31.7	ACC	30.3	AAC	31.5	AGC	16.0
AUA	3.4	ACA	7.8	AAA	8.3	AGA	6.0
AUG	23.6	ACG	10.9	AAG	46.8	AGG	8.8
GUU	16.1	GCU	22.8	GAU	22.6	GGU	24.7
GUC	34.1	GCC	43.1	GAC	33.4	GGC	32.4
GUA	4.0	GCA	8.3	GAA	15.7	GGA	9.7
GUG	11.8	GCG	13.4	GAG	46.2	GGG	6.6

GC content of genes = 56.5%
GC frequency for first nucleotide = 57.3%, for second nucleotide = 43.0%, for third nucleotide = 69.1%

Codon usage for MICROBIAL EUKARYOTES *Pichia pastoris*

Frequencies per thousand codons — Based on 24 genes containing 14 360 codons

UUU	24.4	UCU	22.6	UAU	17.0	UGU	10.1
UUC	16.6	UCC	14.9	UAC	16.1	UGC	5.8
UUA	16.0	UCA	15.5	UAA	0.6	UGA	0.5
UUG	32.5	UCG	6.1	UAG	0.6	UGG	9.1
CUU	17.1	CCU	16.5	CAU	9.6	CGU	6.9
CUC	7.5	CCC	7.9	CAC	6.6	CGC	3.2
CUA	11.6	CCA	16.2	CAA	27.6	CGA	4.2
CUG	16.4	CCG	4.5	CAG	16.1	CGG	2.2
AUU	29.2	ACU	22.9	AAU	25.3	AGU	12.6
AUC	17.1	ACC	12.9	AAC	23.0	AGC	6.5
AUA	12.3	ACA	15.1	AAA	31.3	AGA	17.8
AUG	21.0	ACG	7.0	AAG	30.3	AGG	8.1
GUU	29.2	GCU	32.0	GAU	36.6	GGU	25.3
GUC	13.6	GCC	16.6	GAC	22.8	GGC	9.2
GUA	10.7	GCA	18.8	GAA	40.3	GGA	20.8
GUG	13.0	GCG	3.7	GAG	26.6	GGG	5.8

GC content of genes = 42.8%
GC frequency for first nucleotide = 49.9%, for second nucleotide = 38.1%, for third nucleotide = 40.3%

Codon usage for MICROBIAL EUKARYOTES *Plasmodium falciparum*

Frequencies per thousand codons — Based on 281 genes containing 176 163 codons

UUU	29.6	UCU	15.6	UAU	38.7	UGU	15.0
UUC	7.4	UCC	5.4	UAC	5.0	UGC	2.5
UUA	52.7	UCA	19.4	UAA	1.2	UGA	0.2
UUG	9.2	UCG	1.9	UAG	0.2	UGG	5.9
CUU	8.5	CCU	11.5	CAU	17.3	CGU	3.9
CUC	1.6	CCC	3.7	CAC	4.9	CGC	0.4
CUA	4.1	CCA	22.0	CAA	26.9	CGA	2.5
CUG	0.8	CCG	1.0	CAG	3.2	CGG	0.2
AUU	32.0	ACU	15.7	AAU	87.4	AGU	22.6
AUC	5.2	ACC	6.3	AAC	19.0	AGC	3.8
AUA	35.6	ACA	23.4	AAA	84.8	AGA	17.8
AUG	17.9	ACG	3.1	AAG	15.9	AGG	3.2
GUU	20.5	GCU	19.3	GAU	53.8	GGU	23.3
GUC	2.8	GCC	5.1	GAC	8.2	GGC	2.1
GUA	19.9	GCA	21.2	GAA	69.1	GGA	22.0
GUG	3.7	GCG	1.0	GAG	9.3	GGG	2.8

GC content of genes = 28.8%
GC frequency for first nucleotide = 39.7%, for second nucleotide = 30.4%, for third nucleotide = 16.3%

Codon usage for MICROBIAL EUKARYOTES *Saccharomyces cerevisiae*

Frequencies per thousand codons — Based on 9327 genes containing 4 601 350 codons

UUU	25.9	UCU	23.6	UAU	18.7	UGU	8.0
UUC	18.3	UCC	14.3	UAC	14.7	UGC	4.6
UUA	26.3	UCA	18.7	UAA	1.0	UGA	0.6
UUG	27.3	UCG	8.5	UAG	0.5	UGG	10.4
CUU	12.1	CCU	13.6	CAU	13.7	CGU	6.5
CUC	5.3	CCC	6.8	CAC	7.9	CGC	2.6
CUA	13.4	CCA	18.3	CAA	27.6	CGA	3.0
CUG	10.4	CCG	5.3	CAG	12.1	CGG	1.7
AUU	30.2	ACU	20.3	AAU	35.8	AGU	14.1
AUC	17.2	ACC	12.6	AAC	25.0	AGC	9.6
AUA	17.6	ACA	17.8	AAA	41.9	AGA	21.3
AUG	20.9	ACG	8.0	AAG	30.8	AGG	9.2
GUU	22.1	GCU	21.3	GAU	37.8	GGU	24.2
GUC	11.7	GCC	12.7	GAC	20.4	GGC	9.7
GUA	11.7	GCA	16.2	GAA	45.9	GGA	10.8
GUG	10.6	GCG	6.1	GAG	19.0	GGG	5.9

GC content of genes = 39.8%
GC frequency for first nucleotide = 44.6%, for second nucleotide = 36.6%, for third nucleotide = 38.0%

Codon usage for MICROBIAL EUKARYOTES *Schizosaccharomyces pombe*

Frequencies per thousand codons — Based on 776 genes containing 368 075 codons

UUU	30.4	UCU	30.1	UAU	21.3	UGU	8.5
UUC	13.2	UCC	12.0	UAC	11.9	UGC	5.6
UUA	25.0	UCA	16.8	UAA	1.2	UGA	0.4
UUG	24.1	UCG	7.3	UAG	0.5	UGG	10.0
CUU	25.6	CCU	21.8	CAU	15.8	CGU	17.9
CUC	7.5	CCC	8.7	CAC	6.4	CGC	6.5
CUA	7.8	CCA	12.1	CAA	27.5	CGA	7.6
CUG	6.1	CCG	4.4	CAG	10.7	CGG	2.8
AUU	35.7	ACU	23.3	AAU	33.1	AGU	14.1
AUC	13.8	ACC	11.4	AAC	18.6	AGC	9.3
AUA	11.7	ACA	13.5	AAA	37.8	AGA	10.7
AUG	21.5	ACG	6.1	AAG	26.5	AGG	4.7
GUU	29.7	GCU	31.2	GAU	38.1	GGU	25.4
GUC	12.1	GCC	12.8	GAC	16.5	GGC	9.2
GUA	11.7	GCA	15.3	GAA	43.6	GGA	16.1
GUG	8.1	GCG	5.2	GAG	22.1	GGG	4.0

GC content of genes = 40.5%
GC frequency for first nucleotide = 49.0%, for second nucleotide = 38.5%, for third nucleotide = 33.9%

Codon usage for MICROBIAL EUKARYOTES *Toxoplasma gondii*

Frequencies per thousand codons Based on 48 genes containing 17 039 codons

UUU	16.4	UCU	13.3	UAU	6.8	UGU	8.0
UUC	24.4	UCC	13.7	UAC	13.6	UGC	13.9
UUA	5.9	UCA	10.5	UAA	1.2	UGA	1.2
UUG	17.5	UCG	15.4	UAG	0.4	UGG	8.8
CUU	15.2	CCU	13.8	CAU	6.2	CGU	10.4
CUC	20.2	CCC	15.2	CAC	15.0	CGC	11.6
CUA	5.2	CCA	15.3	CAA	12.4	CGA	6.9
CUG	19.6	CCG	15.4	CAG	21.5	CGG	5.9
AUU	20.2	ACU	13.0	AAU	12.9	AGU	11.5
AUC	20.9	ACC	15.3	AAC	22.5	AGC	13.6
AUA	4.3	ACA	17.7	AAA	21.4	AGA	11.1
AUG	21.8	ACG	15.3	AAG	30.2	AGG	8.3
GUU	17.5	GCU	19.6	GAU	19.7	GGU	19.7
GUC	29.0	GCC	20.3	GAC	30.7	GGC	22.8
GUA	8.9	GCA	21.4	GAA	34.0	GGA	22.8
GUG	23.9	GCG	22.9	GAG	33.2	GGG	12.4

GC content of genes = 53.0%
GC frequency for first nucleotide = 56.9%, for second nucleotide = 44.7%, for third nucleotide = 57.5%

Codon usage for MICROBIAL EUKARYOTES *Trypanosoma brucei*

Frequencies per thousand codons Based on 121 genes containing 50 350 codons

UUU	18.0	UCU	9.9	UAU	11.0	UGU	7.6
UUC	18.7	UCC	11.7	UAC	14.9	UGC	12.0
UUA	6.5	UCA	10.6	UAA	1.4	UGA	0.5
UUG	15.6	UCG	8.3	UAG	0.5	UGG	9.9
CUU	20.2	CCU	11.2	CAU	8.5	CGU	12.9
CUC	17.1	CCC	10.9	CAC	12.8	CGC	13.6
CUA	9.1	CCA	12.1	CAA	16.0	CGA	5.6
CUG	18.2	CCG	9.4	CAG	19.5	CGG	8.6
AUU	17.8	ACU	12.8	AAU	18.5	AGU	12.3
AUC	15.2	ACC	13.6	AAC	24.3	AGC	14.5
AUA	12.5	ACA	19.5	AAA	30.3	AGA	6.9
AUG	24.3	ACG	15.5	AAG	34.6	AGG	8.3
GUU	18.5	GCU	20.8	GAU	25.8	GGU	22.5
GUC	12.2	GCC	20.7	GAC	28.2	GGC	19.4
GUA	12.5	GCA	27.0	GAA	35.0	GGA	17.8
GUG	27.1	GCG	20.6	GAG	36.6	GGG	12.1

GC content of genes = 50.3%
GC frequency for first nucleotide = 56.2%, for second nucleotide = 41.9%, for third nucleotide = 52.9%

Codon usage for ANIMALS — *Bos taurus*

Frequencies per thousand codons — Based on 1059 genes containing 417 154 codons

UUU	15.9	UCU	12.5	UAU	11.7	UGU	9.8
UUC	24.3	UCC	17.6	UAC	19.7	UGC	13.9
UUA	5.1	UCA	9.0	UAA	0.7	UGA	1.3
UUG	11.1	UCG	4.7	UAG	0.5	UGG	13.8
CUU	10.9	CCU	14.6	CAU	7.8	CGU	4.0
CUC	20.7	CCC	20.2	CAC	14.5	CGC	10.9
CUA	5.3	CCA	14.2	CAA	9.8	CGA	5.6
CUG	43.2	CCG	7.5	CAG	33.0	CGG	11.2
AUU	14.9	ACU	11.3	AAU	15.0	AGU	9.6
AUC	26.2	ACC	21.7	AAC	23.2	AGC	18.6
AUA	6.3	ACA	13.0	AAA	22.4	AGA	10.3
AUG	22.7	ACG	7.7	AAG	35.9	AGG	11.1
GUU	10.1	GCU	17.5	GAU	21.2	GGU	11.2
GUC	16.9	GCC	31.1	GAC	29.6	GGC	25.4
GUA	5.9	GCA	13.3	GAA	26.5	GGA	16.5
GUG	32.0	GCG	8.2	GAG	42.1	GGG	17.5

GC content of genes = 53.7%
GC frequency for first nucleotide = 55.8%, for second nucleotide = 41.5%, for third nucleotide = 63.7%

Codon usage for ANIMALS — *Caenorhabditis elegans*

Frequencies per thousand codons — Based on 8299 genes containing 3 659 520 codons

UUU	23.0	UCU	17.8	UAU	17.5	UGU	11.4
UUC	23.7	UCC	10.3	UAC	13.8	UGC	8.9
UUA	10.0	UCA	21.0	UAA	1.1	UGA	0.7
UUG	19.8	UCG	11.1	UAG	0.4	UGG	10.7
CUU	22.4	CCU	9.3	CAU	14.3	CGU	11.8
CUC	14.5	CCC	4.1	CAC	9.0	CGC	5.1
CUA	7.4	CCA	28.1	CAA	28.3	CGA	11.5
CUG	10.9	CCG	8.4	CAG	13.4	CGG	4.0
AUU	32.2	ACU	19.7	AAU	30.3	AGU	12.2
AUC	18.6	ACC	10.3	AAC	18.7	AGC	7.9
AUA	9.2	ACA	20.4	AAA	37.7	AGA	16.0
AUG	26.1	ACG	8.1	AAG	26.0	AGG	3.5
GUU	25.4	GCU	23.5	GAU	36.2	GGU	11.3
GUC	13.4	GCC	12.2	GAC	16.9	GGC	5.9
GUA	10.0	GCA	20.4	GAA	41.8	GGA	33.9
GUG	13.3	GCG	7.2	GAG	23.5	GGG	4.1

GC content of genes = 42.6%
GC frequency for first nucleotide = 50.2%, for second nucleotide = 39.1%, for third nucleotide = 38.4%

Codon usage for ANIMALS — *Drosophila melanogaster*

Frequencies per thousand codons — Based on 2049 genes containing 1 173 401 codons

UUU	11.3	UCU	6.3	UAU	10.0	UGU	5.2
UUC	22.4	UCC	20.2	UAC	18.5	UGC	14.0
UUA	4.1	UCA	7.1	UAA	0.9	UGA	0.5
UUG	14.5	UCG	17.3	UAG	0.5	UGG	9.7
CUU	7.8	CCU	6.6	CAU	10.2	CGU	9.0
CUC	13.4	CCC	19.0	CAC	16.9	CGC	19.0
CUA	7.0	CCA	13.2	CAA	15.1	CGA	7.5
CUG	37.8	CCG	16.7	CAG	38.5	CGG	7.6
AUU	15.6	ACU	8.8	AAU	20.5	AGU	10.7
AUC	24.0	ACC	22.5	AAC	28.6	AGC	20.3
AUA	8.2	ACA	10.8	AAA	15.7	AGA	4.9
AUG	23.3	ACG	14.1	AAG	41.1	AGG	5.8
GUU	10.2	GCU	14.6	GAU	27.1	GGU	14.5
GUC	14.7	GCC	35.9	GAC	25.6	GGC	30.0
GUA	5.7	GCA	12.7	GAA	18.7	GGA	18.5
GUG	27.0	GCG	13.9	GAG	43.7	GGG	4.5

GC content of genes = 54.8%
GC frequency for first nucleotide = 56.3%, for second nucleotide = 42.1%, for third nucleotide = 66.1%

Codon usage for ANIMALS — *Gallus gallus*

Frequencies per thousand codons — Based on 1166 genes containing 564 241 codons

UUU	14.6	UCU	12.6	UAU	10.8	UGU	7.8
UUC	20.5	UCC	16.6	UAC	18.9	UGC	13.6
UUA	5.5	UCA	10.2	UAA	0.8	UGA	1.0
UUG	11.3	UCG	5.5	UAG	0.5	UGG	11.5
CUU	10.6	CCU	14.3	CAU	8.3	CGU	5.6
CUC	16.4	CCC	18.9	CAC	14.4	CGC	12.2
CUA	5.2	CCA	15.3	CAA	11.2	CGA	5.0
CUG	38.5	CCG	8.1	CAG	31.6	CGG	9.4
AUU	16.0	ACU	13.1	AAU	15.5	AGU	9.5
AUC	24.0	ACC	18.8	AAC	24.0	AGC	20.6
AUA	7.4	ACA	15.9	AAA	26.4	AGA	11.1
AUG	23.1	ACG	8.0	AAG	37.7	AGG	11.1
GUU	12.2	GCU	21.1	GAU	25.0	GGU	11.2
GUC	14.7	GCC	25.1	GAC	26.8	GGC	21.7
GUA	7.1	GCA	17.5	GAA	31.4	GGA	17.3
GUG	29.1	GCG	9.0	GAG	45.5	GGG	16.4

GC content of genes = 52.5%
GC frequency for first nucleotide = 55.6%, for second nucleotide = 41.5%, for third nucleotide = 60.3%

Codon usage for ANIMALS *Homo sapiens*

Frequencies per thousand codons — Based on 10 491 genes containing 5 066 851 codons

UUU	16.3	UCU	14.2	UAU	12.3	UGU	9.6
UUC	21.2	UCC	17.7	UAC	17.0	UGC	12.7
UUA	6.5	UCA	10.9	UAA	0.7	UGA	1.2
UUG	11.6	UCG	4.3	UAG	0.5	UGG	13.3
CUU	11.8	CCU	16.8	CAU	9.7	CGU	4.7
CUC	19.5	CCC	19.9	CAC	14.6	CGC	11.0
CUA	6.4	CCA	16.3	CAA	11.5	CGA	6.0
CUG	40.2	CCG	6.9	CAG	33.7	CGG	11.3
AUU	15.8	ACU	12.8	AAU	16.7	AGU	11.3
AUC	23.5	ACC	20.9	AAC	20.9	AGC	19.1
AUA	6.8	ACA	14.8	AAA	23.4	AGA	10.9
AUG	22.6	ACG	6.6	AAG	33.9	AGG	10.9
GUU	10.7	GCU	18.6	GAU	22.2	GGU	11.1
GUC	15.4	GCC	28.9	GAC	26.9	GGC	23.7
GUA	6.6	GCA	15.3	GAA	28.1	GGA	16.9
GUG	29.7	GCG	7.5	GAG	40.6	GGG	16.6

GC content of genes = 52.8%
GC frequency for first nucleotide = 55.9%, for second nucleotide = 42.3%, for third nucleotide = 60.3%

Codon usage for ANIMALS *Mus musculus*

Frequencies per thousand codons — Based on 5367 genes containing 2 415 882 codons

UUU	15.6	UCU	15.5	UAU	12.0	UGU	10.9
UUC	21.5	UCC	17.9	UAC	17.5	UGC	13.0
UUA	5.7	UCA	11.0	UAA	0.6	UGA	1.2
UUG	12.0	UCG	4.5	UAG	0.5	UGG	13.1
CUU	12.0	CCU	18.6	CAU	9.6	CGU	4.8
CUC	19.2	CCC	19.0	CAC	15.0	CGC	9.9
CUA	7.2	CCA	17.2	CAA	11.8	CGA	6.5
CUG	38.5	CCG	6.8	CAG	34.0	CGG	10.0
AUU	14.8	ACU	13.4	AAU	15.7	AGU	11.8
AUC	23.2	ACC	20.2	AAC	22.1	AGC	19.7
AUA	6.6	ACA	15.8	AAA	21.7	AGA	11.5
AUG	22.4	ACG	6.2	AAG	35.2	AGG	11.4
GUU	10.1	GCU	19.9	GAU	21.5	GGU	12.2
GUC	15.9	GCC	26.6	GAC	27.7	GGC	23.3
GUA	6.9	GCA	15.1	GAA	26.8	GGA	18.0
GUG	29.0	GCG	7.1	GAG	39.7	GGG	16.0

GC content of genes = 52.7%
GC frequency for first nucleotide = 55.6%, for second nucleotide = 42.8%, for third nucleotide = 59.8%

Codon usage for ANIMALS ***Rattus norvegicus***

Frequencies per thousand codons — Based on 3626 genes containing 1 684 507 codons

UUU	16.6	UCU	13.9	UAU	11.7	UGU	9.4
UUC	25.0	UCC	18.1	UAC	18.3	UGC	12.2
UUA	5.1	UCA	9.8	UAA	0.6	UGA	1.1
UUG	12.2	UCG	4.3	UAG	0.5	UGG	13.5
CUU	11.7	CCU	16.2	CAU	8.7	CGU	5.0
CUC	20.6	CCC	17.9	CAC	14.5	CGC	10.4
CUA	6.9	CCA	14.8	CAA	10.3	CGA	6.3
CUG	41.8	CCG	6.1	CAG	32.0	CGG	10.2
AUU	16.0	ACU	12.8	AAU	15.2	AGU	10.7
AUC	27.5	ACC	21.2	AAC	23.4	AGC	18.5
AUA	6.5	ACA	14.9	AAA	20.7	AGA	10.1
AUG	24.6	ACG	6.4	AAG	36.5	AGG	11.1
GUU	10.1	GCU	19.6	GAU	20.8	GGU	11.2
GUC	17.2	GCC	28.5	GAC	28.8	GGC	23.2
GUA	6.9	GCA	14.7	GAA	25.6	GGA	16.4
GUG	31.6	GCG	6.9	GAG	40.9	GGG	15.6

GC content of genes = 52.7%
GC frequency for first nucleotide = 55.2%, for second nucleotide = 41.1%, for third nucleotide = 61.9%

Codon usage for ANIMALS ***Xenopus laevis***

Frequencies per thousand codons — Based on 999 genes containing 426 313 codons

UUU	19.2	UCU	19.2	UAU	14.9	UGU	10.8
UUC	16.2	UCC	16.0	UAC	15.4	UGC	11.8
UUA	8.2	UCA	13.3	UAA	1.0	UGA	1.0
UUG	13.7	UCG	3.8	UAG	0.4	UGG	10.2
CUU	15.4	CCU	17.4	CAU	11.9	CGU	7.2
CUC	12.5	CCC	13.8	CAC	13.4	CGC	7.2
CUA	8.3	CCA	19.4	CAA	16.2	CGA	6.1
CUG	26.4	CCG	4.7	CAG	28.7	CGG	6.0
AUU	19.9	ACU	16.8	AAU	21.6	AGU	14.2
AUC	18.1	ACC	16.1	AAC	21.8	AGC	17.1
AUA	10.2	ACA	18.8	AAA	33.0	AGA	14.9
AUG	25.1	ACG	4.7	AAG	34.0	AGG	12.6
GUU	15.4	GCU	21.3	GAU	29.6	GGU	14.3
GUC	12.3	GCC	18.2	GAC	23.6	GGC	15.4
GUA	10.1	GCA	19.4	GAA	36.0	GGA	22.0
GUG	21.1	GCG	4.5	GAG	34.9	GGG	13.1

GC content of genes = 47.7%
GC frequency for first nucleotide = 52.6%, for second nucleotide = 41.1%, for third nuclcotide = 49.3%

Codon usage for PLANTS — *Arabidopsis thaliana*

Frequencies per thousand codons — Based on 1509 genes containing 578 930 codons

UUU	19.7	UCU	21.9	UAU	13.4	UGU	9.2
UUC	22.5	UCC	10.5	UAC	15.6	UGC	6.9
UUA	10.6	UCA	14.8	UAA	1.0	UGA	1.1
UUG	19.8	UCG	8.1	UAG	0.4	UGG	11.5
CUU	25.1	CCU	18.3	CAU	12.7	CGU	9.8
CUC	17.1	CCC	5.7	CAC	9.5	CGC	3.6
CUA	9.1	CCA	15.9	CAA	18.5	CGA	5.4
CUG	9.5	CCG	8.0	CAG	16.5	CGG	4.1
AUU	22.5	ACU	18.9	AAU	19.3	AGU	12.3
AUC	20.8	ACC	11.7	AAC	22.3	AGC	11.0
AUA	10.9	ACA	14.6	AAA	28.4	AGA	17.3
AUG	25.7	ACG	7.4	AAG	34.9	AGG	11.5
GUU	27.7	GCU	33.1	GAU	35.7	GGU	26.0
GUC	14.4	GCC	12.7	GAC	19.1	GGC	9.7
GUA	8.3	GCA	17.8	GAA	31.5	GGA	27.1
GUG	17.8	GCG	9.4	GAG	34.1	GGG	10.2

GC content of genes = 45.7%
GC frequency for first nucleotide = 52.3%, for second nucleotide = 40.6%, for third nucleotide = 44.2%

Codon usage for PLANTS — *Hordeum vulgare*

Frequencies per thousand codons — Based on 296 genes containing 91 701 codons

UUU	8.5	UCU	7.2	UAU	6.0	UGU	3.8
UUC	30.7	UCC	19.5	UAC	24.1	UGC	14.8
UUA	1.6	UCA	5.9	UAA	1.0	UGA	1.4
UUG	8.7	UCG	10.4	UAG	0.9	UGG	13.0
CUU	11.6	CCU	9.7	CAU	6.2	CGU	6.0
CUC	31.2	CCC	16.8	CAC	16.0	CGC	15.5
CUA	3.9	CCA	12.0	CAA	13.6	CGA	2.0
CUG	22.6	CCG	15.9	CAG	28.4	CGG	9.0
AUU	10.1	ACU	8.0	AAU	7.8	AGU	4.8
AUC	30.6	ACC	25.8	AAC	29.9	AGC	17.4
AUA	5.0	ACA	6.6	AAA	8.8	AGA	3.9
AUG	25.5	ACG	12.8	AAG	45.1	AGG	11.7
GUU	11.6	GCU	47.0	GAU	13.9	GGU	15.2
GUC	26.4	GCC	42.4	GAC	38.6	GGC	44.5
GUA	4.0	GCA	13.5	GAA	11.4	GGA	12.9
GUG	29.9	GCG	24.8	GAG	43.5	GGG	18.6

GC content of genes = 59.2%
GC frequency for first nucleotide = 58.9%, for second nucleotide = 44.3%, for third nucleotide = 74.5%

Codon usage for PLANTS *Lycopersicon esculentum*

Frequencies per thousand codons — Based on 84 genes containing 34 238 codons

UUU	23.2	UCU	20.6	UAU	17.5	UGU	11.0
UUC	16.7	UCC	9.7	UAC	12.2	UGC	6.6
UUA	13.6	UCA	19.8	UAA	0.9	UGA	1.0
UUG	22.3	UCG	6.0	UAG	0.5	UGG	11.6
CUU	13.7	CCU	18.3	CAU	14.8	CGU	7.4
CUC	10.7	CCC	5.3	CAC	7.0	CGC	2.8
CUA	9.6	CCA	20.2	CAA	19.7	CGA	5.3
CUG	8.9	CCG	4.8	CAG	13.7	CGG	2.6
AUU	30.8	ACU	22.4	AAU	31.2	AGU	15.2
AUC	14.8	ACC	9.2	AAC	18.2	AGC	9.0
AUA	12.1	ACA	19.4	AAA	31.3	AGA	14.6
AUG	22.9	ACG	5.8	AAG	29.3	AGG	10.9
GUU	29.8	GCU	33.7	GAU	39.5	GGU	26.7
GUC	9.8	GCC	11.2	GAC	15.2	GGC	11.2
GUA	12.0	GCA	25.1	GAA	32.2	GGA	26.7
GUG	18.5	GCG	6.3	GAG	26.1	GGG	10.6

GC content of genes = 43.0%

GC frequency for first nucleotide = 50.1%, for second nucleotide = 41.1%, for third nucleotide = 37.0%

Codon usage for PLANTS *Nicotiana tabacum*

Frequencies per thousand codons — Based on 381 genes containing 128 097 codons

UUU	24.5	UCU	19.8	UAU	17.9	UGU	10.2
UUC	18.5	UCC	9.7	UAC	14.1	UGC	8.9
UUA	11.4	UCA	15.9	UAA	1.4	UGA	1.0
UUG	22.1	UCG	4.7	UAG	0.6	UGG	12.1
CUU	24.3	CCU	20.4	CAU	12.5	CGU	7.9
CUC	12.7	CCC	7.9	CAC	8.8	CGC	3.9
CUA	8.4	CCA	23.1	CAA	21.4	CGA	5.3
CUG	9.6	CCG	4.7	CAG	14.3	CGG	3.0
AUU	28.8	ACU	21.9	AAU	25.9	AGU	12.4
AUC	14.9	ACC	10.3	AAC	19.7	AGC	9.8
AUA	11.8	ACA	17.2	AAA	28.3	AGA	14.2
AUG	23.6	ACG	4.1	AAG	33.5	AGG	12.2
GUU	28.6	GCU	34.3	GAU	35.5	GGU	26.7
GUC	12.4	GCC	13.8	GAC	16.6	GGC	12.8
GUA	10.6	GCA	22.9	GAA	31.2	GGA	25.5
GUG	17.4	GCG	5.9	GAG	26.7	GGG	9.7

GC content of genes = 44.3%

GC frequency for first nucleotide = 51.9%, for second nucleotide = 41.2%, for third nucleotide = 39.9%

Codon usage for PLANTS *Oryza sativa*

Frequencies per thousand codons — Based on 418 genes containing 146 017 codons

UUU	10.9	UCU	10.1	UAU	9.5	UGU	4.9
UUC	27.8	UCC	16.0	UAC	22.4	UGC	14.1
UUA	4.7	UCA	8.5	UAA	0.9	UGA	1.2
UUG	10.9	UCG	9.8	UAG	0.8	UGG	12.6
CUU	13.8	CCU	11.6	CAU	8.5	CGU	6.6
CUC	28.4	CCC	12.8	CAC	14.0	CGC	14.6
CUA	4.9	CCA	11.4	CAA	13.4	CGA	3.0
CUG	20.8	CCG	15.1	CAG	27.0	CGG	7.4
AUU	14.6	ACU	10.7	AAU	13.4	AGU	7.3
AUC	28.4	ACC	20.5	AAC	27.3	AGC	17.1
AUA	6.8	ACA	9.4	AAA	11.0	AGA	7.4
AUG	25.1	ACG	9.5	AAG	42.6	AGG	14.5
GUU	15.6	GCU	20.3	GAU	21.4	GGU	18.1
GUC	24.0	GCC	31.8	GAC	32.6	GGC	35.1
GUA	5.5	GCA	14.3	GAA	15.2	GGA	15.6
GUG	24.4	GCG	22.3	GAG	43.7	GGG	16.0

GC content of genes = 55.6%
GC frequency for first nucleotide = 56.9%, for second nucleotide = 43.0%, for third nucleotide = 67.0%

Codon usage for PLANTS *Pisum sativum*

Frequencies per thousand codons — Based on 283 genes containing 99 210 codons

UUU	22.6	UCU	21.7	UAU	17.1	UGU	7.7
UUC	17.9	UCC	10.3	UAC	12.5	UGC	5.6
UUA	9.5	UCA	16.5	UAA	1.3	UGA	0.9
UUG	23.1	UCG	4.8	UAG	0.6	UGG	10.3
CUU	25.5	CCU	19.2	CAU	13.1	CGU	9.0
CUC	12.7	CCC	5.4	CAC	7.9	CGC	4.4
CUA	7.7	CCA	17.5	CAA	19.1	CGA	4.0
CUG	8.0	CCG	5.0	CAG	12.8	CGG	2.6
AUU	29.3	ACU	22.5	AAU	23.5	AGU	13.7
AUC	16.5	ACC	13.8	AAC	19.9	AGC	8.8
AUA	11.3	ACA	16.7	AAA	30.3	AGA	15.9
AUG	20.5	ACG	3.8	AAG	38.1	AGG	11.2
GUU	33.3	GCU	36.4	GAU	38.7	GGU	30.1
GUC	10.4	GCC	12.5	GAC	16.7	GGC	10.1
GUA	9.1	GCA	23.0	GAA	37.0	GGA	28.0
GUG	17.5	GCG	5.6	GAG	30.7	GGG	8.6

GC content of genes = 43.9%
GC frequency for first nucleotide = 52.2%, for second nucleotide = 40.6%, for third nucleotide = 38.9%

Codon usage for PLANTS *Triticum aestivum*

Frequencies per thousand codons — Based on 234 genes containing 72 161 codons

UUU	12.4	UCU	13.1	UAU	9.1	UGU	4.9
UUC	21.3	UCC	15.1	UAC	19.9	UGC	13.2
UUA	3.1	UCA	11.4	UAA	0.8	UGA	1.8
UUG	12.0	UCG	9.2	UAG	0.7	UGG	9.5
CUU	12.7	CCU	12.5	CAU	7.7	CGU	5.2
CUC	23.8	CCC	13.9	CAC	12.4	CGC	10.4
CUA	7.1	CCA	30.3	CAA	57.4	CGA	2.9
CUG	19.6	CCG	15.1	CAG	42.2	CGG	6.8
AUU	11.9	ACU	11.4	AAU	10.2	AGU	5.6
AUC	22.9	ACC	20.5	AAC	22.2	AGC	14.8
AUA	6.1	ACA	8.8	AAA	10.3	AGA	5.8
AUG	22.0	ACG	9.1	AAG	40.1	AGG	11.4
GUU	14.1	GCU	17.3	GAU	15.3	GGU	15.0
GUC	19.6	GCC	32.0	GAC	26.7	GGC	32.6
GUA	4.9	GCA	14.3	GAA	15.2	GGA	19.8
GUG	21.5	GCG	19.1	GAG	39.8	GGG	22.0

GC content of genes = 55.5%
GC frequency for first nucleotide = 60.9%, for second nucleotide = 43.5%, for third nucleotide = 62.2%

Codon usage for PLANTS *Zea mays*

Frequencies per thousand codons — Based on 525 genes containing 193 587 codons

UUU	9.7	UCU	10.5	UAU	7.0	UGU	4.3
UUC	26.1	UCC	16.6	UAC	21.5	UGC	12.0
UUA	3.6	UCA	9.5	UAA	0.7	UGA	1.2
UUG	11.6	UCG	9.7	UAG	0.9	UGG	11.9
CUU	15.4	CCU	12.8	CAU	7.7	CGU	5.8
CUC	25.1	CCC	14.8	CAC	14.8	CGC	15.3
CUA	6.4	CCA	14.2	CAA	13.6	CGA	3.4
CUG	25.8	CCG	16.6	CAG	28.0	CGG	8.6
AUU	12.6	ACU	10.8	AAU	11.2	AGU	6.4
AUC	27.2	ACC	20.8	AAC	25.3	AGC	17.0
AUA	5.9	ACA	9.2	AAA	11.1	AGA	6.6
AUG	25.0	ACG	11.6	AAG	43.5	AGG	15.1
GUU	14.7	GCU	23.2	GAU	18.9	GGU	15.8
GUC	22.1	GCC	34.3	GAC	34.5	GGC	33.3
GUA	5.2	GCA	15.7	GAA	15.9	GGA	12.3
GUG	27.7	GCG	23.1	GAG	44.4	GGG	14.9

GC content of genes = 56.7%
GC frequency for first nucleotide = 58.4%, for second nucleotide = 43.7%, for third nucleotide = 67.9%

Codon usage for MITOCHONDRIA — *Arabidopsis thaliana*

Frequencies per thousand codons

Based on 36 genes containing 7985 codons

UUU	35.4	UCU	24.7	UAU	21.2	UGU	11.0
UUC	28.4	UCC	14.4	UAC	11.9	UGC	9.1
UUA	20.0	UCA	17.5	UAA	3.1	UGA	3.0
UUG	20.4	UCG	12.5	UAG	1.3	UGG	17.3
CUU	26.0	CCU	20.5	CAU	18.2	CGU	10.8
CUC	18.8	CCC	8.9	CAC	6.9	CGC	7.9
CUA	13.0	CCA	15.2	CAA	18.4	CGA	12.4
CUG	10.1	CCG	7.4	CAG	13.0	CGG	7.3
AUU	26.9	ACU	18.0	AAU	22.3	AGU	15.3
AUC	17.9	ACC	12.0	AAC	16.0	AGC	14.4
AUA	17.7	ACA	14.7	AAA	29.3	AGA	16.7
AUG	23.9	ACG	8.4	AAG	21.7	AGG	11.9
GUU	19.7	GCU	23.3	GAU	21.8	GGU	17.4
GUC	11.6	GCC	9.6	GAC	12.1	GGC	9.4
GUA	12.8	GCA	15.2	GAA	26.8	GGA	22.3
GUG	10.1	GCG	7.1	GAG	17.9	GGG	9.6

GC content of genes = 43.2%
GC frequency for first nucleotide = 46.2%, for second nucleotide = 42.5%, for third nucleotide = 41.0%

Codon usage for MITOCHONDRIA — *Aspergillus nidulans*

Frequencies per thousand codons

Based on 16 genes containing 4620 codons

UUU	48.1	UCU	20.3	UAU	45.7	UGU	5.0
UUC	19.7	UCC	1.5	UAC	7.8	UGC	0.9
UUA	112.8	UCA	25.8	UAA	3.0	UGA	9.3
UUG	2.2	UCG	2.4	UAG	0.4	UGG	1.3
CUU	8.2	CCU	18.8	CAU	13.0	CGU	4.5
CUC	4.3	CCC	2.2	CAC	5.0	CGC	1.9
CUA	6.9	CCA	10.6	CAA	23.8	CGA	3.0
CUG	3.9	CCG	3.0	CAG	5.0	CGG	1.9
AUU	38.1	ACU	19.5	AAU	47.2	AGU	20.6
AUC	10.6	ACC	1.9	AAC	7.6	AGC	3.0
AUA	62.3	ACA	26.8	AAA	39.2	AGA	19.7
AUG	24.2	ACG	1.7	AAG	5.4	AGG	0.2
GUU	24.7	GCU	27.9	GAU	23.2	GGU	35.9
GUC	1.7	GCC	4.5	GAC	5.0	GGC	3.9
GUA	34.8	GCA	21.0	GAA	27.3	GGA	21.6
GUG	3.0	GCG	4.5	GAG	9.1	GGG	1.3

GC content of genes = 28.1%
GC frequency for first nucleotide = 36.6%, for second nucleotide = 32.7%, for third nucleotide = 15.1%

Codon usage for MITOCHONDRIA *Bos taurus*

Frequencies per thousand codons — Based on 26 genes containing 6938 codons

UUU	26.1	UCU	13.8	UAU	16.9	UGU	1.9
UUC	30.6	UCC	16.6	UAC	18.2	UGC	4.9
UUA	28.0	UCA	25.5	UAA	2.2	UGA	21.9
UUG	7.1	UCG	3.0	UAG	0.4	UGG	4.3
CUU	16.1	CCU	12.5	CAU	8.2	CGU	3.0
CUC	22.5	CCC	16.1	CAC	16.9	CGC	3.7
CUA	65.7	CCA	23.2	CAA	18.6	CGA	11.8
CUG	10.5	CCG	1.2	CAG	3.7	CGG	2.3
AUU	39.5	ACU	11.8	AAU	15.0	AGU	3.5
AUC	41.2	ACC	22.5	AAC	24.6	AGC	11.2
AUA	52.8	ACA	38.6	AAA	23.3	AGA	1.3
AUG	12.8	ACG	5.0	AAG	6.5	AGG	0.6
GUU	11.4	GCU	14.4	GAU	8.9	GGU	8.5
GUC	14.3	GCC	23.8	GAC	12.8	GGC	17.3
GUA	21.8	GCA	26.2	GAA	21.0	GGA	24.9
GUG	7.1	GCG	3.2	GAG	8.1	GGG	8.5

GC content of genes = 41.2%
GC frequency for first nucleotide = 46.8%, for second nucleotide = 38.7%, for third nucleotide = 38.2%

Codon usage for MITOCHONDRIA *Dictyostelium discoideum*

Frequencies per thousand codons — Based on 16 genes containing 5196 codons

UUU	42.1	UCU	4.2	UAU	49.5	UGU	7.3
UUC	3.8	UCC	1.7	UAC	4.2	UGC	0.4
UUA	94.9	UCA	15.4	UAA	2.5	UGA	0.0
UUG	3.7	UCG	3.5	UAG	0.6	UGG	8.9
CUU	3.8	CCU	5.2	CAU	11.9	CGU	3.5
CUC	0.0	CCC	1.3	CAC	1.3	CGC	0.0
CUA	8.1	CCA	19.1	CAA	31.8	CGA	14.0
CUG	1.3	CCG	0.4	CAG	4.8	CGG	0.8
AUU	33.9	ACU	7.3	AAU	47.2	AGU	25.6
AUC	9.6	ACC	4.8	AAC	9.2	AGC	4.0
AUA	52.7	ACA	29.8	AAA	108.5	AGA	38.3
AUG	23.9	ACG	6.2	AAG	14.0	AGG	1.5
GUU	10.0	GCU	7.7	GAU	18.9	GGU	10.2
GUC	4.4	GCC	2.9	GAC	2.3	GGC	1.9
GUA	47.2	GCA	31.8	GAA	43.7	GGA	34.4
GUG	4.4	GCG	5.8	GAG	5.8	GGG	1.9

GC content of genes = 26.0%
GC frequency for first nucleotide = 34.1%, for second nucleotide = 30.0%, for third nucleotide = 14.0%

Codon usage for MITOCHONDRIA *Homo sapiens*

Frequencies per thousand codons — Based on 29 genes containing 7263 codons

UUU	15.8	UCU	8.8	UAU	9.6	UGU	2.2
UUC	33.3	UCC	20.4	UAC	25.3	UGC	5.0
UUA	15.4	UCA	17.8	UAA	1.2	UGA	18.0
UUG	5.1	UCG	1.8	UAG	2.1	UGG	3.7
CUU	19.1	CCU	9.4	CAU	5.5	CGU	4.0
CUC	35.0	CCC	35.9	CAC	16.4	CGC	6.3
CUA	62.2	CCA	11.2	CAA	19.7	CGA	8.3
CUG	18.9	CCG	4.3	CAG	6.3	CGG	1.7
AUU	30.3	ACU	14.0	AAU	9.8	AGU	3.4
AUC	51.1	ACC	39.2	AAC	30.4	AGC	8.0
AUA	39.1	ACA	30.8	AAA	20.0	AGA	0.8
AUG	12.8	ACG	2.8	AAG	7.2	AGG	1.4
GUU	6.9	GCU	14.9	GAU	9.1	GGU	7.7
GUC	19.4	GCC	32.1	GAC	21.6	GGC	20.7
GUA	17.9	GCA	21.9	GAA	22.2	GGA	16.0
GUG	8.9	GCG	3.9	GAG	17.2	GGG	8.9

GC content of genes = 46.8%
GC frequency for first nucleotide = 51.3%, for second nucleotide = 38.5%, for third nucleotide = 50.7%

Codon usage for MITOCHONDRIA *Mus musculus*

Frequencies per thousand codons — Based on 20 genes containing 5647 codons

UUU	27.8	UCU	11.5	UAU	16.3	UGU	2.3
UUC	32.9	UCC	12.4	UAC	15.4	UGC	4.6
UUA	32.8	UCA	37.0	UAA	2.1	UGA	25.0
UUG	3.2	UCG	1.1	UAG	0.7	UGG	1.4
CUU	22.8	CCU	9.2	CAU	9.2	CGU	2.7
CUC	16.6	CCC	9.6	CAC	15.9	CGC	5.1
CUA	68.5	CCA	35.6	CAA	20.7	CGA	9.6
CUG	7.1	CCG	0.5	CAG	1.8	CGG	1.2
AUU	59.3	ACU	15.9	AAU	15.2	AGU	3.2
AUC	35.9	ACC	22.0	AAC	27.3	AGC	8.9
AUA	57.4	ACA	41.6	AAA	26.2	AGA	0.4
AUG	8.3	ACG	1.2	AAG	1.9	AGG	0.5
GUU	12.9	GCU	13.3	GAU	9.2	GGU	9.2
GUC	9.2	GCC	23.6	GAC	10.8	GGC	10.6
GUA	20.2	GCA	25.5	GAA	21.6	GGA	29.4
GUG	3.5	GCG	1.8	GAG	3.7	GGG	7.4

GC content of genes = 37.9%
GC frequency for first nucleotide = 44.8%, for second nucleotide = 38.3%, for third nucleotide = 30.6%

CODON USAGE for MITOCHONDRIA *Neurospora crassa*

Frequencies per thousand codons — Based on 27 genes containing 7554 codons

UUU	35.7	UCU	21.8	UAU	29.0	UGU	4.4
UUC	25.3	UCC	11.5	UAC	16.5	UGC	3.0
UUA	60.4	UCA	10.2	UAA	2.4	UGA	8.6
UUG	10.6	UCG	3.8	UAG	0.8	UGG	4.2
CUU	18.3	CCU	20.7	CAU	14.3	CGU	9.1
CUC	17.5	CCC	13.1	CAC	7.5	CGC	7.9
CUA	9.3	CCA	6.2	CAA	14.2	CGA	1.2
CUG	3.2	CCG	2.4	CAG	15.1	CGG	0.5
AUU	36.9	ACU	21.7	AAU	33.9	AGU	20.3
AUC	18.8	ACC	16.4	AAC	18.8	AGC	7.7
AUA	37.3	ACA	16.0	AAA	27.0	AGA	17.6
AUG	21.8	ACG	2.1	AAG	24.2	AGG	3.6
GUU	22.9	GCU	26.2	GAU	22.5	GGU	29.5
GUC	15.6	GCC	25.0	GAC	13.8	GGC	12.6
GUA	16.5	GCA	10.9	GAA	18.9	GGA	15.5
GUG	5.7	GCG	4.4	GAG	22.8	GGG	4.2

GC content of genes = 38.4%
GC frequency for first nucleotide = 42.8%, for second nucleotide = 36.3%, for third nucleotide = 36.1%

Codon usage for MITOCHONDRIA *Nicotiana tabacum*

Frequencies per thousand codons — Based on 12 genes containing 2920 codons

UUU	34.6	UCU	20.2	UAU	21.9	UGU	6.8
UUC	19.9	UCC	18.5	UAC	6.8	UGC	9.2
UUA	25.0	UCA	19.2	UAA	1.7	UGA	1.7
UUG	21.9	UCG	6.8	UAG	0.7	UGG	12.7
CUU	26.0	CCU	22.6	CAU	18.8	CGU	12.0
CUC	10.6	CCC	8.6	CAC	7.5	CGC	8.9
CUA	11.3	CCA	20.5	CAA	18.5	CGA	10.3
CUG	10.3	CCG	6.5	CAG	8.6	CGG	2.4
AUU	35.6	ACU	18.5	AAU	24.3	AGU	18.2
AUC	14.7	ACC	10.6	AAC	11.0	AGC	11.3
AUA	14.7	ACA	13.4	AAA	20.9	AGA	11.0
AUG	24.0	ACG	4.1	AAG	19.9	AGG	10.3
GUU	25.0	GCU	32.9	GAU	31.5	GGU	28.8
GUC	14.4	GCC	19.2	GAC	11.6	GGC	10.3
GUA	14.7	GCA	20.2	GAA	26.7	GGA	19.9
GUG	16.4	GCG	7.5	GAG	15.8	GGG	11.6

GC content of genes = 43.9%
GC frequency for first nucleotide = 51.0%, for second nucleotide = 43.5%, for third nucleotide = 37.3%

Codon usage for MITOCHONDRIA *Oryza sativa*

Frequencies per thousand codons — Based on 24 genes containing 6007 codons

UUU	44.9	UCU	17.3	UAU	25.1	UGU	8.3
UUC	24.6	UCC	11.3	UAC	7.8	UGC	4.2
UUA	25.5	UCA	14.8	UAA	1.3	UGA	0.8
UUG	20.8	UCG	11.3	UAG	1.8	UGG	18.1
CUU	20.3	CCU	19.3	CAU	22.8	CGU	13.0
CUC	13.5	CCC	10.5	CAC	6.0	CGC	5.8
CUA	13.0	CCA	19.6	CAA	21.6	CGA	10.5
CUG	10.0	CCG	8.2	CAG	8.8	CGG	6.0
AUU	35.6	ACU	14.0	AAU	23.5	AGU	16.1
AUC	15.5	ACC	11.7	AAC	13.5	AGC	9.8
AUA	25.0	ACA	12.0	AAA	24.0	AGA	14.0
AUG	24.0	ACG	7.7	AAG	12.5	AGG	8.2
GUU	21.3	GCU	26.0	GAU	27.1	GGU	24.3
GUC	10.5	GCC	12.3	GAC	9.2	GGC	10.7
GUA	19.8	GCA	16.3	GAA	27.1	GGA	29.3
GUG	17.5	GCG	9.3	GAG	11.7	GGG	13.7

GC content of genes = 42.5%
GC frequency for first nucleotide = 49.5%, for second nucleotide = 41.4%, for third nucleotide = 36.6%

Codon usage for MITOCHONDRIA *Paramecium aurelia*

Frequencies per thousand codons — Based on 40 genes containing 9036 codons

UUU	76.3	UCU	18.5	UAU	18.9	UGU	4.6
UUC	45.0	UCC	14.4	UAC	29.1	UGC	13.2
UUA	20.6	UCA	8.6	UAA	2.5	UGA	4.5
UUG	14.3	UCG	8.3	UAG	1.7	UGG	7.4
CUU	25.3	CCU	7.2	CAU	7.3	CGU	3.3
CUC	36.2	CCC	14.4	CAC	13.1	CGC	6.2
CUA	18.7	CCA	5.1	CAA	7.7	CGA	5.3
CUG	9.0	CCG	0.8	CAG	10.7	CGG	1.3
AUU	20.7	ACU	12.3	AAU	19.5	AGU	5.3
AUC	27.4	ACC	19.0	AAC	32.6	AGC	19.7
AUA	12.7	ACA	10.3	AAA	29.0	AGA	6.9
AUG	20.5	ACG	10.1	AAG	42.4	AGG	20.6
GUU	19.5	GCU	19.9	GAU	11.7	GGU	7.5
GUC	15.6	GCC	20.7	GAC	20.3	GGC	16.9
GUA	13.5	GCA	12.7	GAA	10.8	GGA	11.7
GUG	9.5	GCG	3.9	GAG	23.0	GGG	11.1

GC content of genes = 42.4%
GC frequency for first nucleotide = 40.0%, for second nucleotide = 33.5%, for third nucleotide = 53.8%

Codon usage for MITOCHONDRIA *Saccharomyces cerevisiae*

Frequencies per thousand codons

Based on 59 genes containing 12 917 codons

UUU	34.9	UCU	14.9	UAU	44.2	UGU	7.7
UUC	20.9	UCC	4.9	UAC	7.9	UGC	1.3
UUA	95.2	UCA	24.8	UAA	4.0	UGA	11.3
UUG	8.2	UCG	2.0	UAG	0.2	UGG	3.7
CUU	5.4	CCU	19.3	CAU	16.7	CGU	2.9
CUC	1.9	CCC	3.9	CAC	2.5	CGC	1.1
CUA	7.7	CCA	13.2	CAA	20.9	CGA	0.5
CUG	4.3	CCG	2.0	CAG	3.6	CGG	0.5
AUU	74.7	ACU	18.3	AAU	78.3	AGU	12.6
AUC	12.4	ACC	4.7	AAC	10.6	AGC	2.3
AUA	25.6	ACA	20.7	AAA	51.0	AGA	19.8
AUG	24.2	ACG	2.2	AAG	13.2	AGG	2.8
GUU	20.8	GCU	25.4	GAU	28.6	GGU	37.3
GUC	4.8	GCC	4.9	GAC	7.3	GGC	3.6
GUA	26.9	GCA	18.0	GAA	27.4	GGA	12.0
GUG	3.1	GCG	3.1	GAG	5.3	GGG	5.0

GC content of genes = 27.5%

GC frequency for first nucleotide = 34.0%, for second nucleotide = 30.7%, for third nucleotide = 17.9%

Codon usage for MITOCHONDRIA *Schizosaccharomyces pombe*

Frequencies per thousand codons

Based on 7 genes containing 2767 codons

UUU	30.7	UCU	27.5	UAU	32.9	UGU	8.7
UUC	37.9	UCC	0.4	UAC	12.3	UGC	0.0
UUA	79.1	UCA	32.5	UAA	1.1	UGA	1.4
UUG	4.7	UCG	0.0	UAG	0.0	UGG	19.5
CUU	20.6	CCU	25.7	CAU	21.0	CGU	1.8
CUC	0.4	CCC	1.1	CAC	6.1	CGC	0.4
CUA	14.5	CCA	23.1	CAA	17.0	CGA	4.7
CUG	1.8	CCG	1.1	CAG	2.2	CGG	0.0
AUU	56.4	ACU	26.7	AAU	49.5	AGU	20.6
AUC	15.2	ACC	1.4	AAC	6.1	AGC	3.3
AUA	26.4	ACA	20.6	AAA	38.7	AGA	24.9
AUG	19.3	ACG	2.5	AAG	4.3	AGG	0.4
GUU	26.0	GCU	40.1	GAU	18.2	GGU	44.8
GUC	2.2	GCC	2.2	GAC	3.6	GGC	0.4
GUA	24.9	GCA	16.3	GAA	18.8	GGA	25.3
GUG	2.2	GCG	1.9	GAG	3.6	GGG	2.2

GC content of genes = 31.2%

GC frequency for first nucleotide = 38.5%, for second nucleotide = 38.2%, for third nucleotide = 17.0%

Codon usage for MITOCHONDRIA *Triticum aestivum*

Frequencies per thousand codons — Based on 30 genes containing 8596 codons

UUU	39.9	UCU	18.7	UAU	22.3	UGU	8.0
UUC	27.0	UCC	15.4	UAC	7.9	UGC	5.9
UUA	27.0	UCA	15.9	UAA	1.5	UGA	2.8
UUG	20.2	UCG	11.2	UAG	1.2	UGG	15.8
CUU	22.9	CCU	17.8	CAU	20.7	CGU	11.7
CUC	11.3	CCC	10.7	CAC	6.0	CGC	5.2
CUA	15.0	CCA	14.7	CAA	19.5	CGA	13.5
CUG	11.3	CCG	6.2	CAG	7.4	CGG	6.5
AUU	37.2	ACU	18.0	AAU	20.7	AGU	15.4
AUC	19.8	ACC	16.8	AAC	10.4	AGC	11.9
AUA	20.1	ACA	11.5	AAA	22.3	AGA	14.8
AUG	26.5	ACG	6.4	AAG	16.9	AGG	6.9
GUU	20.2	GCU	29.3	GAU	23.4	GGU	24.9
GUC	12.2	GCC	19.1	GAC	9.8	GGC	11.9
GUA	17.9	GCA	15.9	GAA	25.0	GGA	25.7
GUG	15.2	GCG	9.2	GAG	12.4	GGG	10.9

GC content of genes = 43.3%
GC frequency for first nucleotide = 48.4%, for second nucleotide = 42.9%, for third nucleotide = 38.5%

Codon usage for MITOCHONDRIA *Zea mays*

Frequencies per thousand codons — Based on 35 genes containing 8334 codons

UUU	20.9	UCU	14.6	UAU	21.1	UGU	7.2
UUC	20.6	UCC	13.4	UAC	10.0	UGC	4.0
UUA	27.4	UCA	17.3	UAA	1.6	UGA	1.7
UUG	19.3	UCG	8.9	UAG	1.0	UGG	12.7
CUU	17.0	CCU	15.4	CAU	17.5	CGU	11.8
CUC	16.1	CCC	7.9	CAC	6.8	CGC	7.7
CUA	12.8	CCA	17.3	CAA	23.8	CGA	9.5
CUG	13.7	CCG	8.2	CAG	10.3	CGG	5.3
AUU	35.3	ACU	17.3	AAU	22.0	AGU	13.9
AUC	20.4	ACC	11.0	AAC	14.9	AGC	8.3
AUA	16.7	ACA	12.5	AAA	24.2	AGA	12.1
AUG	23.4	ACG	6.7	AAG	19.7	AGG	7.8
GUU	20.0	GCU	32.0	GAU	27.2	GGU	25.0
GUC	15.0	GCC	17.3	GAC	13.6	GGC	13.4
GUA	17.0	GCA	17.4	GAA	32.5	GGA	25.9
GUG	21.1	GCG	11.8	GAG	17.5	GGG	15.5

GC content of genes = 44.6%
GC frequency for first nucleotide = 52.3%, for second nucleotide = 41.1%, for third nucleotide = 40.3%

Codon usage for CHLOROPLASTS — *Chlamydomonas reinhardtii*

Frequencies per thousand codons

Based on 68 genes containing 18 856 codons

UUU	33.4	UCU	17.3	UAU	23.5	UGU	7.4
UUC	17.9	UCC	3.3	UAC	11.4	UGC	1.9
UUA	75.8	UCA	22.3	UAA	2.8	UGA	0.2
UUG	4.9	UCG	4.2	UAG	0.5	UGG	14.0
CUU	14.6	CCU	15.1	CAU	10.0	CGU	31.4
CUC	1.1	CCC	3.9	CAC	8.3	CGC	4.1
CUA	6.6	CCA	24.1	CAA	38.1	CGA	3.1
CUG	4.8	CCG	2.6	CAG	4.1	CGG	0.6
AUU	47.7	ACU	23.8	AAU	41.1	AGU	16.3
AUC	9.5	ACC	5.7	AAC	19.9	AGC	5.7
AUA	7.3	ACA	31.9	AAA	63.7	AGA	5.3
AUG	22.1	ACG	4.0	AAG	7.5	AGG	1.0
GUU	29.4	GCU	35.5	GAU	24.0	GGU	43.4
GUC	3.0	GCC	6.9	GAC	11.1	GGC	6.2
GUA	27.3	GCA	20.8	GAA	40.7	GGA	7.9
GUG	5.3	GCG	3.6	GAG	5.9	GGG	3.6

GC content of genes = 34.4%

GC frequency for first nucleotide = 44.7%, for second nucleotide = 37.7%, for third nucleotide = 20.9%

Codon usage for CHLOROPLASTS — *Cyanophora paradoxa* (cyanelle)

Frequencies per thousand codons

Based on 181 genes containing 36 351 codons

UUU	34.1	UCU	24.4	UAU	26.6	UGU	8.6
UUC	10.0	UCC	5.3	UAC	5.2	UGC	1.0
UUA	81.8	UCA	11.3	UAA	4.2	UGA	0.1
UUG	3.7	UCG	1.7	UAG	0.6	UGG	10.4
CUU	14.8	CCU	14.7	CAU	12.8	CGU	27.0
CUC	0.8	CCC	1.3	CAC	3.3	CGC	3.8
CUA	5.5	CCA	19.9	CAA	39.0	CGA	9.4
CUG	0.8	CCG	2.8	CAG	2.3	CGG	0.7
AUU	67.8	ACU	29.0	AAU	40.4	AGU	17.3
AUC	7.8	ACC	5.3	AAC	7.3	AGC	2.9
AUA	11.3	ACA	19.8	AAA	59.7	AGA	9.3
AUG	19.7	ACG	3.3	AAG	5.2	AGG	0.5
GUU	34.2	GCU	33.1	GAU	35.3	GGU	36.4
GUC	1.8	GCC	3.1	GAC	5.9	GGC	4.8
GUA	24.9	GCA	32.4	GAA	57.1	GGA	22.0
GUG	2.6	GCG	4.2	GAG	4.4	GGG	3.5

GC content of genes = 32.3%

GC frequency for first nucleotide = 46.5%, for second nucleotide = 36.9%, for third nucleotide = 13.6%

Codon usage for CHLOROPLASTS *Euglena gracilis*

Frequencies per thousand codons — Based on 112 genes containing 27 161 codons

UUU	60.0	UCU	26.4	UAU	31.1	UGU	8.0
UUC	8.0	UCC	3.4	UAC	5.2	UGC	2.5
UUA	56.0	UCA	16.8	UAA	3.2	UGA	0.3
UUG	17.7	UCG	4.9	UAG	0.6	UGG	15.3
CUU	19.1	CCU	21.0	CAU	20.3	CGU	15.5
CUC	0.6	CCC	2.4	CAC	2.5	CGC	3.3
CUA	6.8	CCA	10.8	CAA	25.5	CGA	6.8
CUG	1.0	CCG	2.2	CAG	3.8	CGG	0.8
AUU	48.8	ACU	22.3	AAU	45.7	AGU	13.7
AUC	4.9	ACC	2.0	AAC	9.4	AGC	2.2
AUA	31.8	ACA	22.2	AAA	67.8	AGA	18.4
AUG	22.2	ACG	4.8	AAG	10.5	AGG	4.2
GUU	35.3	GCU	28.9	GAU	30.9	GGU	33.5
GUC	2.1	GCC	3.1	GAC	6.5	GGC	4.8
GUA	19.2	GCA	18.7	GAA	38.4	GGA	25.1
GUG	3.5	GCG	5.0	GAG	7.8	GGG	4.5

GC content of genes = 31.2%
GC frequency for first nucleotide = 41.0%, for second nucleotide = 35.4%, for third nucleotide = 17.2%

Codon usage for CHLOROPLASTS *Oryza sativa*

Frequencies per thousand codons — Based on 86 genes containing 19 711 codons

UUU	34.5	UCU	17.9	UAU	26.1	UGU	7.1
UUC	19.8	UCC	12.2	UAC	8.8	UGC	3.3
UUA	32.6	UCA	10.4	UAA	2.0	UGA	1.2
UUG	20.4	UCG	5.9	UAG	1.3	UGG	17.0
CUU	20.7	CCU	18.5	CAU	16.7	CGU	15.0
CUC	7.1	CCC	10.1	CAC	6.4	CGC	6.5
CUA	13.6	CCA	10.9	CAA	26.3	CGA	13.2
CUG	6.5	CCG	6.3	CAG	8.7	CGG	4.8
AUU	38.8	ACU	24.3	AAU	27.6	AGU	15.0
AUC	16.0	ACC	11.4	AAC	12.0	AGC	4.5
AUA	19.1	ACA	13.7	AAA	36.4	AGA	17.0
AUG	23.4	ACG	5.3	AAG	15.5	AGG	6.0
GUU	24.1	GCU	29.7	GAU	28.8	GGU	26.0
GUC	7.3	GCC	9.3	GAC	9.8	GGC	8.4
GUA	23.8	GCA	19.4	GAA	41.3	GGA	26.1
GUG	9.4	GCG	7.9	GAG	15.8	GGG	15.0

GC content of genes = 40.5%
GC frequency for first nucleotide = 49.4%, for second nucleotide = 39.9%, for third nucleotide = 32.2%

Codon usage for CHLOROPLASTS *Spinacea oleracea*

Frequencies per thousand codons — Based on 62 genes containing 21 590 codons

UUU	34.3	UCU	19.8	UAU	23.9	UGU	7.5
UUC	18.9	UCC	10.5	UAC	7.5	UGC	3.2
UUA	30.5	UCA	12.5	UAA	1.3	UGA	0.8
UUG	21.1	UCG	6.3	UAG	0.7	UGG	17.9
CUU	21.9	CCU	19.8	CAU	17.4	CGU	14.5
CUC	7.1	CCC	8.3	CAC	6.5	CGC	5.1
CUA	13.9	CCA	13.3	CAA	26.3	CGA	11.7
CUG	6.5	CCG	5.6	CAG	8.8	CGG	3.9
AUU	34.8	ACU	21.3	AAU	30.0	AGU	14.8
AUC	15.1	ACC	11.3	AAC	12.8	AGC	5.2
AUA	21.2	ACA	15.5	AAA	34.4	AGA	15.3
AUG	22.7	ACG	5.7	AAG	14.5	AGG	7.3
GUU	24.9	GCU	31.6	GAU	31.9	GGU	27.8
GUC	7.7	GCC	10.7	GAC	10.9	GGC	9.6
GUA	21.9	GCA	19.0	GAA	40.5	GGA	29.4
GUG	8.6	GCG	7.5	GAG	15.2	GGG	13.5

GC content of genes = 40.8%
GC frequency for first nucleotide = 50.1%, for second nucleotide = 40.6%, for third nucleotide = 31.6%

Codon usage for CHLOROPLASTS *Triticum aestivum*

Frequencies per thousand codons — Based on 28 genes containing 5271 codons

UUU	26.9	UCU	15.9	UAU	22.2	UGU	9.9
UUC	20.9	UCC	15.2	UAC	8.5	UGC	5.5
UUA	28.3	UCA	9.3	UAA	2.7	UGA	1.1
UUG	16.7	UCG	5.9	UAG	1.5	UGG	11.6
CUU	18.4	CCU	20.9	CAU	12.0	CGU	12.1
CUC	11.4	CCC	9.9	CAC	4.6	CGC	5.9
CUA	10.2	CCA	12.1	CAA	20.9	CGA	10.2
CUG	8.5	CCG	8.9	CAG	10.8	CGG	3.2
AUU	36.6	ACU	23.5	AAU	23.0	AGU	14.6
AUC	15.4	ACC	13.3	AAC	9.5	AGC	6.5
AUA	18.4	ACA	17.3	AAA	28.8	AGA	16.3
AUG	23.3	ACG	9.3	AAG	18.2	AGG	6.1
GUU	24.1	GCU	30.7	GAU	30.2	GGU	25.6
GUC	12.7	GCC	18.6	GAC	15.6	GGC	12.7
GUA	21.8	GCA	19.4	GAA	41.2	GGA	27.7
GUG	12.3	GCG	12.5	GAG	18.0	GGG	14.8

GC content of genes = 43.7%
GC frequency for first nucleotide = 51.8%, for second nucleotide = 42.7%, for third nucleotide = 36.8%

Codon usage for CHLOROPLASTS *Zea mays*

Frequencies per thousand codons — Based on 56 genes containing 14 599 codons

UUU	32.3	UCU	16.9	UAU	29.0	UGU	7.7
UUC	15.9	UCC	14.5	UAC	7.8	UGC	3.2
UUA	37.3	UCA	11.6	UAA	1.9	UGA	0.9
UUG	19.6	UCG	6.0	UAG	1.0	UGG	16.1
CUU	18.7	CCU	15.7	CAU	17.1	CGU	15.0
CUC	6.4	CCC	10.2	CAC	6.3	CGC	7.0
CUA	14.2	CCA	11.1	CAA	30.1	CGA	14.2
CUG	5.8	CCG	6.4	CAG	9.0	CGG	3.8
AUU	41.0	ACU	22.5	AAU	25.8	AGU	14.2
AUC	13.8	ACC	9.9	AAC	9.1	AGC	4.5
AUA	23.3	ACA	16.1	AAA	39.5	AGA	18.1
AUG	23.2	ACG	5.3	AAG	14.2	AGG	5.4
GUU	22.5	GCU	30.6	GAU	31.2	GGU	25.8
GUC	6.4	GCC	9.9	GAC	8.4	GGC	8.4
GUA	22.3	GCA	20.3	GAA	43.2	GGA	29.0
GUG	8.2	GCG	8.0	GAG	14.0	GGG	13.6

GC content of genes = 39.8%
GC frequency for first nucleotide = 49.3%, for second nucleotide = 40.2%, for third nucleotide = 30.1%

4. REFERENCES

1. John, B. and Miklos, G. (1988) *The Eukaryote Genome in Development and Evolution.* Unwin Hyman, London.

2. Bennett, M.D. and Smith, J.B. (1976) *Phil. Trans. R. Soc. Ser. B*, **274**, 227.

3. Bult, C.J., White, O., Olsen, G.J., Zhou, L., Fleischmann, R.D., Sutton, G.G., Blake, J.A., FitzGerald, L.M., Clayton, R.A., Gocayne, J.D., Kerlavage, A.R., Dougherty, B.A., Tomb, J.-F., Adams, M.D., Reich, C.I., Overbeek, R., Kirkness, E.F., Weinstock, K.G., Merrick, J.M., Glodek, A., Scott, J.L., Geoghagen, N.S.M., Weidman, J.F., Fuhrmann, J.L., Nguyen, D., Utterback, T.R., Kelley, J.M., Peterson, J.D., Sadow, P.W., Hanna, M.C., Cotton, M.D., Roberts, K.M., Hurst, M.A., Kaine, B.P., Borodovsky, M., Klenk, H.P., Fraser, C.M., Smith, H.O., Woese, C.R. and Venter, J.C. (1996) *Science*, **273**, 1058.

4. Brown, T.A. (1987) in *Gene Structure in Eukaryotic Microbes* (J.R. Kinghorn ed.). IRL Press, Oxford, p.141.

5. Grossman, L.I. and Hudspeth, M.E.S. (1985) in *Gene Manipulation in Fungi* (J.W. Bennett and L.L. Lasure, eds). Academic Press, London, p. 65.

6. Palmer, J.D. (1985). *Annu. Rev. Genet.*, **19**, 325.

7. Thiery, J.P., Macaya, G. and Bernardi, G. (1976) *J. Mol. Biol.*, **108**, 219.

8. De Ley, J. (1970) *J. Bacteriol.*, **101**, 738.

9. http://www.tigr.org/tdb/mdb/mdb.html

10. http://www.genetics.wisc.edu:80/index.html

11. http://genome4.aist-nara.ac.jp/

12. http://susi.bio.uni-giessen.de/usr/local/www/html/ecdc.html

13. http://mol.genes.nig.ac.jp/ecoli/

14. http://www.ai.sri.com/ecocyc/ecocyc.html

15. http://www.mbl.edu/html/ecoli.html

16. Kröger, M. and Wahl, R. (1997) *Nucl. Acids Res.*, **25**, 39.

17. Karp, P.D., Riley, M., Paley, S.M., Pellegrini-Toole, A. and Krummenacker, M. (1997) *Nucl. Acids Res.*, **25**, 43.

18. Riley, M. (1997) *Nucl. Acids Res.*, **25**, 51.

19. http://www.tigr.org/tdb/mdb/hidb/hidb.html

20. http://susi.bio.uni-giessen.de/usr/local/www/html/hidc.html

21. Fleischmann, R.D., Adams, M.D., White, O., Clayton, R.A., Kirkness, E.F., Kerlavage, A.R., Bult, C.J., Tomb, J.-F., Dougherty, B.A., Merrick,

J.M., McKenney, K., Sutton, G., FitzHugh, W., Fields, C., Gocayne, J.D., Scott, J., Shirley, R., Liu, L.-I., Glodek, A., Kelley, J.M., Weidman, J.F., Phillips, C.A., Spriggs, T., Hedblom, E., Cotton, M.D., Utterback, T.R., Hanna, M.C., Nguyen, D.T., Saudek, D.M., Brandon, R.C., Fine, L.D., Fritchman, J.L., Fuhrmann, J.L., Geoghagen, N.S.M., Gnehm, C.L., McDonald, L.A., Small, K.V., Fraser, C.M., Smith, H.O. and Venter, J.C. (1995) *Science*, **269**, 496.

22. Tomb, J.-F., White, O., Kerlavage, A.R., Clayton, R.A., Sutton, G.G., Fleischmann, R.D., Ketchum, K.A., Klenk, H.P., Gill, S., Dougherty, B.A., Nelson, K., Quakenbush, J., Zhou, L., Kirkness, E.F., Peterson, S., Loftus, B., Richardson, D., Dodson, R., Khalak, H.G., Glodek, A., McKenney, K., Fitzegerald, L.M., Lee, N., Adams, M., Hickey, E.K., Berg, D.E., Gocayne, J.D., Utterback, T.R., Peterson, J.D., Kelley, J.M., Cotton, M.D., Weidman, J.M., Fujii, C., Bowman, C., Watthey, L., Wallin, E., Hayes, W.S., Borodovsky, M., Karp, P.D., Smith, H.O., Fraser, C.M. and Venter, J.C. (1997) *Nature*, **388**, 539.

23. http://www.tigr.org/tdb/mdb/mgdb/mgdb.html

24. Fraser, C.M., Gocayne, J.D., White, O., Adams, M.D., Clayton, R.A., Fleischmann, R.D., Bult, C.J., Kerlavage, A.R., Sutton, G., Kelley, J.M., Fritchman, J.L., Weidman, J.F., Small, K.V., Sandusky, M., Fuhrmann, J., Nguyen, D., Utterback, T.R., Saudek, D.M., Phillips, C.A., Merrick, J.M., Tomb, J.-F., Dougherty, B.A., Bott, K.F., Hu, P.-C., Lucier, T.S., Peterson, S.N., Smith, H.O., Hutchison, C.A. and Venter, J.C. (1995) *Science*, **270**, 397.

25. http://www.zmbh.uni-heidelberg.de/M_pneumoniae/MP_Home.html

26. Himmelreich, PR., Hilbert, H., Plagens, H., Pirkl, E., Li, B.C. and Herrmann, R. (1996) *Nucl. Acids Res.*, **24**, 4420.

27. http://www.tigr.org/tdb/mdb/mjdb/mjdb.html

28. http://www.kazusa.or.jp/cyano/cyano.html

29. Kaneko, T., Sato, S., Kotani, H., Tanaka, A., Asamizu, E., Nakamura, Y., Miyajima, N., Hirosawa, M., Sugiura, M., Sasamoto, S., Kimura, T., Hosouchi, T., Matsuno, A., Muraki, A., Nakazaki, N., Naruo, K., Okumura, S., Shimpo, S., Takeuchi, C., Wada, T., Watanabe, A., Yamada, M., Yasuda, M. and Tabata, S. (1996) *DNA Res.*, **3**, 109.

30. Kaneko, T., Sato, S., Kotani, H., Tanaka, A., Asamizu, E., Nakamura, Y., Miyajima, N., Hirosawa, M., Sugiura, M., Sasamoto, S., Kimura, T., Hosouchi, T., Matsuno, A., Muraki, A., Nakazaki, N., Naruo, K., Okumura, S., Shimpo, S., Takeuchi, C., Wada, T., Watanabe, A., Yamada, M., Yasuda, M. and Tabata, S. (1996) *DNA Res.*, **3**, 185.

31. http://genome-www.stanford.edu/Saccharomyces/

32. http://www.mips.biochem.mpg.de/

33. http://www.sanger.ac.uk/Projects/

34. Mewes, H.W., Albermann, K., Heumann, K., Liebl, S. and Pfeiffer, F. (1997) *Nucl. Acids Res.*, **25**, 28.

35. Dujon, B. (1996) *Trends Genet.*, **12**, 263.

36. ftp://ftp.tigr.org/pub/data/a_fulgidus/

37. http://www.pasteur.fr/Bio/SubtiList.html

38. http://bacillus.tokyo-center.genome.ad.jp:8008/BSORF-DB.html

39. ftp://ftp.tigr.org/pub/data/b_burgdorferi/

40. http://chlamydia-www.berkeley.edu:4231/

41. http://www.genomecorp.com/htdocs/sequences/clostridium/clospage.html

42. ftp://ftp.tigr.org/pub/data/d_radiodurans/

43. ftp://ftp.tigr.org/pub/data/e_faecalis/

44. http://www.genomecorp.com/htdocs/methanobacter/abstract.html

45. http://kiev.physchem.kth.se/MycDB.html

46. http://www.cric.com/htdocs/leprae/index.html

47. http://dna1.chem.uoknor.edu/gono.html

48. ftp://ftp.tigr.org/pub/data/n_meningitidis/

49. http://www.cmcb.uq.edu.au/aeruginosa/

50. http://dna1.chem.uoknor.edu/strep.html

51. http://www.imb.nrc.ca/imb/sulfolob/sulhom_e.html

52. http://utmmg.med.uth.tmc.edu/treponema/tpall.html

53. ftp://ftp.tigr.org/pub/data/v_cholerae/

54. http://gdbwww.gdb.org/

55. Fasman, K.H., Letovsky, S.I., Li, P., Cottingham, R.W. and Kingsbury, D.T. (1997) *Nucl. Acids Res.*, **25**, 72.

56. http://www.ncbi.nih.gov/dbEST/index.html

57. http://www.informatics.jax.org/

58. Blake, J.A., Richardson, J.E., Davisson, M.T. and Eppig, J.T. (1997) *Nucl. Acids Res.*, **25**, 85.

59. http://www.genethon.fr/

60. http://www.chlc.org/

61. http://www.cephb.fr/bio/ceph-genethon-map.html

62. http://www.genome.mit.edu/

63. http://www.ncbi.nlm.nih.gov

64. Benson, D.A., Boguski, M.S., Lipman, D.J. and Ostell, J. (1997) *Nucl. Acids Res.*, **25**, 1.

65. http://www.ebi.ac.uk

66. Stoesser, G., Sterk, P., Tuli, M.A., Stoehr, P.J. and Cameron, G.N. (1997) *Nucl. Acids Res.*, **25**, 7.

67. http://www.nig.oc.jp/home.html

68. Tateno, Y. and Gojobori, T. (1997) *Nucl. Acids Res.*, **25**, 14.

69. George, D.G., Dodson, R.J., Garavelli, J.S., Haft, D.H., Hunt, L.T., Marzec, C.R., Orcutt, B.C., Sidman, K.E., Srinivasarao, G.Y., Yeh, L.-S.L., Arminski, L.M., Ledley, R.S., Tsugita, A. and Barker, W.C. (1997) *Nucl. Acids Res.*, **25**, 24.

70. Bairoch, A. and Apweiler, R. (1997) *Nucl. Acids Res.*, **25**, 31.

71. http://alces.med.umn.edu/Candida.html

72. http://glamdring.ucsd.edu/others/dsmith/dictydb.html

73. http://ben.vub.ac.be/malaria/mad.html

74. http://www.wehi.edu.au/biology/malaria/who.html

75. http://parasite.arf.ufl.edu/malaria.html

76. http://www.ab.a.u-tokyo.ac.jp/sericulture/shimada.html

77. http://moulon.inra.fr/acedb/acedb.html

78. http://www.ddbj.nig.ac.jp/htmls/c-elegans/html/CE_INDEX.html

79. http://flybase.bio.indiana.edu/

80. The Flybase Consortium (1997) *Nucl. Acids Res.*, **25**, 63.

81. http://klab.agsci.colostate.edu/

82. http://fugu.hgmp.mrc.ac.uk/

83. http://niol1.bio.nagoya-u.ac.jp:8000/

84. http://zfish.uoregon.edu/

85. http://www.ri.bbsrc.ac.uk/chickmap/chickgbase/manager.html

86. http://locus.jouy.inra.fr/cgi-bin/bovmap/intro2.pl

87. http://mendel.berkeley.edu/dog.html

88. http://www-seq.wi.mit.edu/public_release/

89. http://www.informatics.jax.org/mgd.html

90. http://www.ri.bbsrc.ac.uk/pigmap/pigbase/pigbase.html

91. http://ratmap.gen.gu.se/

92. http://dirk.invermay.cri.nz/

93. http://lenti.med.umn.edu/arabidopsis/arab_top_page.html

94. http://cbil.humgen.upenn.edu/~atgc/ATGCUP.html

95. http://www.arabidopsis.com/bb/b950928z.html

96. http://nasc.nott.ac.uk/JIC-contigs/JIC-contigs.html

97. http://genome-www.stanford.edu/Arabidopsis/

98. http://www.tigr.org/tdb/at/at.html

99. http://scaffold.biologie.uni-kl.de/Beanref/

100. http://algodon.tamu.edu/

101. http://s27w007.pswfs.gov/

102. http://teosinte.agron.missouri.edu/

103. http://moulon.moulon.inra/imgd

104. http://www.staff.or.jp/

105. http://mendel.agron.iastate.edu:8000/main.html

106. Low, S.C. and Berry, M.J. (1996) *Trends Biochem. Sci.*, **21**, 203.

107. Yamao, F., Muto, A., Kawauchi, Y., Iwami, M., Iwagami, S., Azumi, Y. and Osawa, S. (1985) *Proc. Natl. Acad. Sci. USA*, **82**, 2306.

108. Anderson, S., Bankier, A.T., Barrell, B.G., de Bruijn, M.H.L., Coulson, A.R., Drouin, J., Eperon, I.C., Nierlich, D.P., Roe, B.A., Sanger, F., Schreier, P.H., Smith, A.J.H., Staden, R. and Young, I.G. (1981) *Nature*, **290**, 457.

109. Clary, D.O., Wahleithner, J.A. and Wolstenholme, D.R. (1984). *Nucl. Acids Res.*, **12**, 3747.

110. Sibler, A.P., Dirheimer, G. and Martin, R.P. (1981) *FEBS Lett.*, **132**, 344.

111. Waring, R.B., Davies, R.W., Lee, S., Grisi, E., Berks, M.M. and Scazzocchio, C. (1981) *Cell*, **27**, 4.

112. Fox, T.D. and Leaver, C.J. (1981) *Cell*, **26**, 315.

113. Caron, F. and Meyer, E. (1985) *Nature*, **314**, 185.

114. Preer, J., Preer, L.B., Rudman, B.M. and Barnett, A.J. (1985) *Nature*, **314**, 188.

115. Horowitz, S. and Gorovsky, M.A. (1985) *Proc. Natl. Acad. Sci. USA*, **82**, 2452.

116. Kuchino, Y., Hanyu, N., Tashiro, F. and Nishimura, S. (1985) *Proc. Natl. Acad. Sci. USA*, **82**, 4758.

117. Kawaguchi, Y., Honda, H., Taniguchi-Morimura, J. and Iwasaki, S. (1969) *Nature*, **341**, 164.

118. Gualberto, J.M., Lamattina, L., Bonnard, G., Weil, J.-H. and Grienenberger, J.-M. (1989) *Nature*, **341**, 660.

119. Covello, P.S. and Gray, M.W. (1989) *Nature*, **341**, 662.

120. Nakamura, Y., Gojobori, T. and Ikemura, T. (1997) *Nucl. Acids Res.*, **25**, 244.

121. http://www.dna.affrc.go.jp/~nakamura/codon.html

CHAPTER 5
CLONING VECTORS

An enormous number of cloning vectors have been developed over the last 25 years. In the early days of gene cloning, vectors were constructed by individual researchers and distributed freely to other groups. Several of these vectors are still regularly used but many have disappeared, at least from common usage. If you encounter one and need to find out more about it then consult ref. 1, which provides descriptions of virtually all cloning vectors in use up to the late 1980s. In more recent years, most of the new vectors that have appeared have been developed by commercial companies whose catalogues and product information sheets are the best sources of detailed information on individual types.

This chapter describes the most frequently used *E. coli* cloning vectors, including most of the commercially available ones and several from the 'pre-commercial' era.

1. *E. COLI* PLASMID CLONING VECTORS

1.1 Basic features

All plasmid cloning vectors contain three basic components.

(i) One or more origins of replication, also called replicators (*Table 1*), most frequently the one from plasmid ColE1 or the closely related pMB1.

(ii) One or more selectable markers (*Table 2*), either a gene conferring resistance to an antibiotic or other inhibitor, or a gene such as *lacZ′* that enables transformants to be distinguished by a histochemical test.

(iii) One or more unique restriction sites into which new DNA can be inserted.

Table 1. Replicators found in *E. coli* plasmid cloning vectors

Replicator	Type	Copy number[a]	Reference
ColE1	Plasmid	10–15	2
f1[b]	Single-stranded phage	-[c]	3
M13[b]	Single-stranded phage	-[c]	3
p15A	Plasmid	15–20	4
pMB1	Plasmid	15–20	5
R6K	Plasmid	10–15	2

[a]The copy numbers refer to the original plasmids from which the replicators were obtained. Several cloning vectors carry mutations that result in a higher copy number, and the copy numbers of most vectors can be increased by suitable treatment such as chloramphenicol amplification (6).
[b]The f1 and M13 replicators are not identical (f1 and M13 are different phages). It is possible that some commercial vectors that are described as having the M13 replicator in fact have the f1 replicator.
[c]The copy number of a phagemid is determined by the plasmid replicator.

Table 2. Selectable markers found in *E. coli* plasmid cloning vectors

Marker	Source	Gene product	Phenotype[a]
amp	Tn*3*	ß-Lactamase	Ampicillin resistance
cml or cat	Tn*9*	Acetyltransferase	Chloramphenicol resistance
kan	Tn*903*	Phosphotransferase	Kanamycin resistance
lacZ′	*E. coli lac* operon	ß-Galactosidase	Lac^+ (histochemical test)
neo	Tn*5*	Phosphotransferase	Kanamycin resistance
tet	pSC101	Membrane protein	Tetracycline resistance

[a]See Chapter 1, *Table 15* for the amounts of antibiotic to use in selective media.

1.2 Types of plasmid vector

Most *E. coli* plasmid vectors fall into one of eight categories.

(i) *Basic cloning vectors*, which simply comprise an origin of replication and one or more selectable markers.

(ii) *Phagemids*, which also have an origin of replication from a single-stranded phage (usually f1 or M13) and which can therefore be obtained as single-stranded DNA when the cells harbouring the vector are infected with a suitable helper phage (e.g. M13KO7; ref. 7).

(iii) *RNA expression vectors*, which include one or more bacteriophage promoter sequences (almost exclusively the promoters from SP6, T3 and/or T7 phages; *Table 3*) and which can therefore be used for *in vitro* synthesis of large quantities of RNA.

(iv) *Protein expression vectors*, which contain a strong *E. coli* promoter (*Table 4*) and direct synthesis of the protein coded by a cloned gene, often as a fusion protein.

(v) *Promoter probe vectors*, which contain a marker gene that lacks a promoter. Insertion of DNA containing a promoter sequence into a cloning site upstream of the marker gene switches on its expression.

(vi) *Cosmids*, which contain *cos* sequences from bacteriophage λ, and which can therefore be packaged in λ heads. The major use of cosmids is in cloning large pieces of DNA, up to 40 kb in size.

(vii) *Cloning vectors for PCR products*, most of which are modified versions of phagemids.

(viii) *Mutagenesis vectors*, which are specifically designed for use in *in vitro* mutagenesis procedures.

Table 3. Promoters for RNA synthesis found in *E. coli* plasmid cloning vectors

Promoter	Source	For further details see page
SP6	SP6 bacteriophage	199
T3	T3 bacteriophage	202
T7	T7 bacteriophage	212

Table 4. Promoters for protein expression found in *E. coli* plasmid cloning vectors

Promoter	Source	Induction[a]
lac	*E. coli lac* operon	IPTG
lac–tac	Hybrid	IPTG
P_L	Bacteriophage λ	*cIts*857 repressor
P_R	Bacteriophage λ	*cIts*857 repressor
tac	*trp–lac* hybrid	IPTG
trc	*trp–lac* hybrid	IPTG
trp	*E. coli trp* operon	IAA

[a]Abbreviations: IAA, 3-ß-indoleacrylic acid; IPTG, isopropyl-ß-D-thiogalactoside.
[b]The *cIts*857 repressor is temperature sensitive. At 42°C and above the repressor is inactive, resulting in induction of the promoter.

1.3 Individual vectors

Table 5 lists the most frequently used *E. coli* plasmid cloning vectors. Note that many vectors are multifunctional and fit into more than one of the categories described in Section 1.2.

All of the vectors listed in *Table 5* are described in more detail on pp. 295–338. These pages give maps and the additional details needed to choose a suitable plasmid for a particular application. For many plasmids, the name of a commercial supplier is given. In several cases the named company holds a trademark for the vector and is the only supplier, in other cases the plasmid is not trademarked and may also be available from other suppliers. The list of cloning sites given for each plasmid is taken from the product information provided by the supplier or from the original published description of the vector. These lists have been checked wherever possible but there are still some uncertainties: for example, in most vectors a *Sal*I cloning site is also a unique site for *Acc*I and/or *Hinc*II, but this is not always clear from the available information. When in doubt I have not included a site in the list.

Plasmids c2RB, pcos1EMBL, pEMBL8/9 and pJB8 are redrawn from ref. 59; pBR325 and pUC118/119 are redrawn from ref. 1; pBR327 and pGB2 are reproduced with permission from Pouwels, P.H. (1991) in *Essential Molecular Biology: A Practical Approach* (T.A. Brown, ed.). IRL Press, Oxford, Vol. 1; pMOS*Blue* and the pTZ vectors are reproduced with permission from Amersham International; pBR328 and pHelix are reproduced with permission from Boehringer Mannheim; pDIRECT, pKK233-2, pKK388-1, pKT vectors and pT3/T7 vectors are reproduced with permission from Clontech; pSPORT vectors and pT712/13 are reproduced with permission from Life Technologies; pACYC177, pACY184, pLITMUS, pMAL vectors, pNEB193, pNO1523) – source: copyright © 1996/7 New England Biolabs Catalog, reprinted with permission; pBR322, pEZZ 18, pGEX series, pKK175-6, pKK223-3, pKK232-8, pMC1871, pRIT2T, pRIT5, pSL1180, pT7/T3-18U/19U, pTrc99A, pUC vectors except 118/119 and pYEJ001 are reproduced with permission from Pharmacia Biotech; pALTER vectors, pGEM vectors, pGEMEX vectors, PinPoint vectors, pSELECT-1 and pSP vectors are reproduced with permission from Promega Corporation; pBC, pBluescript II, pBluescript SK, pBS+/–, pCAL, pCR-Script vectors, pSPUTK, pWE15, pWE16 and SuperCos 1 are reproduced with permission from Stratagene; pT7-1/2 is reproduced with permission from United States Biochemicals.

Table 5. *E. coli* plasmid cloning vectors

Type	Vectors[a]
Basic cloning vectors	pACYC177, pACYC184, pBR322, pBR325, pBR327, pBR328, pGB2, pNEB193, pNO1523, pUC7, pUC8, pUC9, pUC12, pUC13, pUC18, pUC19
Phagemids	pEMBL8, pEMBL9, pSL1180, pUC118, pUC119
RNA expression vectors	pBS, pGEM-3Z, pGEM-4Z, pSP64, pSP64 Poly(A), pSP65, pSP70, pSP71, pSP72, pSP73, pT3/T7-1, pT3/T7-2, pT3/T7-3, pT7-1, pT7-2, pT712, pT713
Phagemid/RNA expression vectors	pBC, pBluescript II, pBluescript SK, pBS +/-, pGEM3Zf(+/-), pGEM5Zf(+/-), pGEM7Zf(+/-), pGEM9Zf(-), pGEM11Zf(+/-), pGEM13Zf(+), pHelix, pLITMUS, pSELECT-1, pSPORT1, pSPORT2, pT7/T3-18U, pT7/T3-19U, pTZ18R, pTZ18U, pTZ19R, pTZ19U
Protein expression vectors	pCAL, pET series, pEZZ 18, pGEX series, PinPoint Xa-1, pKK223-3, pKK233-2, pKK388-1, pKT279, pKT280, pKT287, pMAL-c2, pMAL-p2, pMC1871, pRIT2T, pRIT5, pSPUTK, pTrc 99A, pYEJ001
Phagemid/RNA expression/ protein expression vectors	pGEMEX-1, pGEMEX-2
Promoter probe vectors	pKK175-6, pKK232-8
Cosmids	c2RB, pcos1EMBL, pJB8, pWE15, pWE16, SuperCos 1
Vectors for PCR products	pCR-Script Amp, pCR-Script Cam, pCR-Script Direct, pDIRECT, pGEM-T, pGEM-T Easy, pMOS*Blue*
PCR products/protein expression	PinPoint Xa-1 T
Mutagenesis vectors	pALTER-1, pALTER-*Ex*1, pALTER-*Ex*2

[a]For maps and additional details see pp. 295–338.

c2RB

Type Cosmid

Size 6.8 kb

Replicator ColE1

Markers Ampicillin resistance
Kanamycin resistance

Cloning site *Sma*I

Reference 8

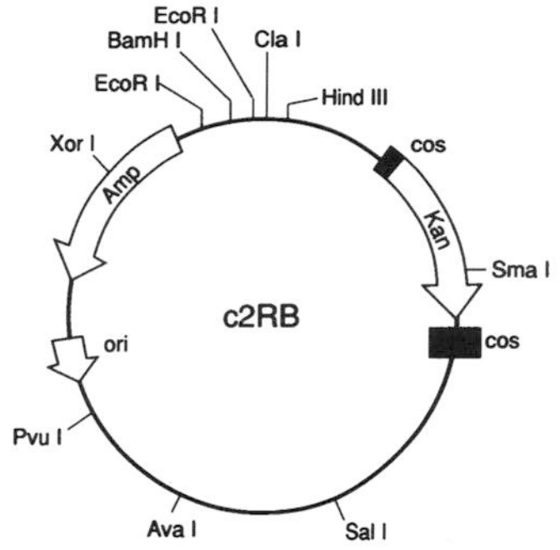

pACYC177

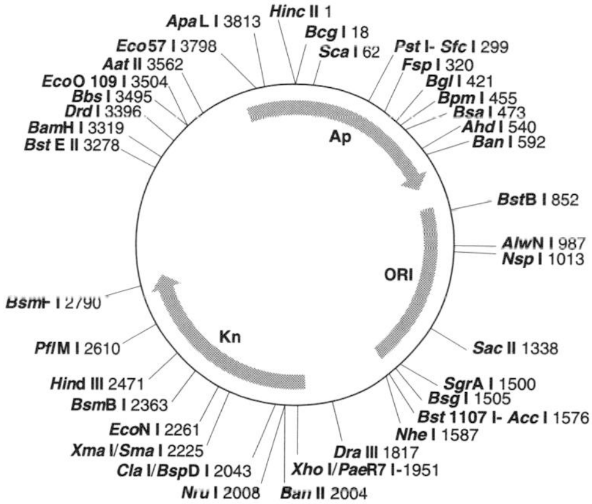

Type Basic cloning vector

Size 3940 bp

Replicator p15A

Markers Ampicillin resistance
Kanamycin resistance

Cloning sites *Ahd*I, *Apa*LI, *Ban*I, *Ban*II, *Bcg*I, *Bgl*I, *Bpm*I, *Bsa*I, *Bsm*BI, *Bsp*DI, *Cla*I, *Eco*NI, *Fsp*I, *Hinc*II, *Hin*dIII, *Nru*I, *Pae*R7I, *Pfl*MI, *Pst*I, *Sca*I, *Sfc*I, *Sma*I, *Xho*I, *Xma*I

Reference 4

pACYC184

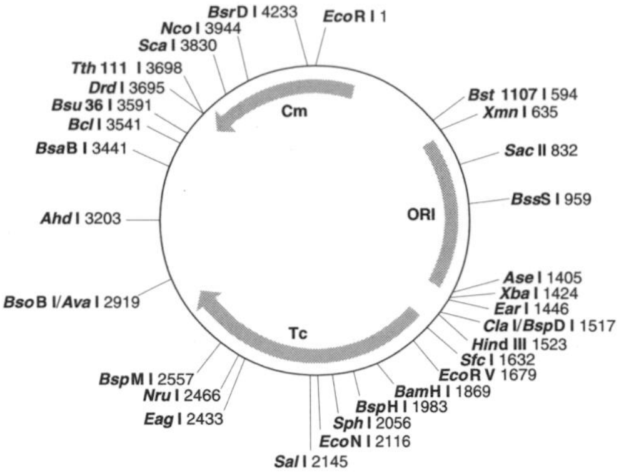

Type Basic cloning vector

Size 4244 bp

Replicator p15A

Markers Chloramphenicol resistance
Tetracycline resistance

Cloning sites *Bam*HI, *Bsp*HI, *Bsp*MI, *Bsr*DI, *Eag*I, *Eco*NI, *Eco*RI, *Eco*RV, *Nco*I, *Nru*I, *Sal*I, *Sca*I, *Sfc*I, *Sph*I

Reference 4

pALTER-1

Marketed by Promega Corporation

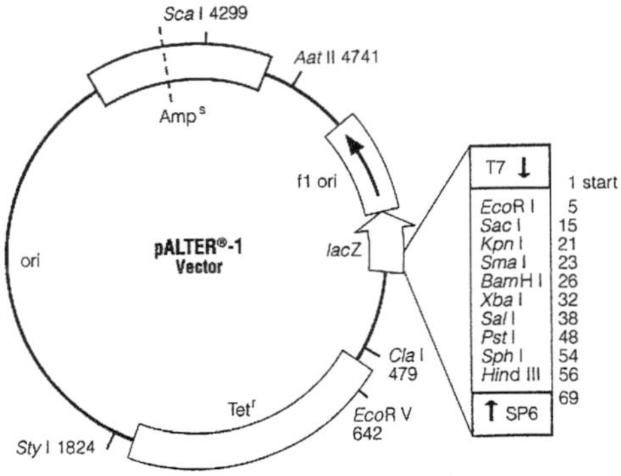

Type Mutagenesis vector

Size 5680 bp

Replicators pMB1, f1

Markers Ampicillin resistance
Tetracycline resistance
lacZ'

Promoters SP6, T7

Cloning sites *Bam*HI, *Eco*RI, *Hin*dIII, *Kpn*I, *Pst*I, *Sac*I, *Sal*I, *Sma*I, *Sph*I, *Xba*I

pALTER-*Ex*1

Marketed by Promega Corporation

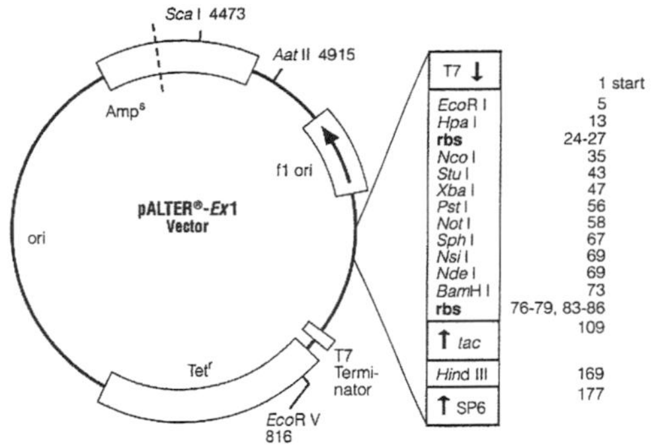

Type Mutagenesis vector

Size 5658 bp

Replicators pMB1, f1

Markers Ampicillin resistance
Tetracycline resistance

Promoters SP6, T7, *tac*

Cloning sites *Bam*HI, *Eco*RI, *Hin*dIII, *Hpa*I, *Nco*I, *Nde*I, *Not*I, *Nsi*I, *Pst*I, *Sph*I, *Stu*I, *Xba*I

pALTER-*Ex*2

Marketed by Promega Corporation

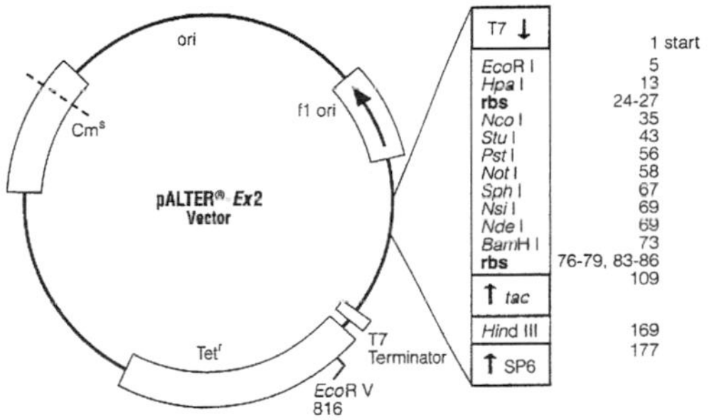

Type Mutagenesis vector

Size 5638 bp

Replicators p15A, f1

Markers Chloramphenicol resistance
Tetracycline resistance

Promoters SP6, T7, *tac*

Cloning sites *Bam*HI, *Eco*RI, *Hin*dIII, *Hpa*I, *Nco*I, *Nde*I, *Not*I, *Nsi*I, *Pst*I, *Sph*I, *Stu*I

Type	Phagemid RNA synthesis
Size	3399 bp
Replicators	ColE1, f1
Markers	Chloramphenicol resistance *lacZ′*
Promoters	T3, T7
Cloning sites	*Acc*I, *Apa*I, *Bam*HI, *Bsp*106I, *Bss*HII, *Bst*XI, *Eag*I, *Eco*O109I, *Eco*RI, *Eco*RV, *Hinc*II, *Hin*dIII, *Kpn*I, *Not*I, *Pst*I, *Sac*I, *Sac*II, *Sal*I, *Sma*I, *Spe*I, *Xba*I, *Xho*I

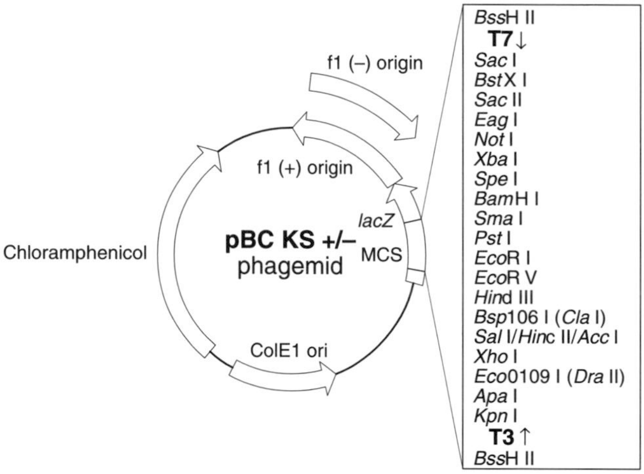

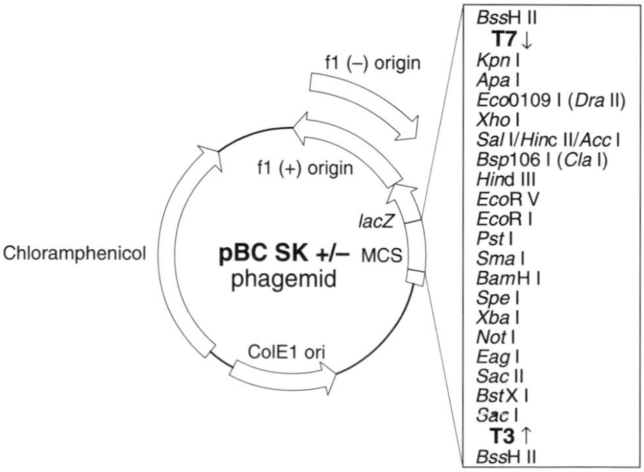

Type	Phagemid RNA synthesis
Size	2961 bp
Replicators	ColE1, f1
Markers	Ampicillin resistance *lacZ′*
Promoters	T3, T7
Cloning sites	*Acc*I, *Apa*I, *Bam*HI, *Bsp*106I, *Bss*HII, *Bst*XI, *Eag*I, *Eco*O109I, *Eco*RI, *Eco*RV, *Hinc*II, *Hin*dIII, *Kpn*I, *Not*I, *Pst*I, *Sac*I, *Sac*II, *Sal*I, *Sma*I, *Spe*I, *Xba*I, *Xho*I
Reference	9

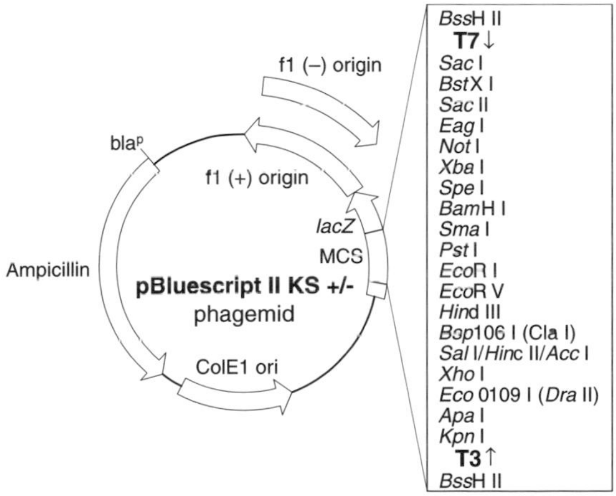

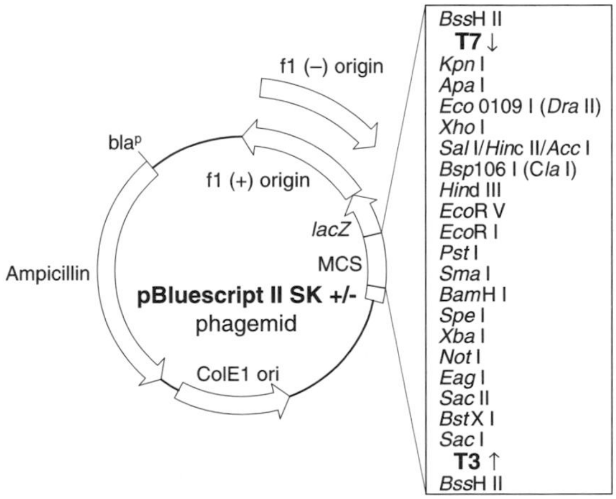

pBluescript SK

Marketed by Stratagene

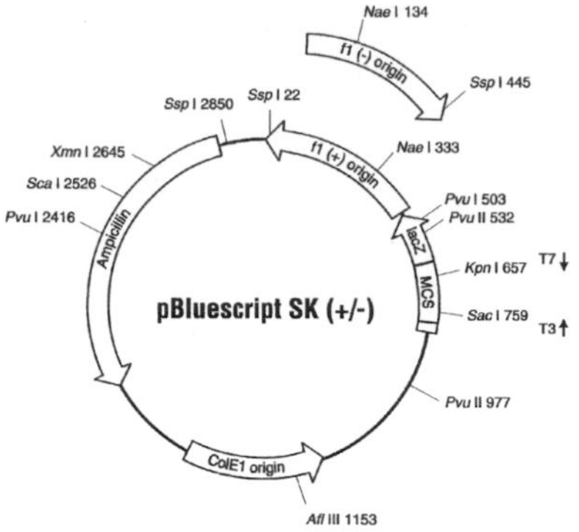

Type Phagemid
RNA synthesis

Size 2958 bp

Replicators ColE1, f1

Markers Ampicillin resistance
lacZ′

Promoters T3, T7

Cloning sites *Acc*I, *Apa*I, *Bam*HI, *Bsp*106I *Bst*XI, *Eag*I, *Eco*O109I, *Eco*RI, *Eco*RV, *Hinc*II, *Hin*dIII, *Kpn*I, *Not*I, *Pst*I, *Sac*I, *Sac*II, *Sal*I, *Sma*I, *Spe*I, *Xba*I, *Xho*I

pBR322

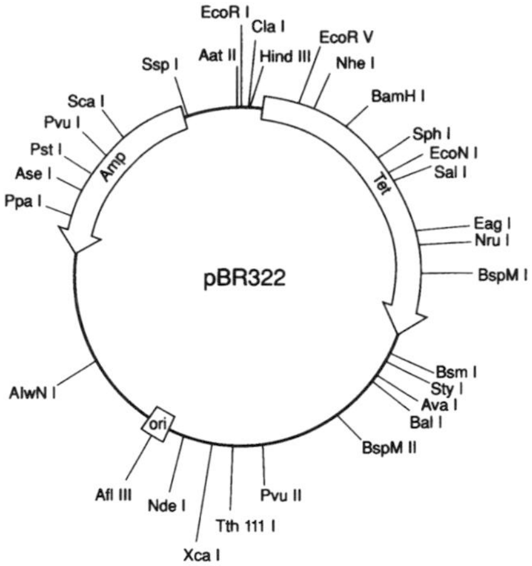

Type Basic cloning vector

Size 4363 bp

Replicator pMB1

Markers Ampicillin resistance
Tetracycline resistance

Cloning sites *Bam*HI, *Bsp*MI, *Eco*31I, *Eco*NI, *Eco*RV, *Nhe*I, *Nru*I, *Pst*I, *Pvu*I, *Sal*I, *Sca*I, *Sph*I, *Xma*III

References 10, 11

pBR325

Type Basic cloning vector

Size 5995 bp

Replicator pMB1

Markers Ampicillin resistance
Chloramphenicol resistance
Tetracycline resistance

Cloning sites *Bam*HI, *Eco*RI, *Eco*RV, *Nco*I, *Nru*I, *Pst*I, *Pvu*I, *Sal*I, *Sph*I, *Xma*III, *Xmn*I

Reference 11

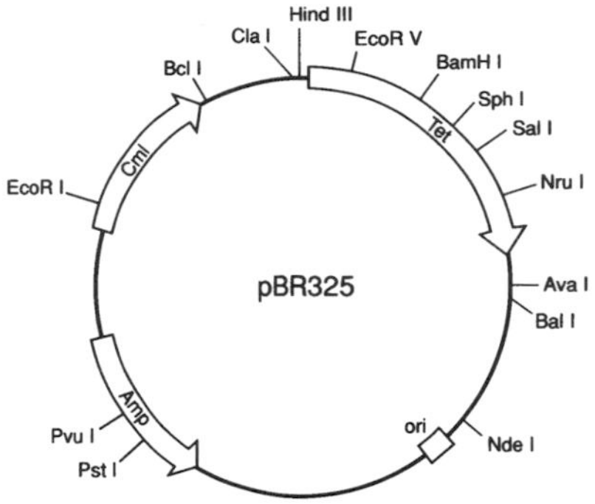

pBR327

Type Basic cloning vector

Size 3274 bp

Replicator pMB1

Markers Ampicillin resistance
Tetracycline resistance

Cloning sites *Acc*I, *Asp*700I, *Bam*HI, *Bsp*MI, *Dsa*I, *Eco*31I, *Eco*NI, *Eco*RV, *Ksp*632I, *Nhe*I, *Nru*I, *Nsp*HI, *Pst*I, *Pvu*I, *Sal*I, *Sca*I, *Sph*I, *Xma*III

Reference 12

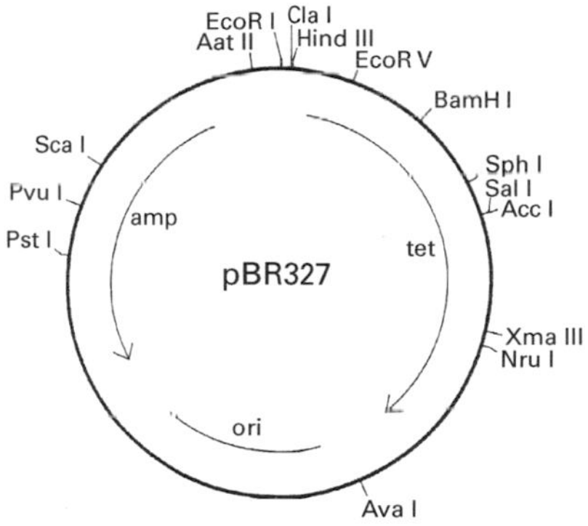

Cloning Vectors

pBR328

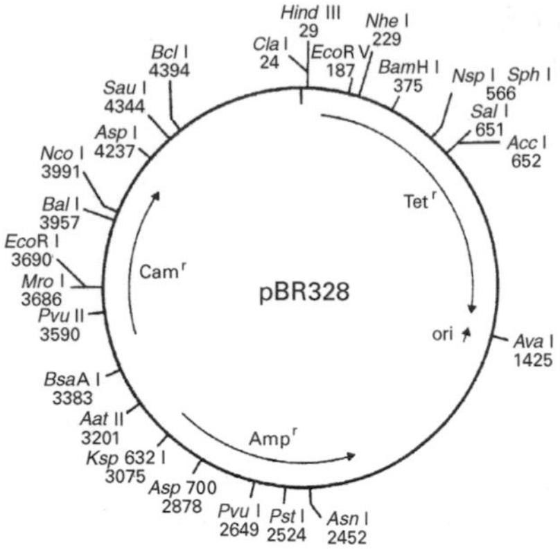

Type Basic cloning vector

Size 4907 bp

Replicator pMB1

Markers Ampicillin resistance
Chloramphenicol resistance
Tetracycline resistance

Cloning sites *Bam*HI, *Eco*RI, *Eco*RV, *Pst*I, *Pvu*I, *Sal*I, *Sph*I

pBS

Type RNA synthesis

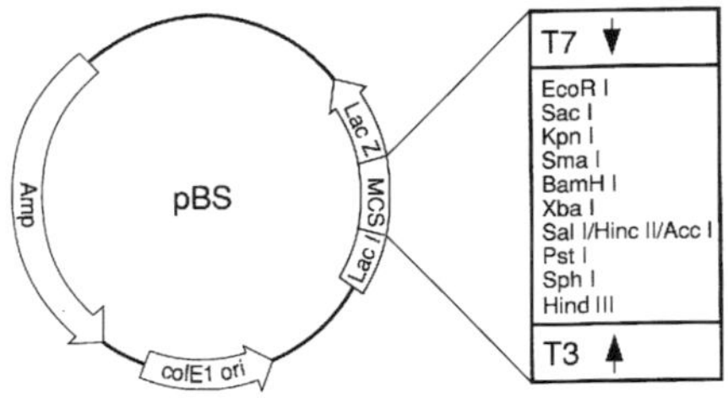

Size 2740 bp

Replicator ColE1

Markers Ampicillin resistance
lacZ′

Promoters T3, T7

Cloning sites *Acc*I, *Bam*HI, *Eco*RI, *Hinc*II, *Hind*III, *Kpn*I, *Pst*I, *Sac*I, *Sal*I, *Sma*I, *Sph*I, *Xba*I

References 5, 13, 14

pBS +/–

Marketed by Stratagene

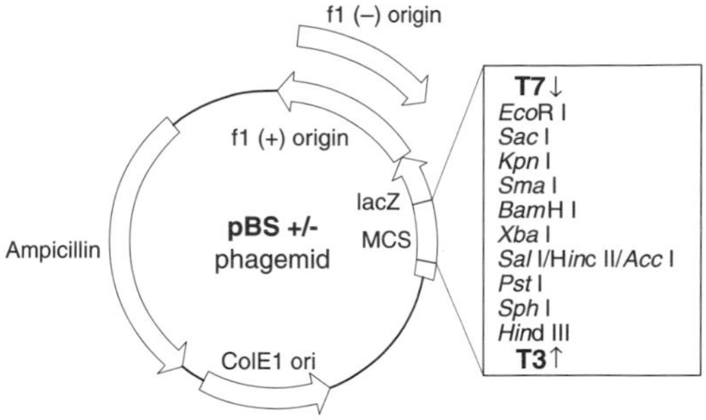

Type	Phagemid RNA synthesis
Size	3.2 kb
Replicators	ColE1, f1
Markers	Ampicillin resistance *lacZ′*
Promoters	T3, T7
Cloning sites	*Acc*I, *Bam*HI, *Eco*RI, *Hinc*II, *Hind*III, *Kpn*I, *Pst*I, *Sac*I, *Sal*I, *Sma*I, *Sph*I, *Xba*I

pCAL

Marketed by Stratagene

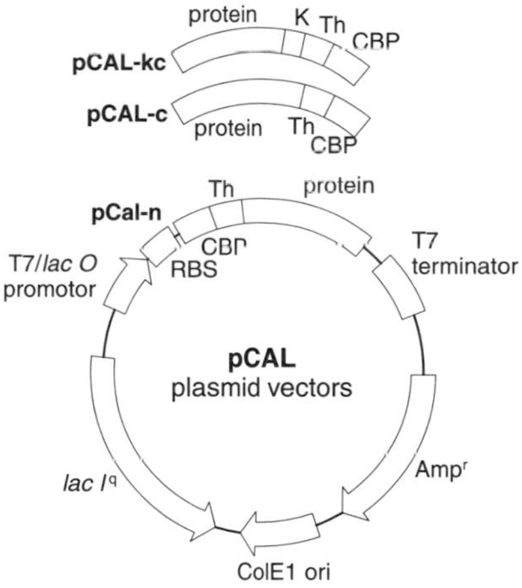

Type	Protein expression
Replicator	ColE1
Marker	Ampicillin resistance
Promoter	T7/*lacO*
Fusion protein	Calmodulin-binding peptide
Cloning sites	pCAL-n: *Bam*HI, *Eco*RI, *Hind*III, *Nco*I, *Sac*I, *Sal*I, *Sma*I, *Xba*I, *Xho*I pCAL-c: *Bam*HI, *Nco*I, *Nhe*I pCAL-kc: *Bam*HI, *Nco*I, *Nhe*I
Note	Abbreviations: CBP, calmodulin-binding peptide; Th, thrombin cleavage site; K, kemptide sequence.

pcos1EMBL

Type Cosmid

Size 6.1 kb

Replicator R6K

Markers Kanamycin resistance
Tetracycline resistance

Cloning sites *Bam*HI, *Sal*I, *Sma*I, *Xho*I

Reference 15

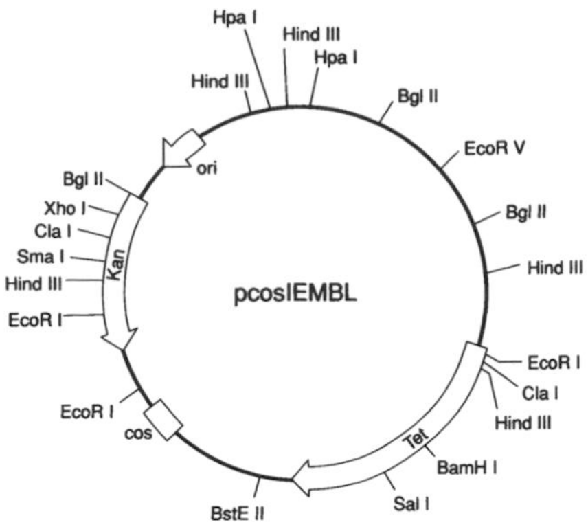

pCR-Script Amp

Marketed by Stratagene

Type Cloning vector for PCR products

Size 2961 bp

Replicators ColE1, f1

Markers Ampicillin resistance
lacZ′

Promoters T3, T7

Cloning site Vector is sold predigested with *Srf*I

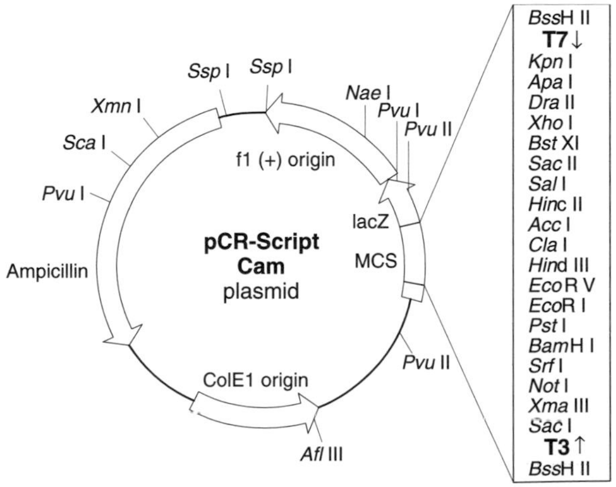

pCR-Script Cam

Marketed by Stratagene

Type	Cloning vector for PCR products
Size	3399 bp
Replicators	ColE1, f1
Markers	Chloramphenicol resistance *lacZ'*
Promoters	T3, T7
Cloning site	Vector is sold predigested with *Srf*I

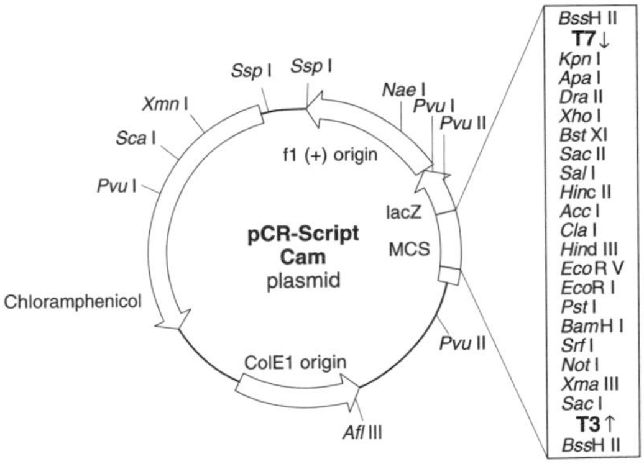

pCR-Script Direct

Marketed by Stratagene

Type	Cloning vector for PCR products
Size	2904 bp
Replicators	ColE1, f1
Markers	Ampicillin resistance *lacZ'*
Promoters	T3, T7
Cloning site	Vector is sold predigested with *Srf*I
Reference	16

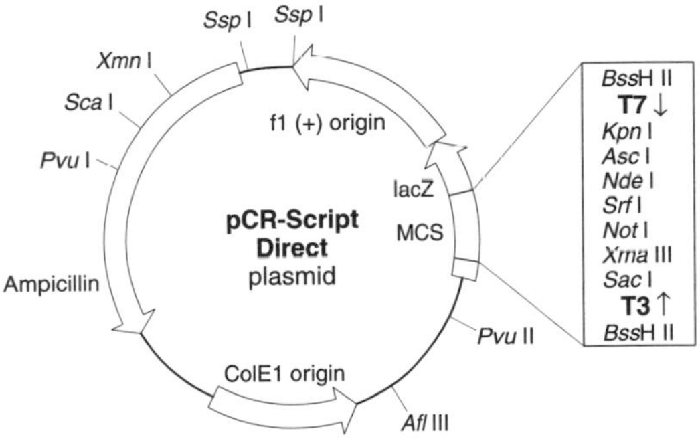

Cloning Vectors

pDIRECT

Marketed by Clontech

Type	Cloning vector for PCR products
Size	3.0 kb
Replicators	pMB1, f1
Markers	Ampicillin resistance *lacZ′*
Promoters	T3, T7
Cloning site	Vector is sold with 11 and 12 bp 5′ tails
References	17, 18

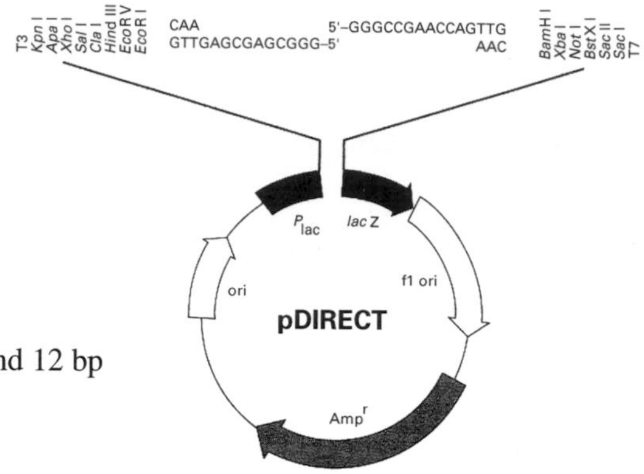

pEMBL8
pEMBL9

Type	Phagemid
Size	4.0 kb
Replicators	pMB1, f1
Markers	Ampicillin resistance *lacZ′*
Cloning sites	*Acc*I, *Bam*HI, *Eco*RI, *Hinc*II, *Hin*dIII, *Pst*I, *Sal*I, *Sma*I, *Xma*I
Reference	19

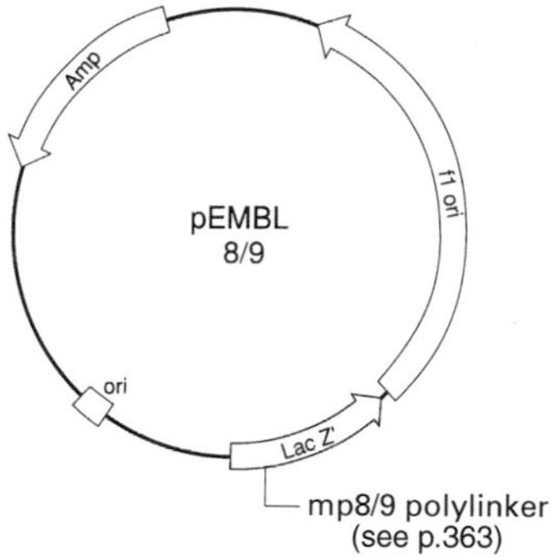

pET series

Type Protein expression

Sizes pET-5: 4134 bp
pET-9: 4341 bp

Replicator pMB1

Markers pET-3, pET-5, pET-11: ampicillin resistance
pET-9: kanamycin resistance

Promoters pET-3, pET-5, pET-9: T7
pET-11: T7/*lacO*

Fusion protein T7 gene 10 protein

Cloning sites Various of *Bam*HI, *Eco*RI, *Nco*I, *Nde*I, *Nhe*I

References 20–22

Note There are a variety of pET vectors with different markers, promoters, restriction sites, and positions of the restriction sites relative to the reading frame. See refs 20–22 for details.

pEZZ 18

Marketed by Pharmacia Biotech

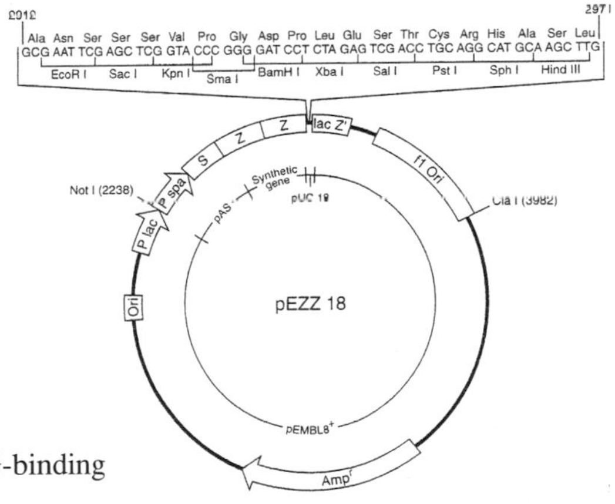

Type Protein expression

Size 4591 bp

Replicators pMB1, f1

Markers Ampicillin resistance
lacZ′

Promoters *lacUV5*
Protein A

Fusion protein Synthetic, based on IgG-binding domain of Protein A

Cloning sites *Bam*HI, *Eco*RI, *Hin*dIII, *Kpn*I, *Pst*I, *Sac*I, *Sal*I, *Sma*I, *Sph*I, *Xba*I

References 23–25

Cloning Vectors

pGB2

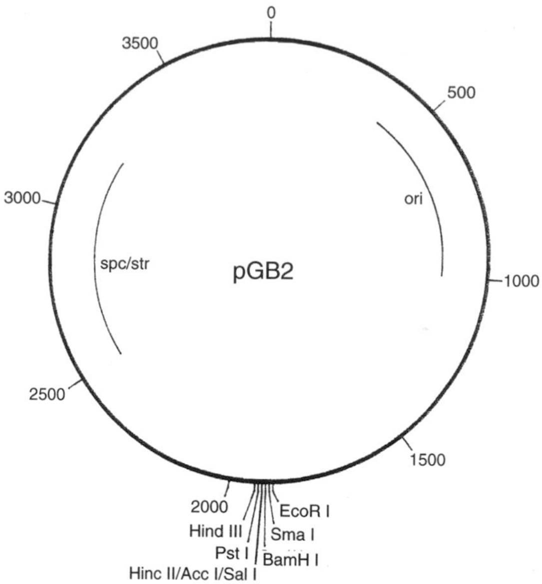

Type Basic cloning vector

Size 3.8 kb

Replicator pSC101

Markers Spectinomycin resistance
Streptomycin resistance

Cloning sites *Acc*I, *Bam*HI, *Eco*RI, *Hinc*II, *Hind*III, *Pst*I, *Sal*I, *Sma*I

Reference 26

pGEM-3Z

Marketed by Promega Corporation

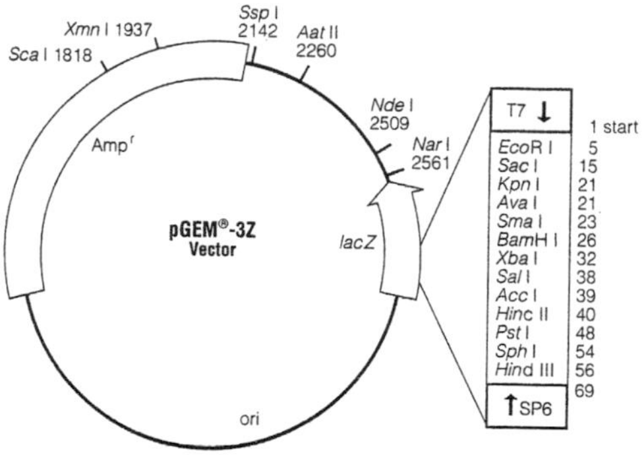

Type RNA synthesis

Size 2743 bp

Replicator pMB1

Markers Ampicillin resistance
lacZ′

Promoters SP6, T7

Cloning sites *Acc*I, *Ava*I, *Bam*HI, *Eco*RI, *Hinc*II, *Hind*III, *Kpn*I, *Pst*I, *Sac*I, *Sal*I, *Sma*I, *Sph*I, *Xba*I

pGEM-3Zf(+/–)

Marketed by Promega Corporation

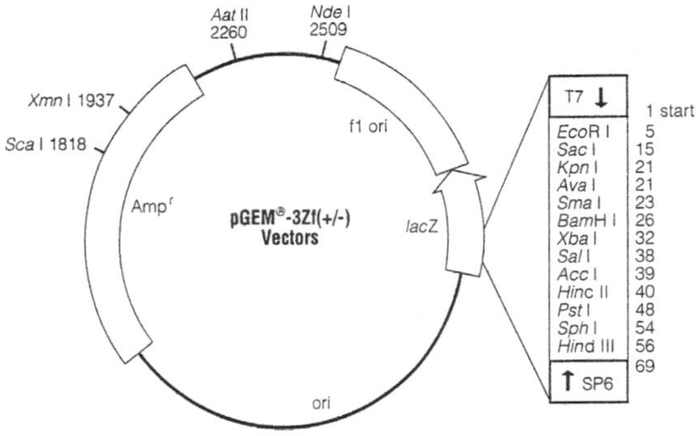

Type Phagemid
RNA synthesis

Size 3199 bp

Replicators pMB1, f1

Markers Ampicillin resistance
lacZ′

Promoters SP6, T7

Cloning sites *Acc*I, *Ava*I, *Bam*HI, *Eco*RI, *Hinc*II, *Hind*III, *Kpn*I, *Pst*I, *Sac*I, *Sal*I, *Sma*I, *Sph*I, *Xba*I

pGEM-4Z

Marketed by Promega Corporation

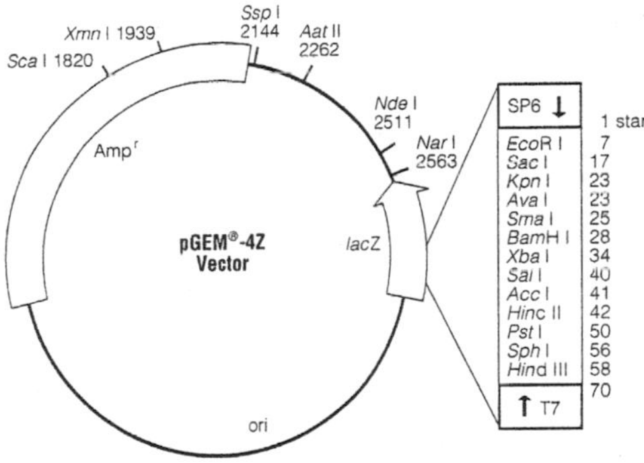

Type RNA synthesis

Size 2746 bp

Replicator pMB1

Markers Ampicillin resistance
lacZ′

Promoters SP6, T7

Cloning sites *Acc*I, *Ava*I, *Bam*HI, *Eco*RI, *Hinc*II, *Hind*III, *Kpn*I, *Pst*I, *Sac*I, *Sal*I, *Sma*I, *Sph*I, *Xba*I

pGEM-5Zf(+/–)

Marketed by Promega Corporation

Type Phagemid
RNA synthesis

Size 3003 bp

Replicators pMB1, f1

Markers Ampicillin resistance
lacZ′

Promoters SP6, T7

Cloning sites *Aat*II, *Apa*I, *Bst*XI, *Eco*RV, *Nco*I, *Nde*I, *Not*I, *Nsi*I, *Pst*I, *Sac*I, *Sac*II, *Sal*I, *Spe*I, *Sph*I

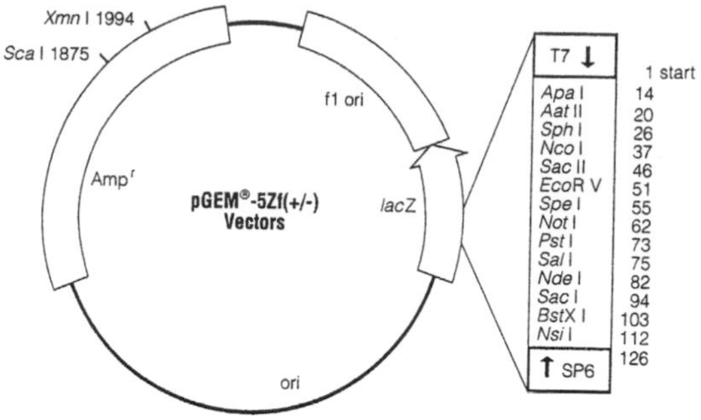

pGEM-7Zf(+/–)

Marketed by Promega Corporation

Type Phagemid
RNA synthesis

Size 3000 bp

Replicators pMB1, f1

Markers Ampicillin resistance
lacZ′

Promoters SP6, T7

Cloning sites *Aat*II, *Apa*I, *Bam*HI, *Bst*XI, *Cla*I, *Csp*45I, *Eco*RI, *Hind*III, *Kpn*I, *Nsi*I, *Sac*I, *Sma*I, *Sph*I, *Xba*I, *Xho*I

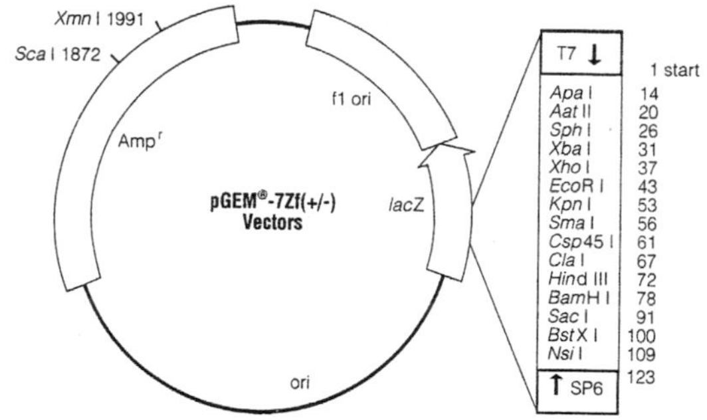

pGEM-9Zf(–)

Marketed by Promega Corporation

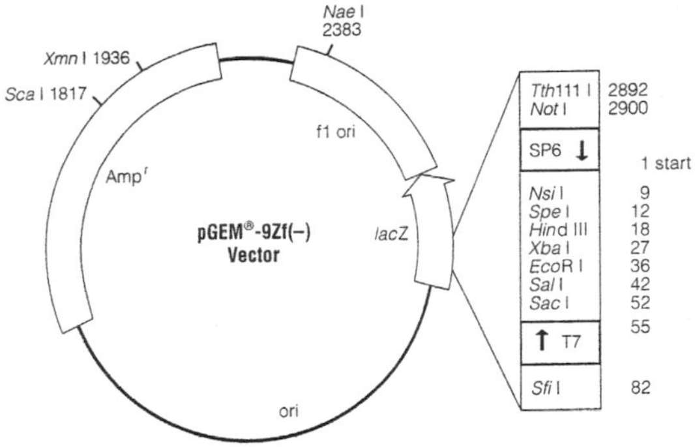

Type	Phagemid RNA synthesis
Size	2925 bp
Replicators	pMB1, f1
Markers	Ampicillin resistance *lacZ′*
Promoters	SP6, T7
Cloning sites	*Eco*RI, *Hin*dIII, *Not*I, *Nsi*I, *Sac*I, *Sal*I, *Sfi*I, *Spe*I, *Tth*111I, *Xba*I

pGEM-11Zf(+/–)

Marketed by Promega Corporation

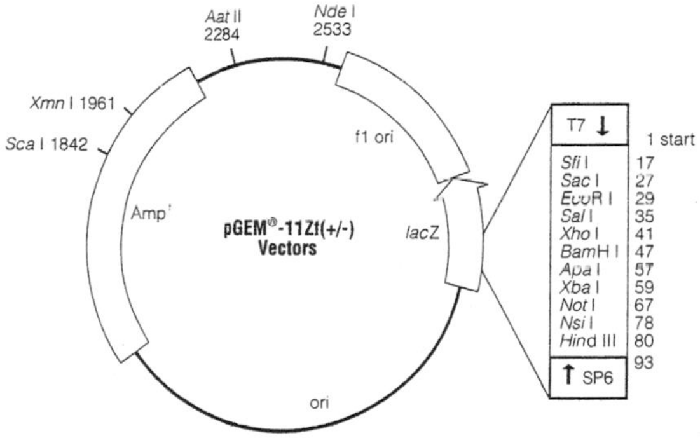

Type	Phagemid RNA synthesis
Size	3223 bp
Replicators	pMB1, f1
Markers	Ampicillin resistance *lacZ′*
Promoters	SP6, T7
Cloning sites	*Apa*I, *Bam*HI, *Eco*RI, *Hin*dIII, *Not*I, *Nsi*I, *Sac*I, *Sal*I, *Sfi*I, *Xba*I, *Xho*I

pGEM-13Zf(+)

Marketed by Promega Corporation

Type	Phagemid RNA synthesis
Size	3181 bp
Replicators	pMB1, f1
Markers	Ampicillin resistance *lacZ′*
Promoters	SP6, T7
Cloning sites	*Hin*dIII, *Not*I, *Sfi*I

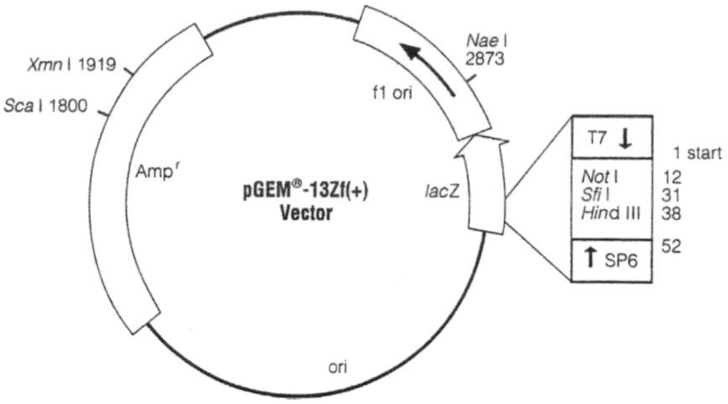

pGEMEX-1
pGEMEX-2

Marketed by Promega Corporation

Type	Phagemid RNA synthesis Protein expression
Sizes	pGEMEX-1: 3995 bp pGEMEX-2: 3997 bp
Replicators	pMB1, f1
Markers	Ampicillin resistance
Promoters	SP6, T3, T7
Fusion protein	T7 gene 10 protein
Cloning sites	*Apa*I, *Bam*HI, *Eco*RI, *Hin*dIII, *Not*I, *Nsi*I, *Sac*I, *Sal*I, *Sfi*I, *Sph*I, *Xho*I
Note	The only difference between the two vectors is the position of the cloning sites relative to the reading frame.

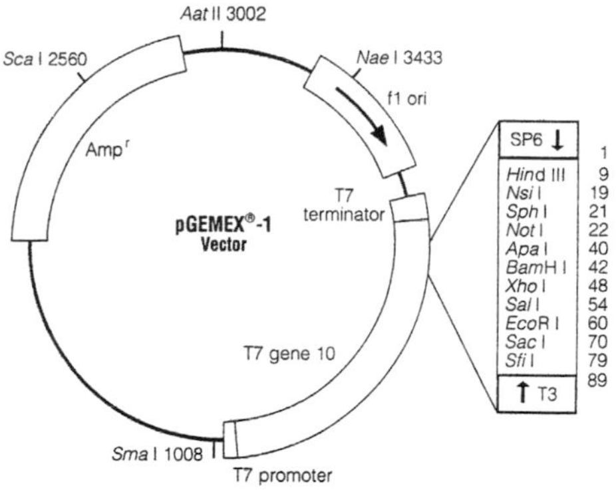

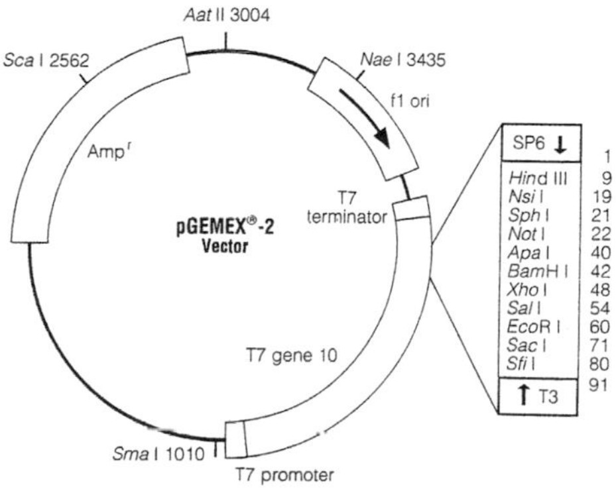

pGEM-T

Marketed by Promega Corporation

Type	Cloning vector for PCR products
Size	3003 bp
Replicators	pMB1, f1
Markers	Ampicillin resistance *lacZ′*
Promoters	SP6, T7
Cloning site	3′ T overhangs

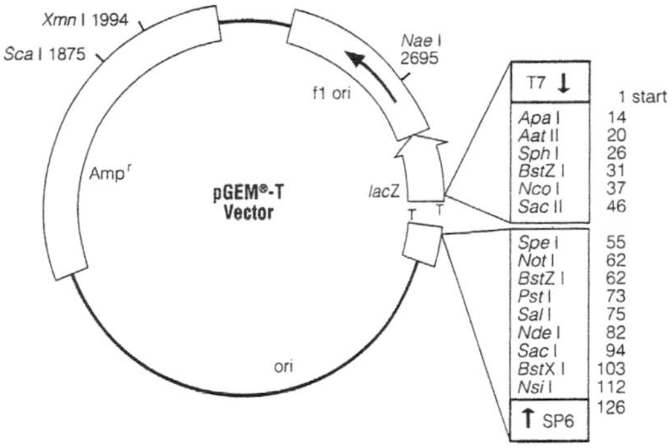

pGEM-T Easy

Marketed by Promega Corporation

Type	Cloning vector for PCR products
Size	3018 bp
Replicators	pMB1, f1
Markers	Ampicillin resistance *lacZ′*
Promoters	SP6, T7
Cloning site	3′ T overhangs

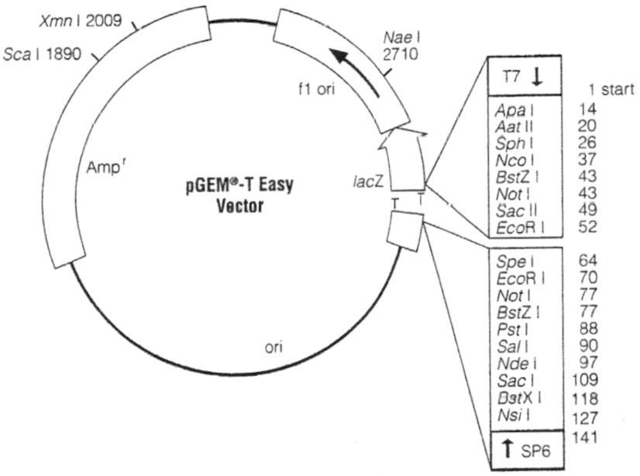

pGEX series

Marketed by Pharmacia Biotech

Type	Protein expression
Size	~4950 bp
Replicator	pMB1
Marker	Ampicillin resistance
Promoter	*tac*
Fusion protein	Glutathione *S*-transferase
Cloning sites	pGEX-1λT: *Bam*HI, *Eco*RI pGEX-2T, 2TK, 3X: *Bam*HI, *Eco*RI, *Sma*I pGEX-4T-1, 4T-2, 4T-3, 5X-1, 5X-2, 5X-3: *Bam*HI, *Eco*RI, *Not*I, *Sal*I, *Sma*I, *Xho*I
References	27, 28

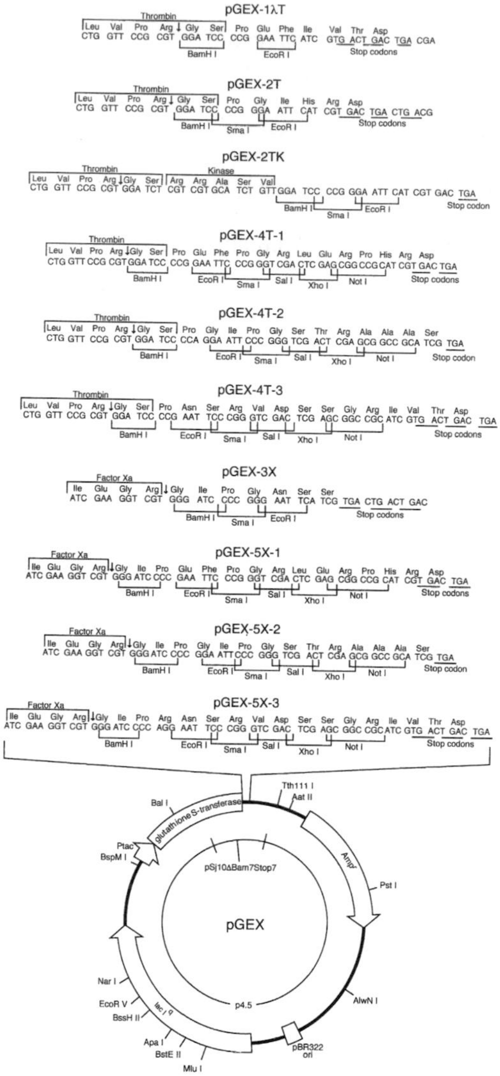

Type Phagemid
RNA synthesis

Size 2967 bp

Replicators pMB1, f1

Markers Ampicillin resistance
lacZ′

Promoters T3, T7

Cloning sites *Acc*I, *Bam*HI, *Bss*HII, *Eco*RI, *Hinc*II, *Hind*III, *Kpn*I, *Not*I, *Pst*I, *Sac*I, *Sal*I, *Sma*I, *Sph*I, *Swa*I, *Xba*I, *Xma*I

Note The 'triple helix-forming site' within the polylinker is used in vector DNA purification. The system uses a biotinylated capture probe that specifically hybridizes to this sequence.

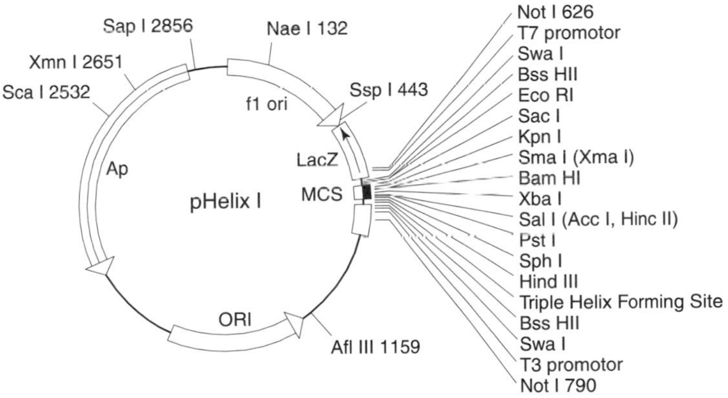

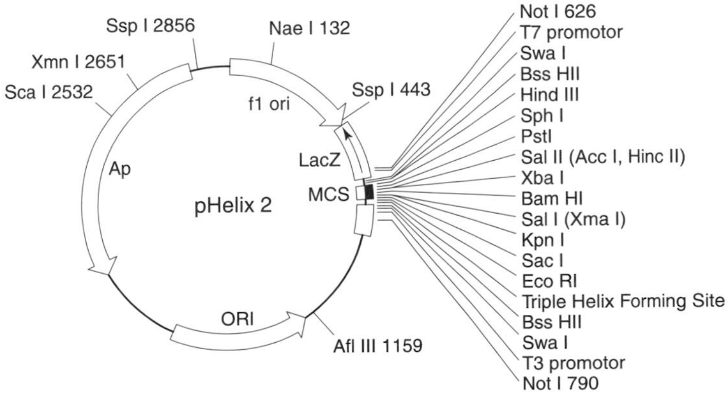

PinPoint Xa-1

Marketed by Promega Corporation

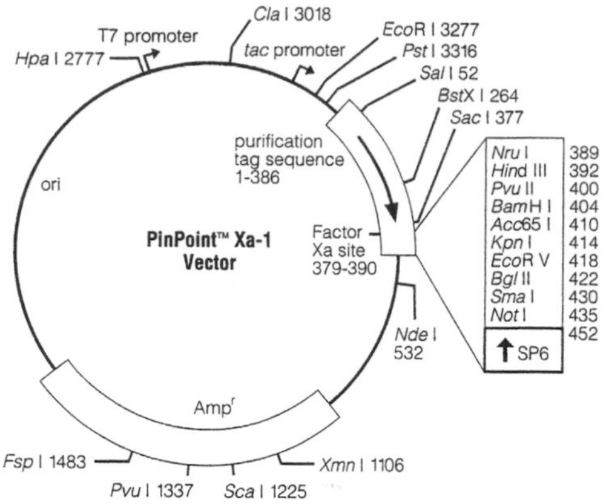

Type Protein expression

Size 3351 bp

Replicator pMB1

Marker Ampicillin resistance

Promoters SP6, T7, *tac*

Fusion protein Biotinylated peptide

Cloning sites *Acc*65I, *Bam*HI, *Bgl*II, *Eco*RV, *Hin*dIII, *Kpn*I, *Not*I, *Nru*I, *Pvu*II, *Sma*I

Notes
1. PinPoint Xa-2 is identical to PinPoint Xa-1 except for one additional adenosine at position 394.
2. PinPoint Xa-3 is identical to PinPoint Xa-1 except for two additional adenosines at position 394.

PinPoint Xa-1 T

Marketed by Promega Corporation

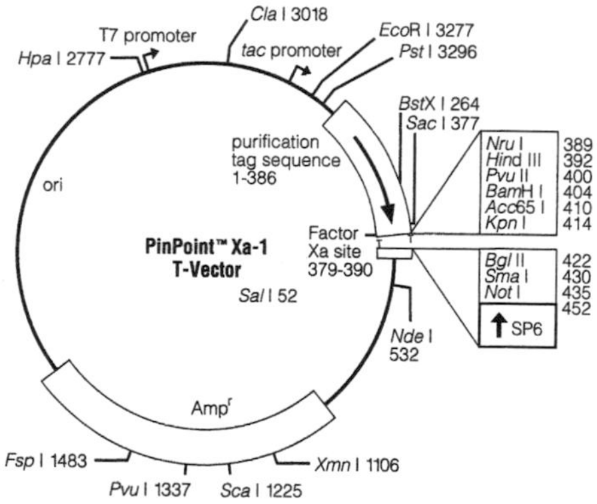

Type Cloning vector for PCR products
Protein expression

Size 3331 bp

Replicator pMB1

Marker Ampicillin resistance

Promoters SP6, T7, *tac*

Fusion protein Biotinylated peptide

Cloning site 3′ T overhangs

pJB8

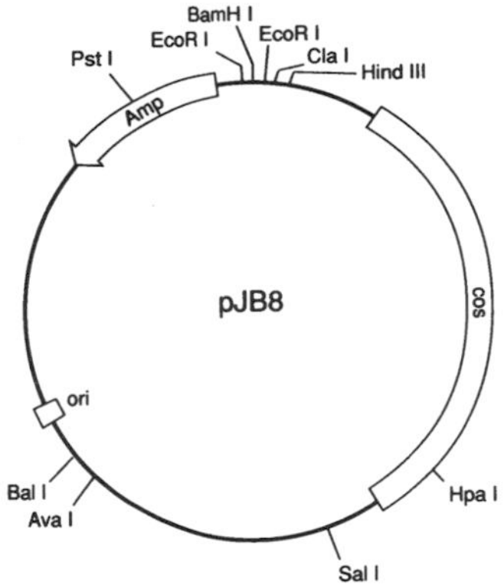

Type	Cosmid
Size	5.4 kb
Replicator	pMB1
Marker	Ampicillin resistance
Cloning sites	*Bam*HI, *Cla*I, *Eco*RI, *Hin*dIII, *Sal*I
Reference	29

pKK175-6

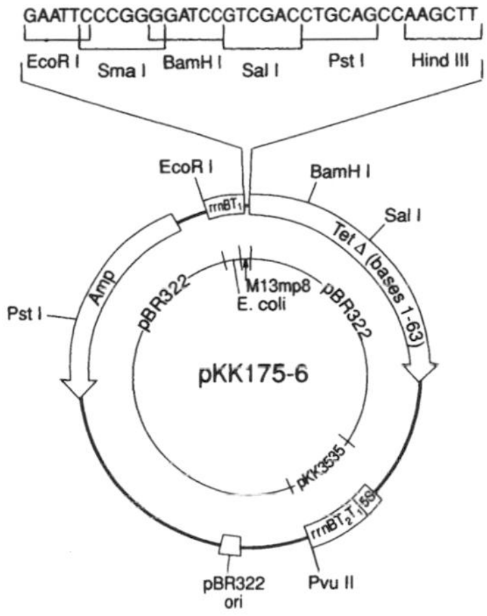

Type	Promoter probe
Size	4.4 kb
Replicator	pMB1
Markers	Ampicillin resistance Tetracycline resistance
Cloning sites	*Acc*I, *Hinc*II, *Hin*dIII, *Sma*I, *Xma*I
References	30, 31

pKK223-3

Marketed by Pharmacia Biotech

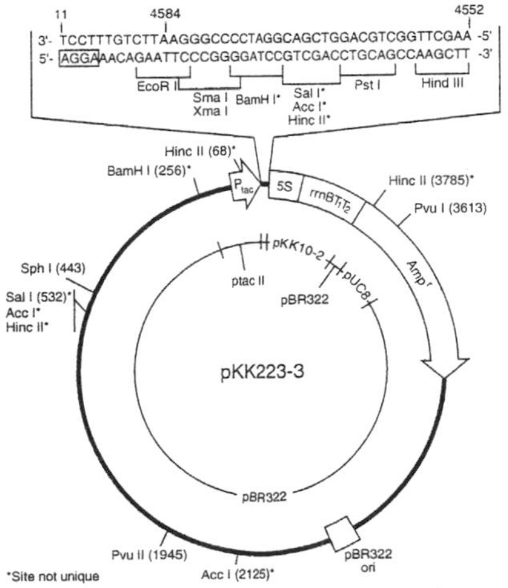

Type	Protein expression
Size	4584 bp
Replicator	pMB1
Marker	Ampicillin resistance
Promoter	*tac*
Fusion protein	None
Cloning sites	*Eco*RI, *Hin*dIII, *Pst*I, *Sma*I, *Xma*I
References	30, 32–34

pKK232-8

Marketed by Pharmacia Biotech

Type	Promoter probe
Size	5094 bp
Replicator	pMB1
Markers	Ampicillin resistance Chloramphenicol resistance
Cloning sites	*Acc*I, *Bam*HI, *Hinc*II, *Hind*III, *Sal*I, *Sma*I, *Xma*I
Reference	32

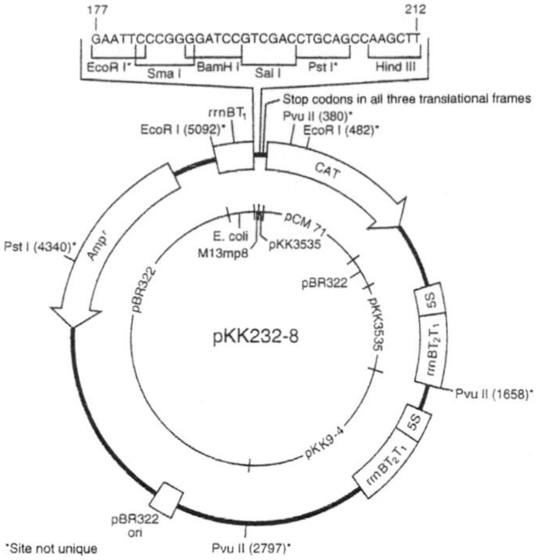

pKK233-2

Marketed by Clontech

Type	Protein expression
Size	4593 bp
Replicator	pMB1
Marker	Ampicillin resistance
Promoter	*trc*
Fusion protein	None
Cloning sites	*Hind*III, *Nco*I, *Pst*I
References	35–37

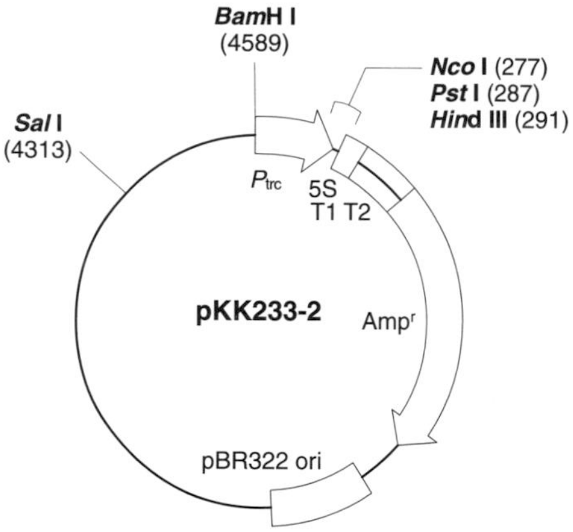

Type	Protein expression
Size	5.1 kb
Replicator	pMB1
Marker	Ampicillin resistance Tetracycline resistance
Promoter	*trc*
Fusion protein	None
Cloning sites	*Eco*RI, *Kpn*I, *Nco*I, *Pst*I, *Sac*I, *Sma*I, *Xba*I
Reference	38

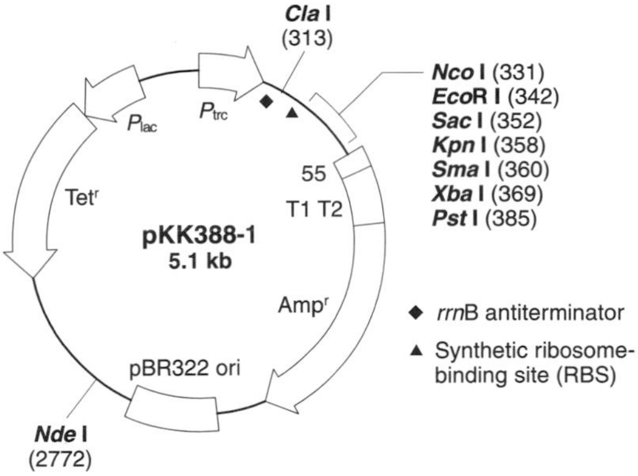

pKT279
pKT280
pKT287

Type	Protein expression
Size	4.3 kb
Replicator	pMB1
Marker	Tetracycline resistance
Promoter	ß-Lactamase
Fusion protein	ß-Lactamase signal peptide
Cloning site	*Pst*I
References	39, 40
Note	The three vectors differ in the position of the cloning site relative to the reading frame.

pKT 279
ATG AGT ATT CAA CAT TTC CGT GTC GCC CTT
ATT CCC TTT TTT GCG GCA TTT TGC CTT
CCT GTT TTT GCT CAC CG**C TGC AG**
pKT 280
ATG AGT ATT CAA CAT TTC CGT GTC GCC CTT
ATT CCC TTT TTT GCG GCA TTT TGC CTT
CCT GTT TTT GCT **CAC CCG CTG CAG**
pKT 287
ATG AGT ATT CAA CAT TTC CGT GTC GCC CTT
ATT CCC TTT TTT GCG GCA TTT TGC CTT
CCT GTT TTT GCT **CAC CCA GAA ACG GCT**
GCA G

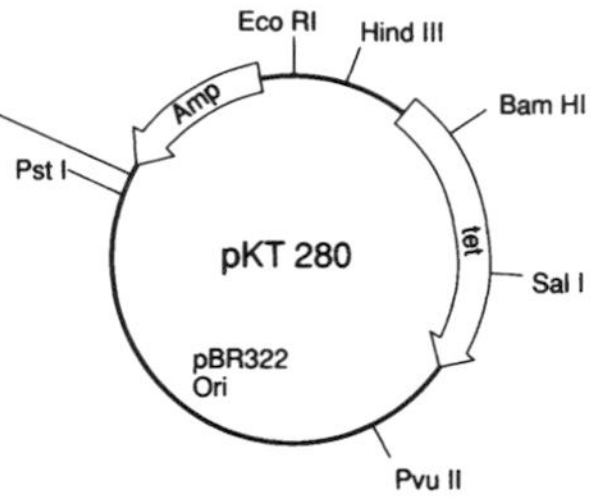

Type Phagemid
RNA synthesis

Size 2.8 kb

Replicators pMB1, M13

Markers Ampicillin resistance
lacZ′

Promoter T7

Cloning sites pLITMUS 28, pLITMUS 29: *Aat*II, *Acc*65I, *Afl*II, *Age*I, *Avr*II, *Bam*HI, *Bgl*II, *Bsi*WI, *Bss*HII, *Eco*RI, *Eco*RV, *Hind*III, *Kpn*I, *Nco*I, *Nsi*I, *Ppu*10I, *Pst*I, *Sac*I, *Spe*I, *Stu*I, *Xba*I, *Xho*I
pLITMUS 38, pLITMUS 39: *Afl*II, *Apa*I, *Bam*HI, *Bsp*120I, *Bsp*EI, *Bsr*GI, *Eag*I, *Eco*RI, *Eco*RV, *Hind*III, *Kas*I, *Mfe*I, *Mlu*I, *Nhe*I, *Ngo*MI, *Pst*I, *Sal*I, *Spe*I, *Sph*I, *Stu*I

Reference 41

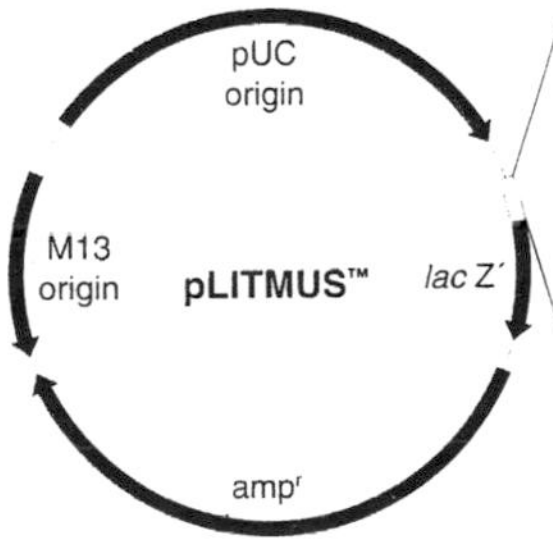

LITMUS 28	LITMUS 29	LITMUS 38	LITMUS 39
Spe I **T7↓**	*Spe* I **T7↓**	*Spe* I **T7↓**	*Spe* I **T7↓**
Bgl II	*Kpn* I-*Acc65* I	*Apa* I-*Bsp*120 I	*Sal* I
Nsi I-*Ppu*10 I	*Sac* I	*Mfe* I	*Sph* I
*Bss*H II	*Avr* II	*Ngo*M I	*Bsr*G I
*Bsi*W I	*Xba* I	*Kas* I	*Bsp*E I
Xho I	*Age* I	*Hind* III	*Mlu* I
*Eco*R I	*Aat* II	*Pst* I	*Eag* I
Pst I	*Nco* I	*Eco*R V	*Nhe* I
*Eco*R V	*Hind* III	*Bam*H I	*Eco*R I
*Bam*H I	*Bam*H I	*Eco*R I	*Bam*H I
Hind III	*Eco*R V	*Nhe* I	*Eco*R V
Nco I	*Pst* I	*Eag* I	*Pst* I
Aat II	*Eco*R I	*Mlu* I	*Hind* III
Age I	*Xho* I	*Bsp*E I	*Kas* I
Xba I	*Bsi*W I	*Bsr*G I	*Ngo*M I
Avr II	*Bss*H II	*Sph* I	*Mfe* I
Sac I	*Nsi* I-*Ppu*10 I	*Sal* I	*Apa* I-*Bsp*120 I
Kpn I-*Acc65* I	*Bgl* II	*Stu* I	*Stu* I
Stu I	*Stu* I		
Afl II **T7↑**	*Afl* II **T7↑**	*Afl* II **T7↑**	*Afl* II **T7↑**

Cloning Vectors

pMAL-c2
pMAL-p2

Marketed by New England Biolabs

Type	Protein expression
Size	pMAL-c2: 6647 bp pMAL-p2: 6721 bp
Replicators	pMB1, M13
Markers	Ampicillin resistance *lacZ′*
Promoter	*tac*
Fusion protein	Maltose-binding protein
Cloning sites	*Bam*HI, *Eco*RI, *Hind*III, *Pst*I, *Sal*I, *Xba*I, *Xmn*I
Reference	42
Note	The only difference between the two vectors is that pMAL-c2 lacks the *malE* signal sequence.

pMAL-c2, p2 polylinker

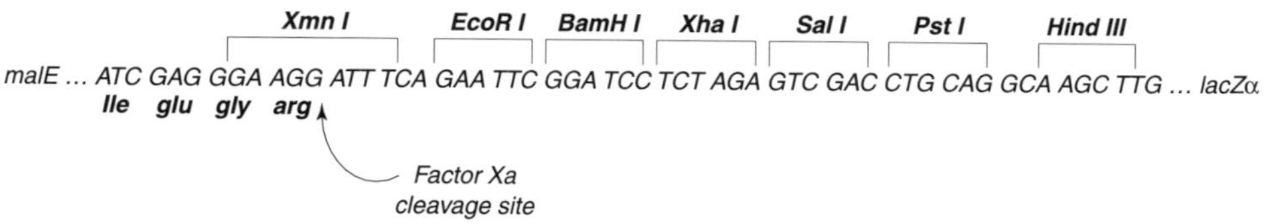

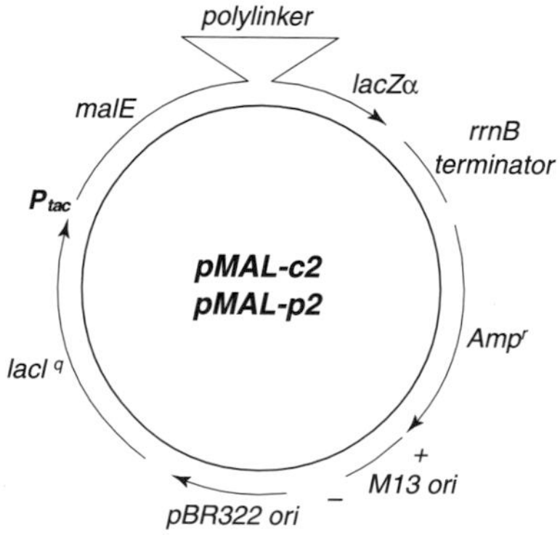

pMC1871

Marketed by Pharmacia Biotech

Type	Protein expression
Size	7476 bp
Replicator	pMB1
Marker	Tetracycline resistance
Promoter	None
Fusion protein	ß-Galactosidase
Cloning site	*Sma*I
References	43, 44

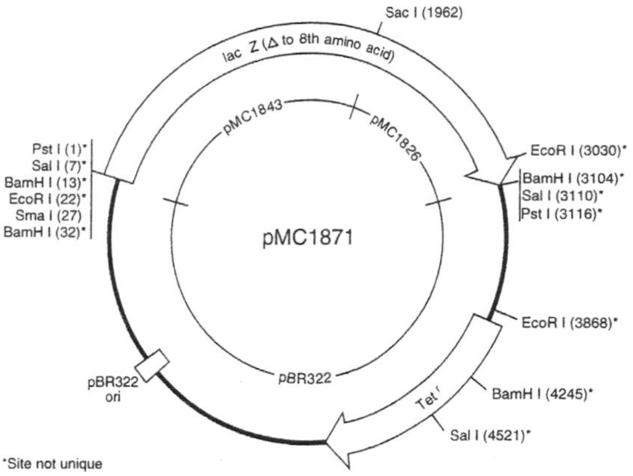

pMOS*Blue*

Marketed by Amersham International

Type products	Cloning vector for PCR
Size	2887 bp
Replicators	pMB1, f1
Markers	Ampicillin resistance *lacZ′*
Cloning site	3′ T overhangs

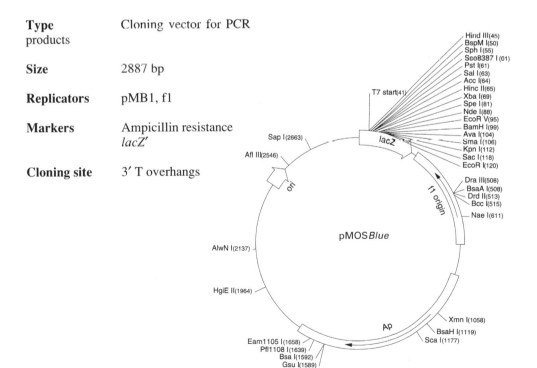

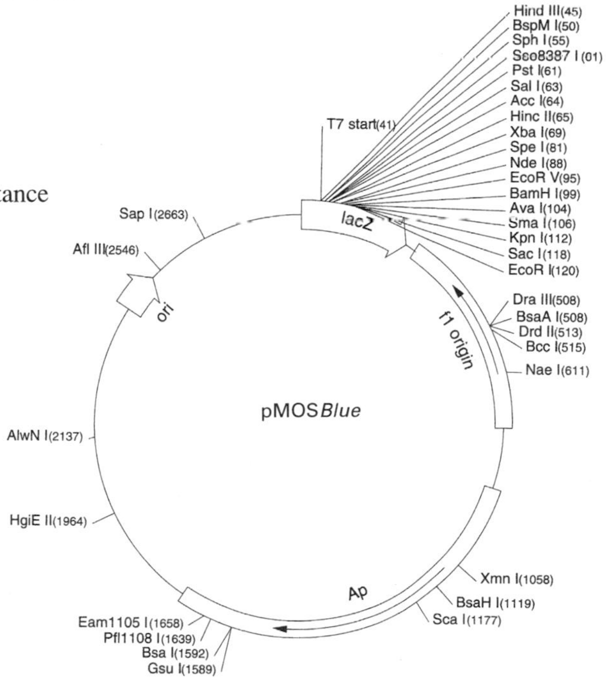

Cloning Vectors

Type	Basic cloning vector
Size	2713 bp
Replicator	pMB1
Marker	Ampicillin resistance
Cloning sites	*Acc*I, *Acc*65I, *Apo*I, *Asc*I, *Ava*I, *Bam*HI, *Ban*I, *Bsp*MI, *Bss*HII, *Ecl*136II, *Eco*RI, *Hinc*II, *Hind*III, *Kpn*I, *Pac*I, *Pme*I, *Pst*I, *Sac*I, *Sal*I, *Sma*I, *Sph*I, *Sse*8387I, *Xba*I, *Xma*I

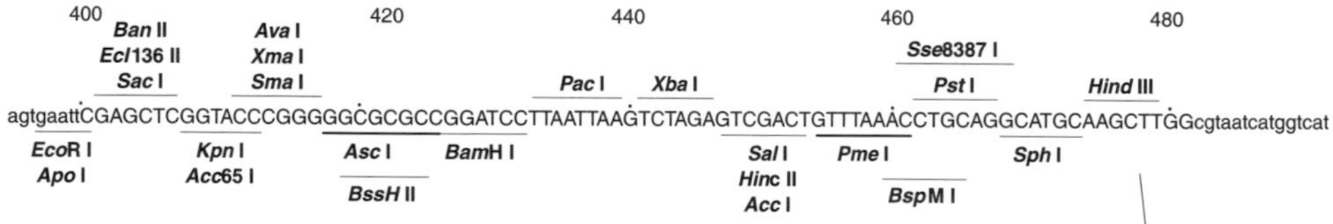

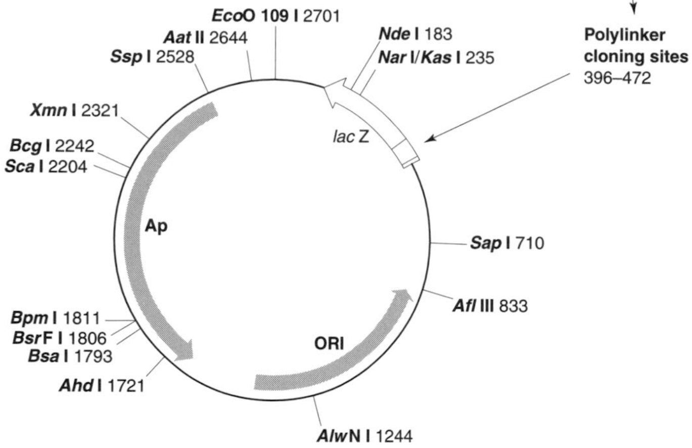

pNO1523

Marketed by New England Biolabs

Type	Basic cloning vector
Size	5.2 kb
Replicator	pMB1
Markers	Ampicillin resistance Streptomycin sensitivity
Cloning sites	*Hpa*I, *Sma*I
Reference	45
Note	The vector must be used in a *rpsL* host so that insertional inactivation of the streptomycin sensitivity gene leads to streptomycin resistance.

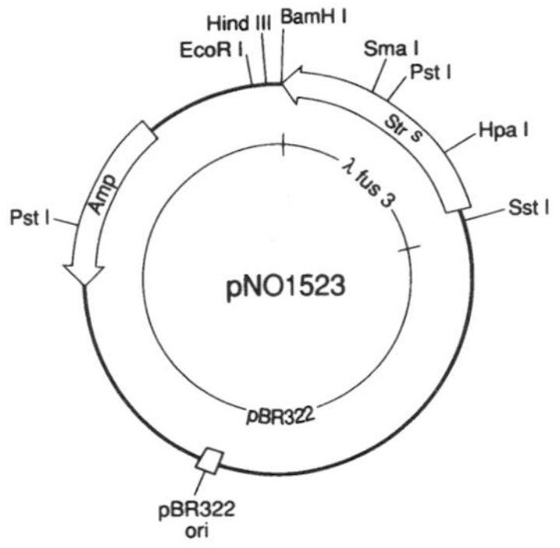

pRIT2T

Marketed by Pharmacia Biotech

Type	Protein expression
Size	4229 bp
Replicator	pMB1
Marker	Ampicillin resistance
Promoter	λP_R
Fusion protein	Protein A
Cloning sites	*Bam*HI, *Eco*RI, *Pst*I, *Sal*I, *Sma*I, *Xcy*I
Reference	46

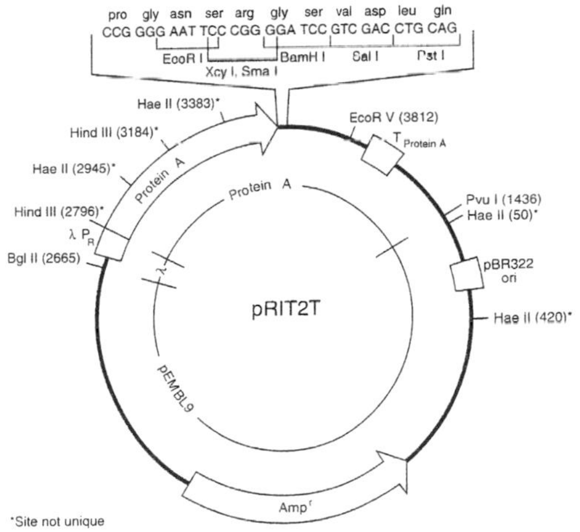

*Site not unique

pRIT5

Type Protein expression

Size 6.9 kb

Replicators pMB1, *Staphylococcus aureus* ori

Markers Ampicillin resistance
Chloramphenicol resistance

Promoter Protein A

Fusion protein Protein A

Cloning sites *Bam*HI, *Eco*RI, *Hinc*II, *Pst*I, *Sal*I, *Sma*I, *Xcy*I

Reference 46

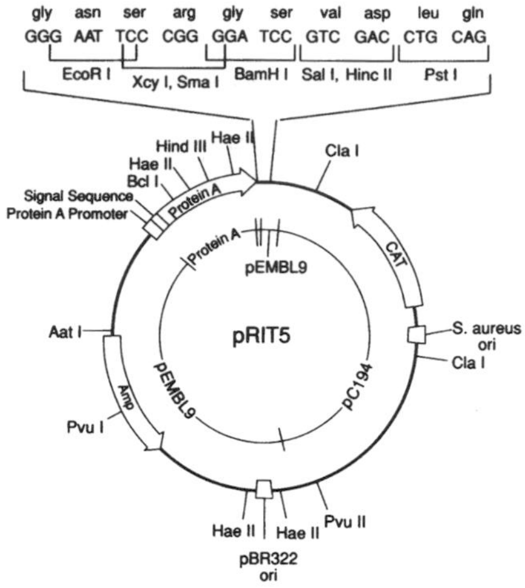

pSELECT-1

Marketed by Promega Corporation

Type Phagemid
RNA synthesis

Size 3422 bp

Replicators pMB1, M13

Markers Tetracycline resistance
lacZ'

Promoters SP6, T7

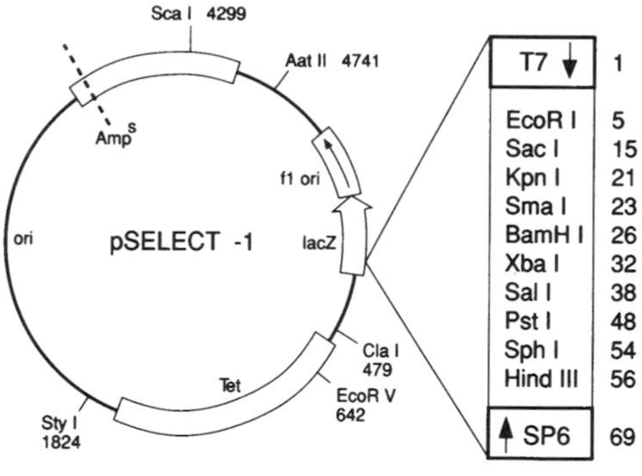

Cloning sites *Bam*HI, *Eco*RI, *Hind*III, *Kpn*I, *Pst*I, *Sac*I, *Sal*I, *Sma*I, *Sph*I, *Xba*I

pSL1180

Marketed by Pharmacia Biotech

Type	Phagemid
Size	3422 bp
Replicators	pMB1, M13
Marker	Ampicillin resistance
Cloning sites	Contains all 64 possible hexamer sequences
Reference	47

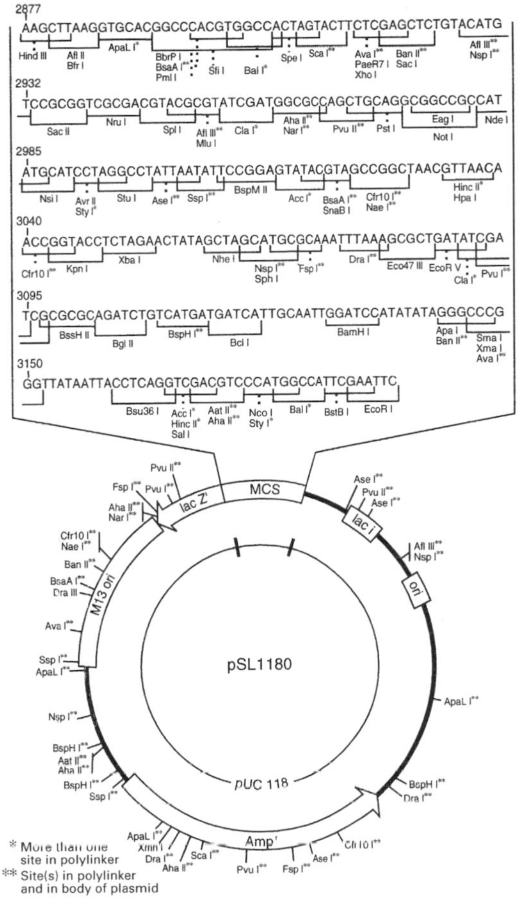

pSP64

Marketed by Promega Corporation

Type	RNA synthesis
Size	2999 bp
Replicator	pMB1
Marker	Ampicillin resistance
Promoter	SP6

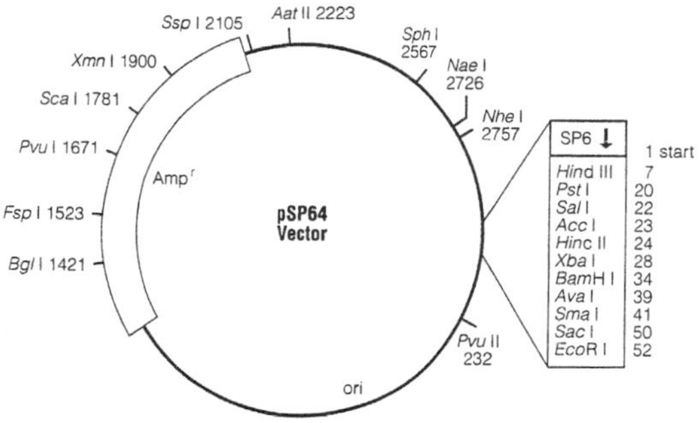

Cloning sites *Acc*I, *Ava*I, *Bam*HI, *Eco*RI, *Hinc*II, *Hind*III, *Pst*I, *Sac*I, *Sal*I, *Sma*I, *Xba*I

pSP64 Poly(A)

Marketed by Promega Corporation

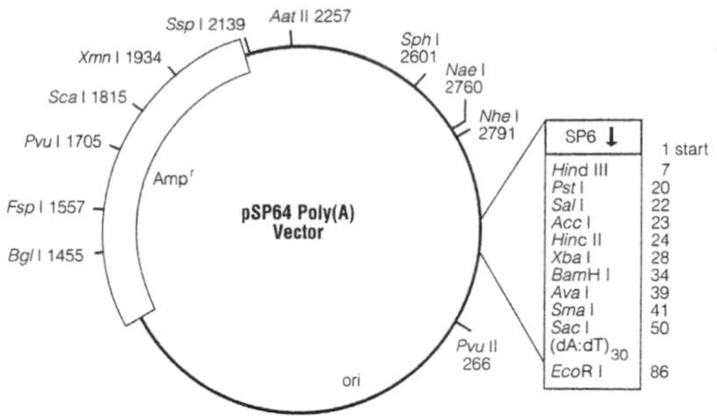

Type	RNA synthesis
Size	3033 bp
Replicator	pMB1
Marker	Ampicillin resistance
Promoter	SP6
Cloning sites	*Acc*I, *Ava*I, *Bam*HI, *Eco*RI, *Hinc*II, *Hin*dIII, *Pst*I, *Sac*I, *Sal*I, *Sma*I, *Xba*I

pSP65

Marketed by Promega Corporation

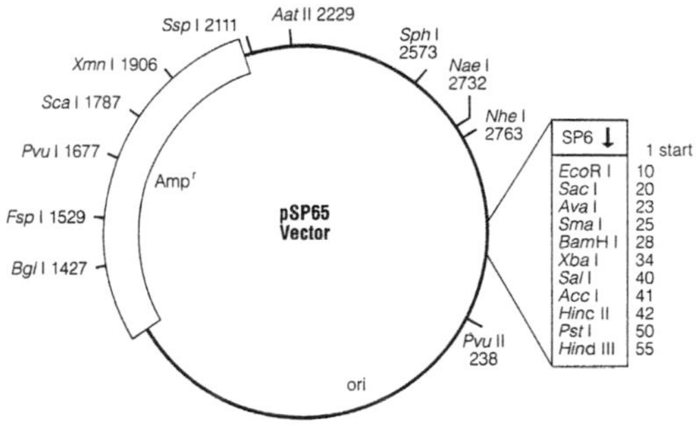

Type	RNA synthesis
Size	3005 bp
Replicator	pMB1
Marker	Ampicillin resistance
Promoter	SP6
Cloning sites	*Acc*I, *Ava*I, *Bam*HI, *Eco*RI, *Hinc*II, *Hin*dIII, *Pst*I, *Sac*I, *Sal*I, *Sma*I, *Xba*I

pSP70

Marketed by Promega Corporation

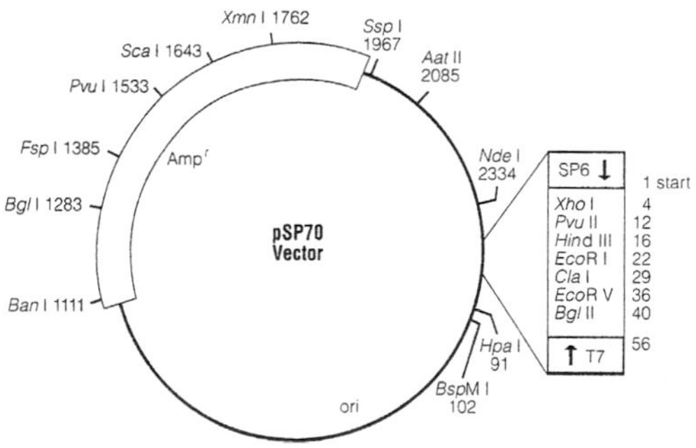

Type	RNA synthesis
Size	2417 bp
Replicator	pMB1
Marker	Ampicillin resistance
Promoters	SP6, T7
Cloning sites	*Bgl*II, *Cla*I, *Eco*RI, *Eco*RV, *Hin*dIII, *Pvu*II, *Xho*I

pSP71

Marketed by Promega Corporation

Type	RNA synthesis
Size	2419 bp
Replicator	pMB1
Marker	Ampicillin resistance
Promoters	SP6, T7
Cloning sites	*Bgl*II, *Cla*I, *Eco*RI, *Eco*RV, *Hin*dIII, *Pvu*II, *Xho*I

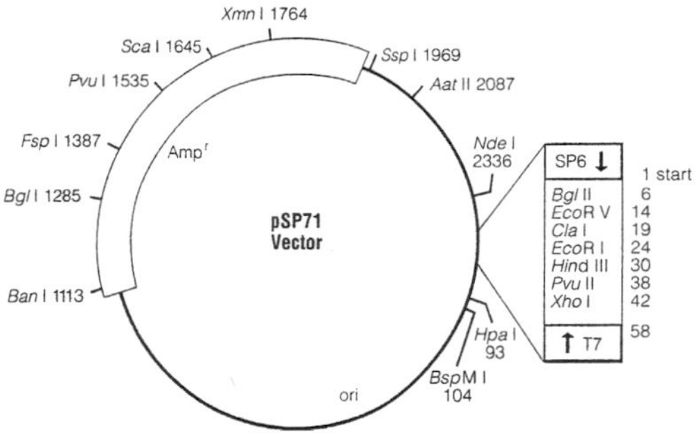

pSP72

Marketed by Promega Corporation

Type	RNA synthesis
Size	2462 bp
Replicator	pMB1
Marker	Ampicillin resistance
Promoters	SP6, T7
Cloning sites	*Acc*I, *Bam*HI, *Bgl*II, *Cla*I, *Eco*RI, *Eco*RV, *Hin*dIII, *Kpn*I, *Pst*I, *Pvu*II, *Sac*I, *Sal*I, *Sma*I, *Sph*I, *Xba*I, *Xho*I

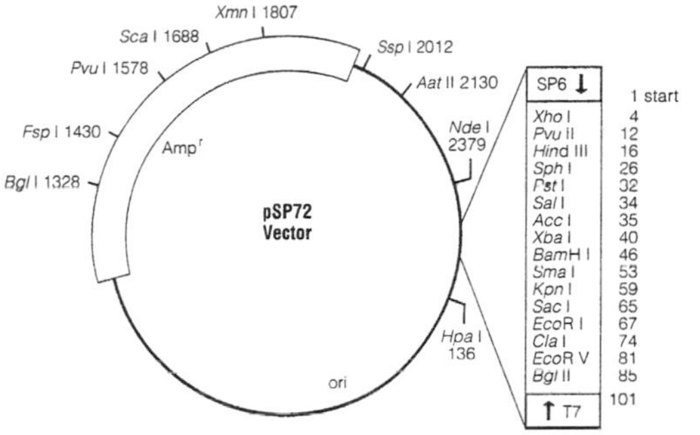

pSP73

Marketed by Promega Corporation

Type	RNA synthesis
Size	2464 bp
Replicator	pMB1
Marker	Ampicillin resistance
Promoters	SP6, T7
Cloning sites	*Acc*I, *Bam*HI, *Bgl*II, *Cla*I, *Eco*RI, *Eco*RV, *Hin*dIII, *Kpn*I, *Pst*I, *Pvu*II, *Sac*I, *Sal*I, *Sma*I, *Sph*I, *Xba*I, *Xho*I

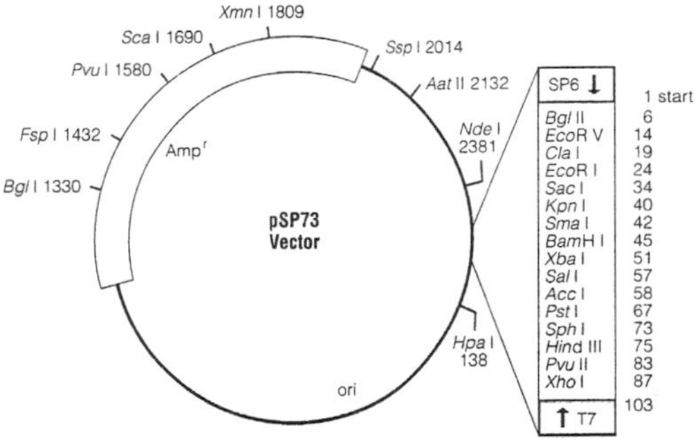

Type	Phagemid RNA synthesis
Sizes	pSPORT1: 4109 bp pSPORT2: 4310 bp
Replicators	pMB1, f1
Markers	Ampicillin resistance *lacZ′*
Promoters	SP6, T7
Cloning sites	*Aat*II, *Bam*HI, *Bsp*MII, *Eco*RI, *Hin*dIII, *Kpn*I, *Mlu*I, *Not*I, *Pst*I, *Rsr*II, *Sal*I, *Sma*I, *Sna*BI, *Spe*I, *Sph*I, *Spl*I, *Sst*I, *Xba*I, *Xma*III

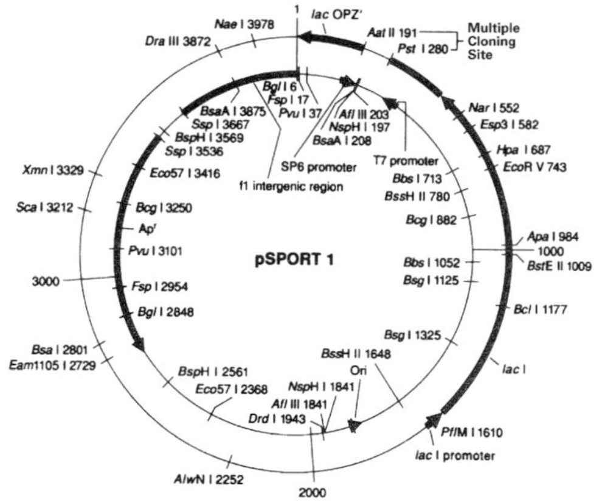

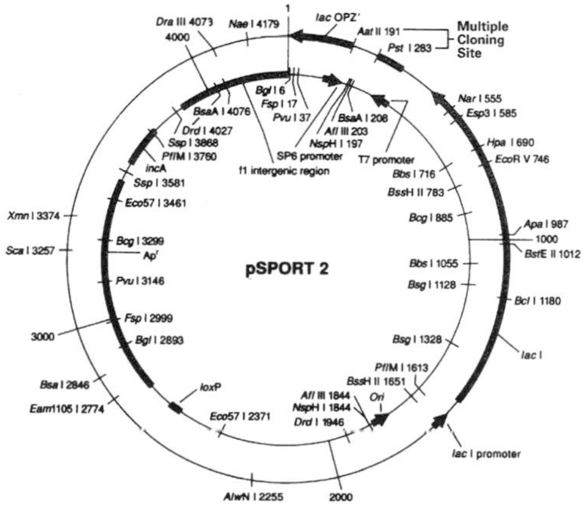

pSPORT 1	pSPORT 2
SP6↓	SP6↓
Aat II	*Aat* II
Sph I	*Sph* I
Mlu I	*Mlu* I
Spl I	*Spl* I
*Sna*B I	*Sna*B I
Hind III	*Hin*d III
*Bam*H I	*Bam*H I
Xba I	*Xba* I
Not I	*Sal* I
Xma III	*Sst* I
Spe I	*Spe* I
Sst I	*Xma* III
Sal I	*Not* I
Sma I	*Sma* I
*Eco*R I	*Eco*R I
*Bsp*M II	*Bsp*M II
Rsr II	*Rsr* II
Kpn I	*Kpn* I
Pst I	*Pst* I
T7↑	T7↑

Type	Protein expression
Size	2957 bp
Replicator	ColE1
Marker	Ampicillin resistance
Promoter	SP6
Fusion protein	None
Cloning sites	*Acc*I, *Apa*I, *Bam*HI, *Bgl*II, *Cla*I, *Eco*RI, *Eco*RV, *Hinc*II, *Kpn*I, *Nco*I, *Pst*I, *Sac*I, *Sal*I, *Sma*I, *Sph*I, *Xba*I
Reference	48
Note	The UTK untranslated leader sequence, containing the *Xenopus* ß-globin 5′ UTR plus the Kozak initiation site, results in high-level translational efficiency.

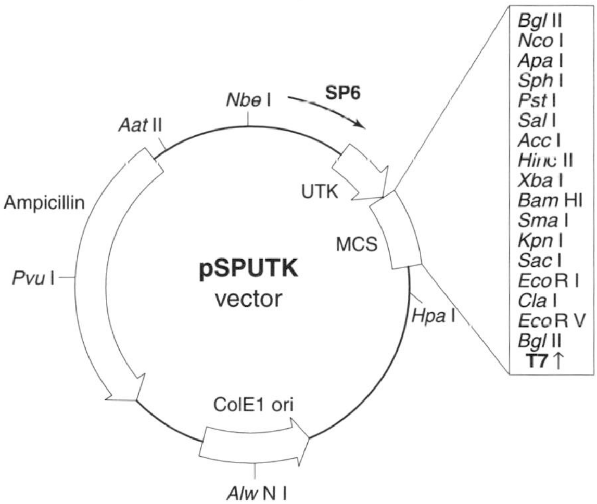

pT3/T7-1
pT3/T7-2
pT3/T7-3

Type RNA expression

Sizes pT3/T7-1, pT3/T7-2: 2.7 kb
pT3/T7-3: 2.95 kb

Replicator ColE1

Markers Ampicillin resistance
lacZ′

Promoters T3, T7

Cloning sites pT3/T7-1: *Apa*I, *Bam*HI, *Cla*I, *Eco*RI, *Eco*RV, *Hin*dIII, *Kpn*I, *Pst*I, *Sal*I, *Sma*I, *Sph*I, *Xba*I, *Xho*I
pT3/T7-2: *Bst*XI, *Eco*RI, *Hin*dIII, *Not*I, *Pst*I, *Sac*I, *Sac*II, *Sal*I, *Sph*I, *Xba*I
pT3/T7-3: *Apa*I, *Bam*HI, *Bst*XI, *Cla*I, *Eco*RI, *Eco*RV, *Hin*dIII, *Kpn*I, *Not*I, *Sac*I, *Sac*II, *Sal*I, *Sma*I, *Xba*I, *Xho*I

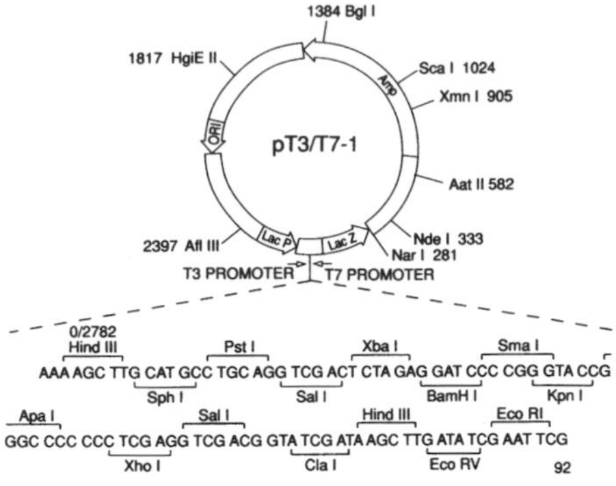

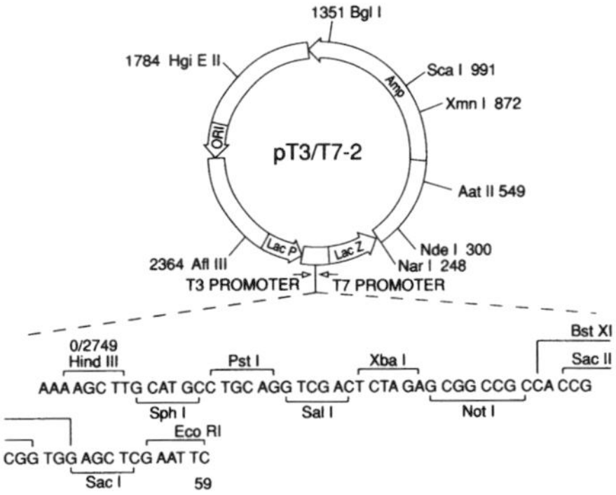

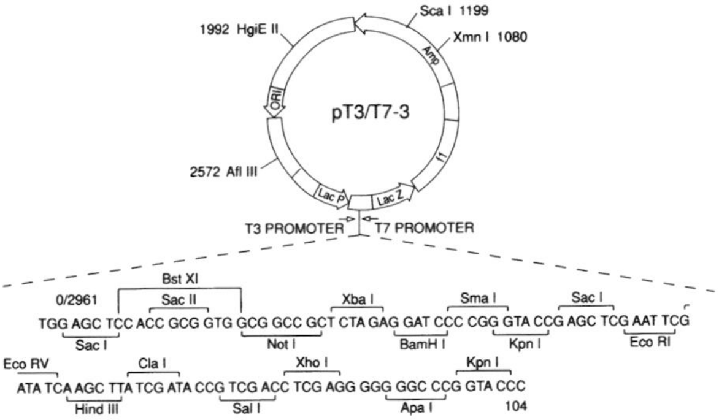

pT7-1
pT7-2

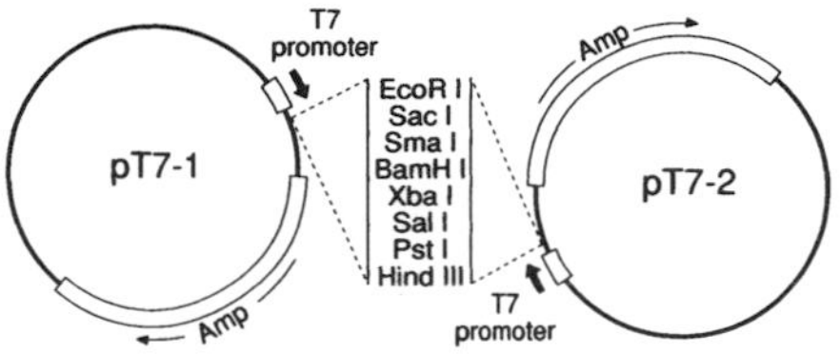

Type RNA expression

Size 2.4 kb

Replicator pMB1

Marker Ampicillin resistance

Promoter T7

Cloning sites *Acc*I, *Bam*HI, *Eco*RI, *Hinc*II, *Hind*III, *Pst*I, *Sac*I, *Sal*I, *Sma*I, *Sst*I, *Xba*I, *Xma*I

Reference 49

pT712
pT713

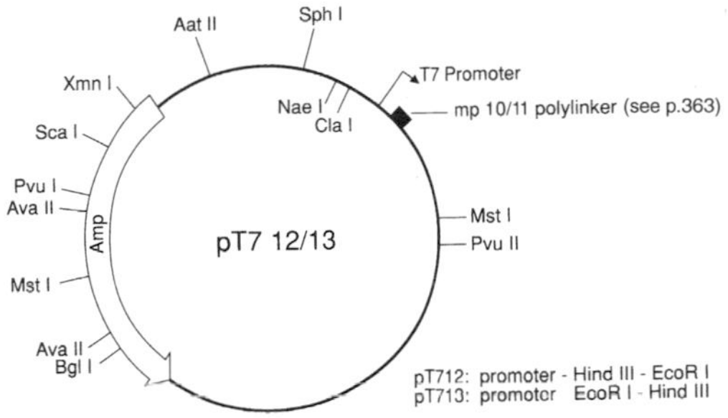

Type RNA expression

Size 2818 bp

Replicator pMB1

Marker Ampicillin resistance

Promoter T7

Cloning sites *Acc*I, *Bam*HI, *Eco*RI, *Hinc*II, *Hind*III, *Pst*I, *Sal*I, *Sma*I, *Sst*I, *Xba*I, *Xma*I

pT7/T3-18U
pT7/T3-19U

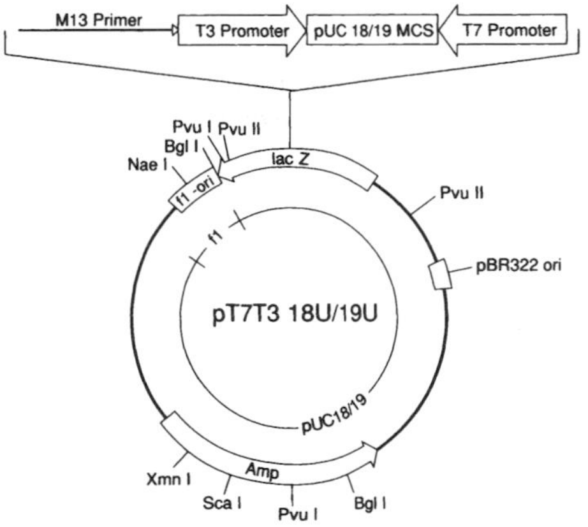

Type Phagemid
RNA expression

Size 2.89 kb

Replicators pMB1, f1

Markers Ampicillin resistance
lacZ′

Promoters T3, T7

Cloning sites *Acc*I, *Bam*HI, *Eco*RI, *Hinc*II, *Hind*III, *Kpn*I, *Pst*I, *Sal*I, *Sma*I, *Sph*I, *Sst*I, *Xba*I, *Xma*I

Reference 50

pTrc 99A

Marketed by Pharmacia Biotech

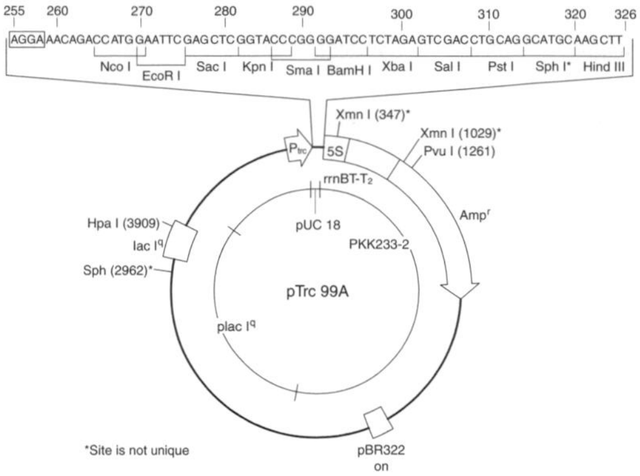

Type	Protein expression
Size	4176 bp
Replicator	pMB1
Marker	Ampicillin resistance
Promoter	*tac*
Fusion protein	None
Cloning sites	*Bam*HI, *Eco*RI, *Hin*dIII, *Kpn*I, *Nco*I, *Pst*I, *Sac*I, *Sal*I, *Sma*I, *Xba*I
Reference	32

pTZ18R
pTZ18U
pTZ19R
pTZ19U

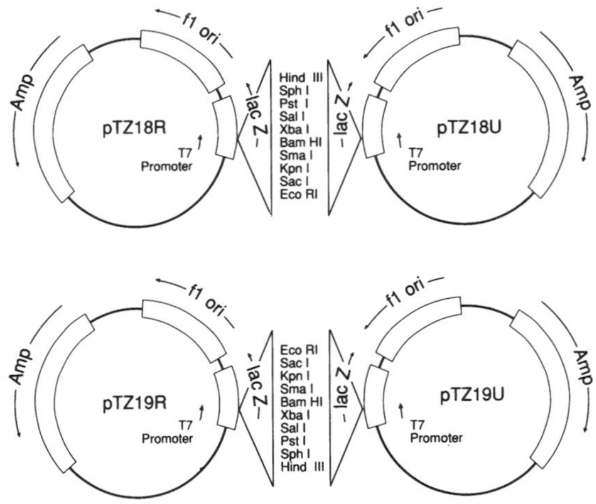

Type	Phagemid RNA expression
Size	2.9 kb
Replicators	pMB1, f1
Markers	Ampicillin resistance *lacZ′*
Promoter	T7
Cloning sites	*Acc*I, *Bam*HI, *Eco*RI, *Hin*cII, *Hin*dIII, *Kpn*I, *Pst*I, *Sac*I, *Sal*I, *Sma*I, *Sph*I, *Xba*I, *Xma*I

pUC7

Type Basic cloning vector

Size 2.7 kb

Replicator pMB1

Markers Ampicillin resistance
lacZ′

Cloning sites *Acc*I, *Bam*HI, *Eco*RI, *Hinc*II, *Pst*I, *Sal*I

References 51–53

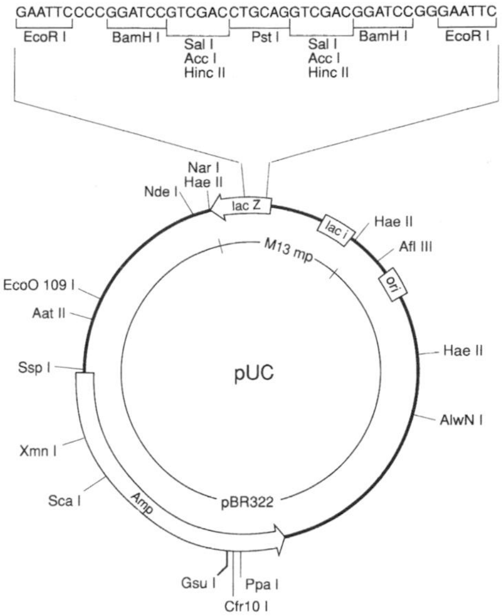

pUC8
pUC9

Type Basic cloning vector

Size 2.7 kb

Replicator pMB1

Markers Ampicillin resistance
lacZ′

Cloning sites *Acc*I, *Bam*HI, *Eco*RI, *Hinc*II, *Hind*III, *Pst*I, *Sal*I, *Sma*I, *Xma*I

References 51–53

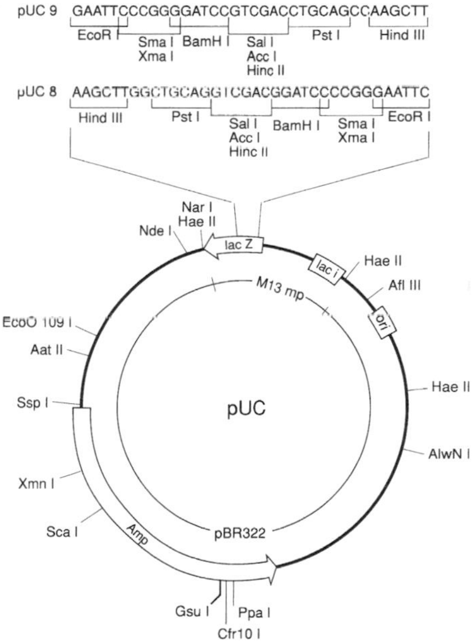

pUC12
pUC13

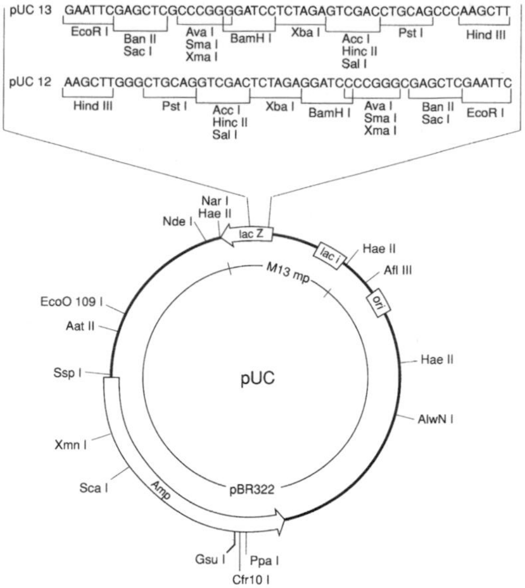

Type Basic cloning vector

Size 2.7 kb

Replicator pMB1

Markers Ampicillin resistance
lacZ'

Cloning sites *Acc*I, *Bam*HI, *Eco*RI, *Hinc*II, *Hind*III, *Pst*I, *Sal*I, *Sma*I, *Sst*I, *Xba*I, *Xma*I

References 51–53

pUC18
pUC19

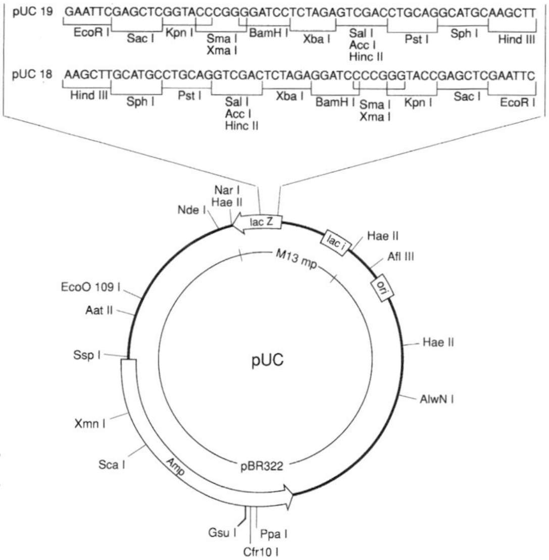

Type Basic cloning vector

Size 2686 bp

Replicator pMB1

Markers Ampicillin resistance
lacZ'

Cloning sites *Acc*I, *Bam*HI, *Eco*RI, *Hinc*II, *Hind*III, *Kpn*I, *Pst*I, *Sal*I, *Sma*I, *Sph*I, *Sst*I, *Xba*I, *Xma*I

References 51–53

Note Beware of a pUC18 variant with a single bp deletion in codon 2 of *lacZ'* (54).

pUC118
pUC119

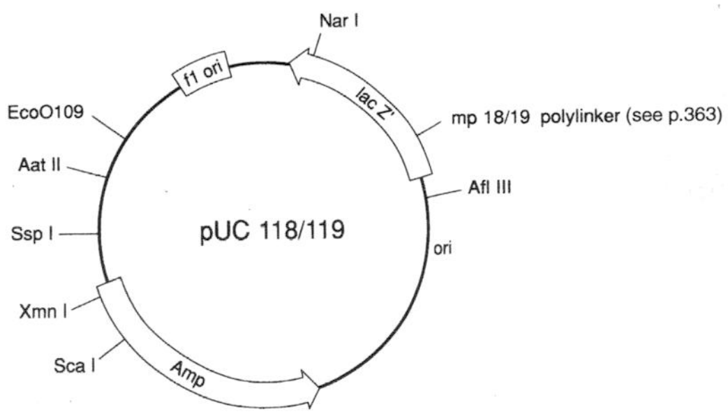

Type Phagemid

Size 3.2 kb

Replicators pMB1, f1

Markers Ampicillin resistance
lacZ'

Cloning sites *Acc*I, *Bam*HI, *Eco*RI, *Hinc*II, *Hin*dIII, *Kpn*I, *Pst*I, *Sal*I, *Sma*I, *Sph*I, *Sst*I, *Xba*I, *Xma*I

Reference 55

pWE15

Type Cosmid

Size 8.2 kb

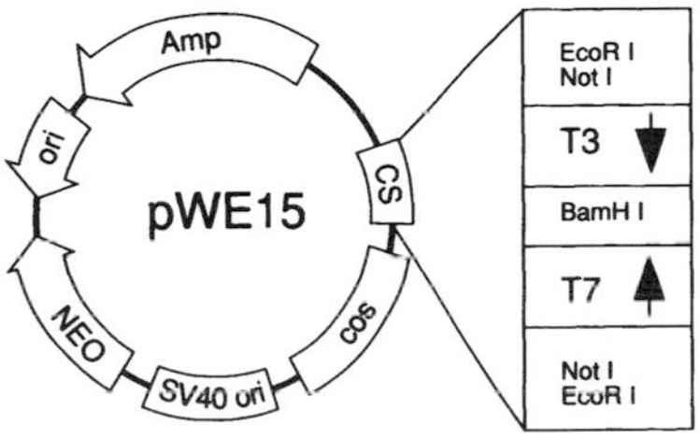

Replicators ColE1, SV40

Markers Ampicillin resistance
Kanamycin resistance

Promoters T3, T7

Cloning site *Bam*HI

References 56, 57

pWE16

Type Cosmid

Size 8.8 kb

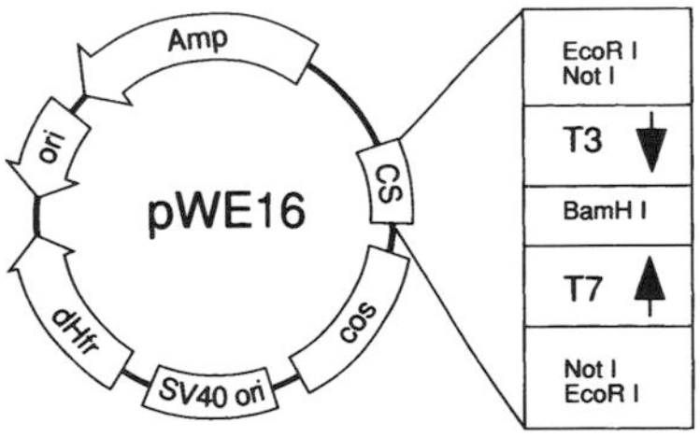

Replicators ColE1, SV40

Markers Ampicillin resistance
Methotrexate resistance

Promoters T3, T7

Cloning site *Bam*HI

References 56, 57

pYEJ001

Type	Protein expression
Size	4.1 kb
Replicator	pMB1
Markers	Ampicillin resistance Chloramphenicol resistance Tetracycline resistance
Promoter	Synthetic
Fusion protein	Chloramphenicol acetyltransferase
Cloning sites	*Eco*RI, *Hin*dIII
Reference	58

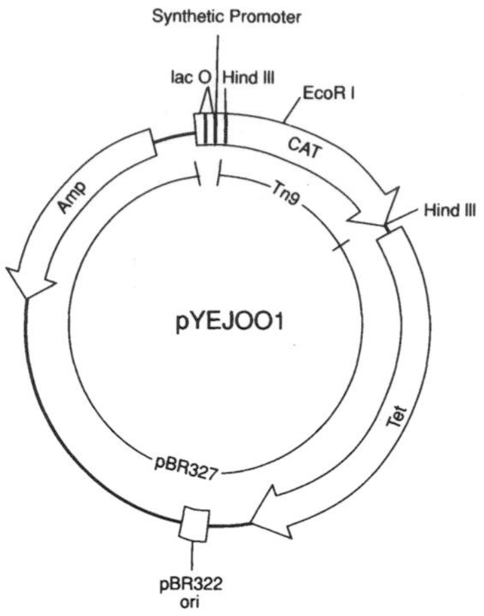

SuperCos 1

Marketed by Stratagene

Type	Cosmid
Size	7.6 kb
Replicators	ColE1, SV40
Markers	Ampicillin resistance Kanamycin resistance
Promoters	T3, T7
Cloning site	*Bam*HI

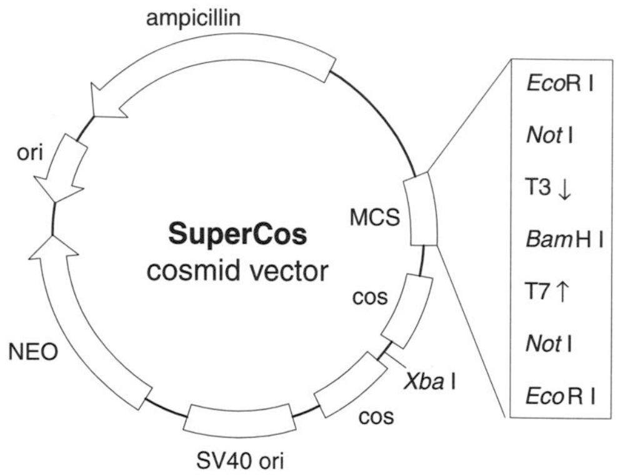

2. PHAGE λ CLONING VECTORS

2.1 Types of λ vector

The λ phage particle can accommodate approximately 52 kb of DNA, which means that the wild-type genome (48.5 kb, see Chapter 1, Section 2) can only be increased by about 3.5 kb. All λ cloning vectors are therefore based on deleted versions of the wild-type genome, the deletions removing regions that are not essential for the lytic growth cycle. The positions of the deletions found in different vectors are shown in *Figure 1*. More details are provided about these deletions, and about other genetic features of λ cloning vectors in *Table 6*.

There are two types of λ cloning vector:

(i) *Insertion vectors*, which are simple deleted versions of the wild-type λ genome, with one or more unique restriction sites into which new DNA can be inserted. The maximum insert size is about 10–12 kb. This limitation is imposed by the *minimum* size for packaging into a λ phage head: if a non-recombinant insertion vector is less than about 38 kb in length then it will not be packaged and cannot be propagated.

(ii) *Replacement vectors*, which have a higher capacity because the non-recombinant version carries a stuffer fragment, flanked by pairs of restriction sites, which is replaced by new DNA during a cloning experiment. This replacement step means that the capacity of the vector is increased to a maximum of about 24 kb, this maximum being set by the amount of the λ genome that can be deleted without affecting essential functions. Note that many replacement vectors also have a minimum capacity, because if the cloned DNA is too short the recombinant vector is below the minimum length for packaging. In some vectors the stuffer fragment is part of the λ genome, in others it is *E. coli* DNA. A recent innovation is to use plasmid DNA as the stuffer fragment, in such a way that sequences cloned in the λ vector can be recovered as double-stranded or single-stranded plasmid DNA by simple manipulations (e.g. by co-infection with an M13 helper phage resulting in synthesis of single-stranded copies of DNA cloned into a stuffer fragment carrying an f1 origin of replication).

Examples of commonly used insertion and replacement vectors are listed in *Table 7*. All of these vectors are described in more detail in pp. 343–362.

2.2 Identification of recombinants

The majority of λ vectors use one of two systems for recombinant identification:

(i) *Lac selection*, with recombinants identified as clear plaques on agar containing X-gal and IPTG.

(ii) *Spi selection*. Spi is 'sensitivity to P2 prophage inhibition' and refers to the fact that wild-type λ phage cannot infect *E. coli* cells that already possess an integrated form of the related phage P2. The Spi phenotype is controlled by the *red* and *gam* genes. Several replacement vectors carry the *red* and *gam* genes on the stuffer fragment, so replacement of the stuffer with the DNA to be cloned converts the vector from Red^+ Gam^+ to Red^- Gam^-, meaning that only recombinant phage give plaques when a P2 lysogenic *E. coli* is used as the host.

In addition to the Lac and Spi systems, a few λ vectors utilize *c*I inactivation as the basis for recombinant selection. Insertional inactivation of *c*I causes a change in plaque morphology

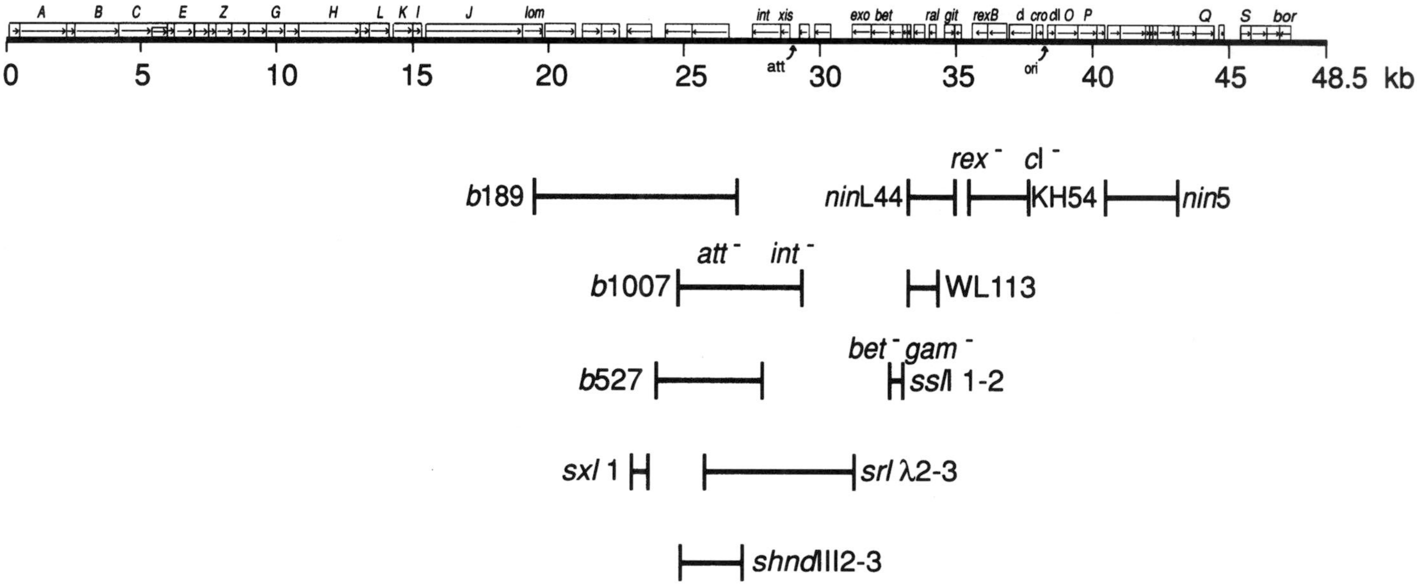

Figure 1. Genetic map of wild-type phage λ, showing the positions of the main deletions carried by λ cloning vectors.

Table 6. Genetic markers carried by phage λ vectors (see also ref. 59)

Marker	Description
*A*am	Amber mutation in gene *A*
*b*2, *b*189, *b*527, *b*1007	Deletions within the *b* region that prevent the phage entering the lysogenic cycle
*B*am	Amber mutation in gene *B*
bio, *bio*252, *bio*256	Substitutions from the *bio* region of *E. coli*
*chi*A131, *chi*C	Directional recombination sites
*c*I*ts*857	A temperature-sensitive mutation in *c*I; lysogens carrying this marker can be induced by heat shock
*cos*2	Defective *cos* sites
*D*am	Amber mutation in gene *D*, resulting in intracellular accumulation of phage λ structural proteins
*E*am	Amber mutation in gene *E*, resulting in intracellular accumulation of phage λ structural proteins
exo bet	Mutations in these genes (the *red* region) result in the Spi^- phenotype, meaning that phage can infect $recA^+$ hosts that are lysogenic for P2
*gam*am	Amber mutation in *gam*. A mutation in *gam* results in the Spi^- phenotype (see *exo bet*)
*imm*434	Substitution from phage ϕ434
*int*29	Substitution from phage ϕ29
*int*am	Amber mutation in *int*
KH54	Deletion that prevents lysogeny
*lac*5, *lac*UV5, *lac*Z, *lac*Z′	Substitutions from the *lac* region of *E. coli*
*nin*5, *nin*L44	Deletions that remove t_{R2}, allowing delayed early transcription independent of the *N* gene product
*QSR*80	Substitution from phage ϕ80
*red*3	See *exo bet*
s....λ.	Designates a deletion: the letters immediately following the *s* and the number(s) following the λ denote which restriction sites are involved (e.g. *ssl*Iλ1-2 = deletion of the region between *Sal*I sites 1 and 2; *shn*dIIIλ2-3 = deletion of the region between *Hin*dIII sites 2 and 3). λ1°, λ2°, etc, indicate exonuclease digestion at the site indicated
*S*am100	Amber mutation in gene *S*, which results in intracellular accumulation of phage particles
*trp*E	Substitution from the *trp* region of *E. coli*
*W*am	Amber mutation in gene *W*
WL113	Deletion that removes *kil*, *c*III, *Ea*10 and *ral*

Table 7. Phage λ vectors

Type	Vectors[a]
Insertion	λCharon 4, λgt10, λgt11, λgt18, λgt19, λgt20, λgt21, λgt22, λgt23, λSurfZAP, λZAP Express
Replacement	λ1059, λ2001, λCharon 4, λCharon 30, λCharon 34, λCharon 35, λCharon 40, λDASH II, λEMBL3, λEMBL4, λFIX II, λGEM-11, λGEM-12, λgtWES.λB′, λZAP II

[a]For maps and additional details see pp. 343–362.

from turbid to clear, which is recognizable to the trained eye. Other replacement vectors rely simply on the supposition that only recombinants are large enough to be packaged, as constructions in which the left and right arms have been ligated without an insert will be too short, so all plaques contain inserted DNA. The drawback with this system is that if the stuffer fragment is reinserted instead of the DNA to be cloned then a non-recombinant vector will give a plaque. Most stuffers in fact contain internal sites for the restriction enzyme used for cloning, and so are cut into two or more small pieces when the vector is restricted, but their use still requires scrupulous technique.

2.3 Individual vectors

The following pages give details for the commonly used λ vectors listed in *Table 7*. If the vector description includes an acknowledgement of the commercial supplier, then the map shown is reproduced (possibly with small modifications for the sake of clarity) with permission from that supplier's product information: λGEM-11 and λGEM-12 are reproduced with permission from Promega Corporation; λDASHII, λFIXII, λSurfZAP, λZAPII and λZAP Express are reproduced with permission from Stratagene. The maps for other vectors are redrawn, with permission, from *Molecular Cloning (2nd edn)* by J. Sambrook, E.F. Fritsch and T. Maniatis (copyright 1989 Cold Spring Harbor Laboratory Press).

λ1059

Type Replacement vector

Capacity 10–20 kb

Left arm 20.0 kb
Right arm 8.8 kb

Properties of recombinants cI^{-} Red^{-} Gam^{-}

Recognition of recombinants Spi

Genotype h$^{\lambda}$*sbh*λ1° *b*189 <*int*29 *nin*L44 cI*ts*857 p*acl*29> Δ[*int*-*c*III] KH54 *sr*Iλ4° *nin*5 *chi*3

Cloning sites and coordinates *Bam*HI 20020, 34440

Reference 60

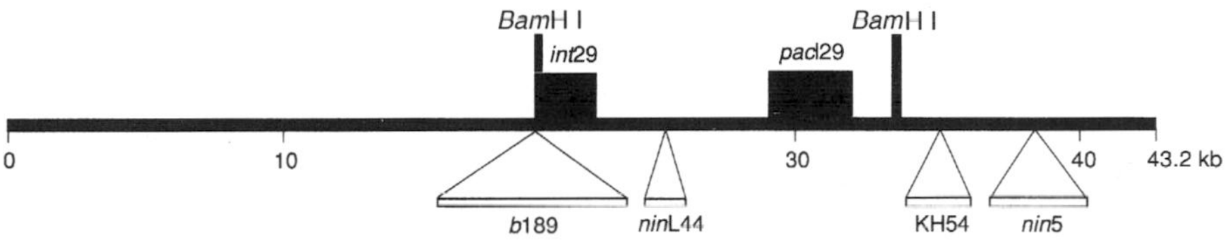

λ2001

Type	Replacement vector
Capacity	9–23 kb
Left arm	20.0 kb
Right arm	8.8 kb
Properties of recombinants	cI$^-$ Red$^-$ Gam$^-$
Recognition of recombinants	Spi
Genotype	λ*sbh*Iλ1° *b*189 <polycloning site *sr*Iλ3° *nin*L44 *bio* polycloning site> KH54 *chi*C *sr*Iλ4° *nin*5 *shnd*IIIλ6° *sr*Iλ5°

Cloning sites and coordinates

Polylinker 1:	*Xba*I	19990, 20020
	*Sac*I	20000
	*Xho*I	20000
	*Bam*HI	20000
	*Hin*dIII	20010
	*Eco*RI	20010
Polylinker 2:	*Xba*I	32760, 32770
	*Sac*I	32770
	*Xho*I	32770
	*Bam*HI	32770
	*Hin*dIII	32760
	*Eco*RI	32760

Reference 61

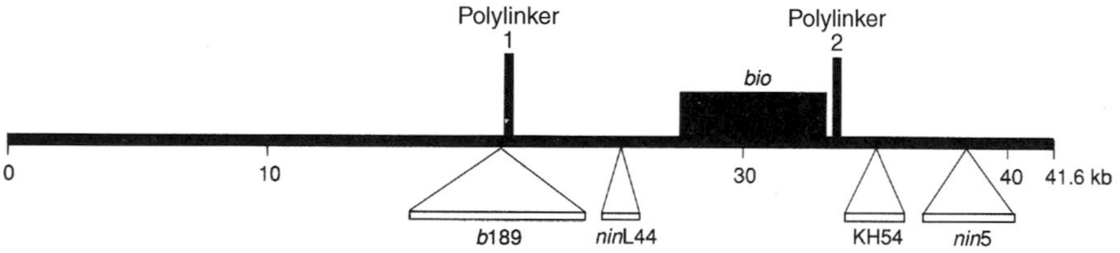

λCharon 4

Type	Replacement and insertion vector
Capacity	Replacement: 7–20 kb Insertion: 0–6 kb
Left arm	Replacement: 19.6 kb Insertion: 24.8 kb
Right arm	Replacement: 11.0 kb Insertion: 20.5 kb
Properties of recombinants	cI^- Red^- Gam^-
Recognition of recombinants	Replacement: Bio^- Lac^- Insertion: Bio^+ Lac^+
Genotype	λ*lac*5 *bio*256 KH54 *sr*Iλ4° *nin*5 *QSR*80
Cloning sites and coordinates	Replacement: *Eco*RI 19600, 26420, 34320 Insertion: *Xba*I 24820
References	62–65
Note	λCharon 4A is the same as λCharon 4 but carries the *A*am32 and *B*am1 mutations.

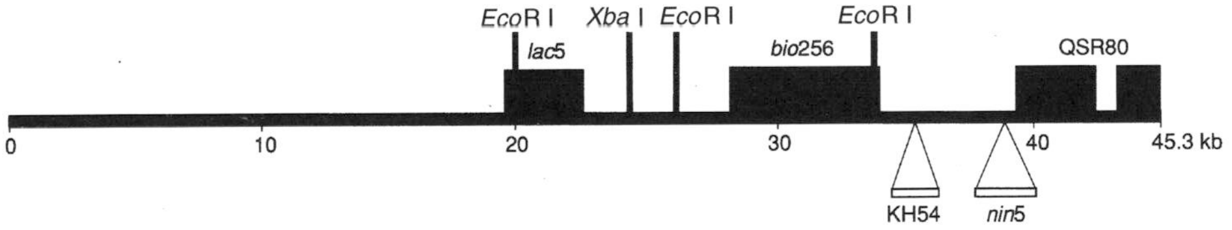

λCharon 30

Type	Replacement vector
Capacity	7–17 kb
Left arm **Right arm**	22.4 kb 8.8 kb
Properties of recombinants	cI^- Red^- Gam^-
Recognition of recombinants	Spi
Genotype	λ*sbh*Iλ1° [*sbh*Iλ2–(*b*1007)–*sbh*Iλ4]^{dup} KH54 *sr*Iλ4° *nin*5 *shn*dIIIλ6° *sr*Iλ5°
Cloning sites and coordinates	*Bam*HI 22350, 29810, 37270
Reference	65

Notes

1. The insertion in λCharon 30 is *Bam*HI fragment 3/4 (coordinates 22350–34500), minus the *b*1007 deletion, from the wild-type λ genome. This region therefore occurs twice in the vector.
2. Other restriction sites can be used for cloning purposes but the properties of recombinants are uncertain and there are no definite means of recognizing recombinant plaques. Details relating to these sites are:
   *Eco*RI (coordinates 21230, 27060, 34390; E1, E2 and E3 on the map below): capacity 5–15 kb, left arm 21.3 kb, right arm 11.7 kb.
   *Hin*dIII (coordinates 23130, 30590; H1, H2): capacity 0–9 kb, left arm 23.1 kb, right arm 15.5 kb.
   *Sal*I (coordinates 28060, 28550, 35520, 36010; S1–S4): capacity 0-10 kb, left arm 28.1 kb, right arm 10.1 kb.
   *Xho*I (coordinates 28810, 36270; X1, X2): capacity 0–9 kb, left arm 28.8 kb, right arm 9.9 kb.

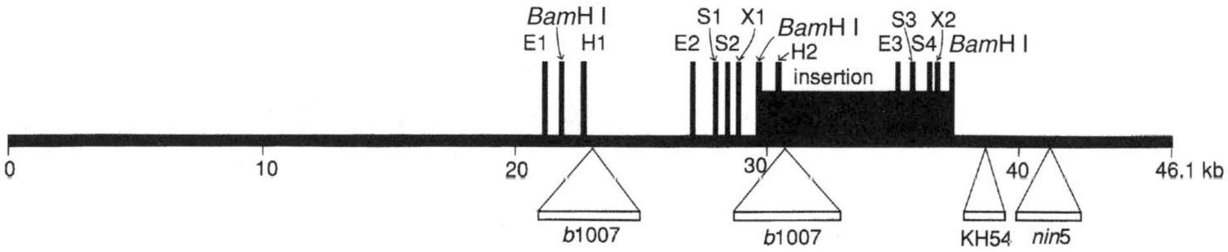

λCharon 34

Type Replacement vector

Capacity 9–20 kb

Left arm 19.5 kb
Right arm 10.6 kb

Properties of recombinants cI^- Red^- Gam^+

Recognition of recombinants None

Genotype λ*sbh*Iλ1° *lac*5 *sr*I *lac*Z <polycloning site–*E. coli* DNA–polycloning site> *sr*Iλ3 WL113 KH54 *nin*5 *shn*dIIIλ6° *sr*Iλ5°

Cloning sites and coordinates

Polylinker 1:	*Eco*RI	19470
	*Sac*I	19480
	*Xba*I	19500
	*Hin*dIII	19520
	*Bam*HI	19520
Polylinker 2:	*Bam*HI	35920
	*Hin*dIII	35930
	*Xba*I	35960
	*Sac*I	35970
	*Eco*RI	35980

Reference 66

Notes

1. λCharon 34 and λCharon 35 are identical except that they contain different pieces of *E. coli* DNA as the stuffer fragment.
2. *Sal*I cuts within the two polylinkers (coordinates 19500, 35950) and at a downstream site (36970). Restriction with *Sal*I therefore results in a greater capacity (10–21 kb; left arm 19.5 kb, right arm 9.6 kb). Recombinants produced in this way are still cI^- Red^- Gam^+.

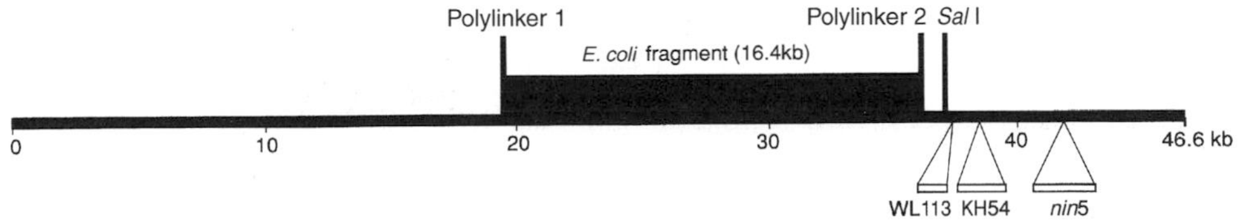

λCharon 35

Type Replacement vector

Capacity 9–20 kb

Left arm 19.5 kb
Right arm 10.6 kb

Properties of recombinants cI^- Red^- Gam^+

Recognition of recombinants None

Genotype λ*sbh*Iλ1° *lac*5 *sr*I *lac*Z <polycloning site–*E. coli* DNA-polycloning site> *sr*Iλ3 WL113 KH54 *nin*5 *shn*dIIIλ6° *sr*Iλ5°

Cloning sites and coordinates

Polylinker 1:	*Eco*RI	19470
	*Sac*I	19480
	*Xba*I	19500
	*Hin*dIII	19520
	*Bam*HI	19520
Polylinker 2:	*Bam*HI	35120
	*Hin*dIII	35130
	*Xba*I	35160
	*Sac*I	35170
	*Eco*RI	35180

Reference 66

Notes

1. λCharon 34 and λCharon 35 are identical except that they contain different pieces of *E. coli* DNA as the stuffer fragment.
2. *Sal*I cuts within the two polylinkers (coordinates 19500, 35150) and at a downstream site (36170). Restriction with *Sal*I therefore results in a greater capacity (10–21 kb; left arm 19.5 kb, right arm 9.6 kb). Recombinants produced in this way are still cI^- Red^- Gam^+.

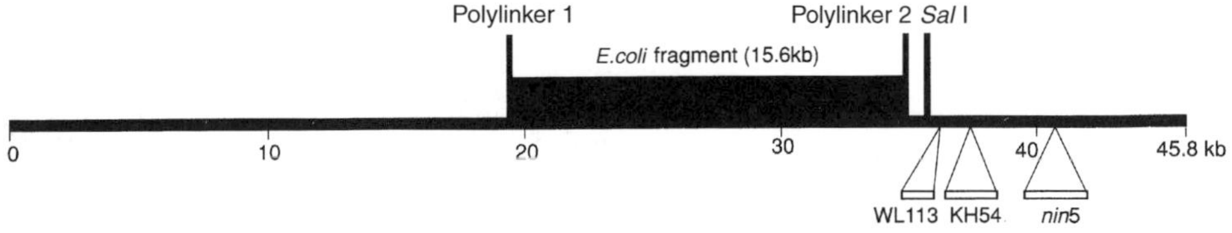

λCharon 40

Type Replacement vector

Capacity 9–24 kb

Left arm 19.2 kb
Right arm 9.6 kb

Properties of recombinants cI$^-$ Red$^-$ Gam$^+$

Recognition of recombinants Lac$^-$

Genotype λ*sbh*Iλ1° *sap*Iλ1° *sk*Iλ1° *sk*Iλ2° *ssm*Iλ1 <polycloning site–polystuffer–polycloning site> *ssl*Iλ2 WL113 KH54 *ssm*Iλ3° *nin*5 *shn*dIIIλ6° *sr*Iλ5°

Cloning sites and coordinates

Polylinker 1:	*Eco*RI	19180
	*Sac*I	19190
	*Kpn*I	19200
	*Sma*I	19200
	*Xba*I	19210
	*Sal*I	19220
	*Hin*dIII	19240
	*Not*I	19240
	*Xma*III	19240
	*Avr*II	19250
	*Spe*II	19260
	*Xho*I	19260
	*Apa*I	19290
	*Bam*HI	19290
	*Sfi*I	19310
	*Nae*I	19320
Polylinker 2:	See note below	

Reference 67

Note The stuffer fragment in λCharon 40 is variable in length, so the co-ordinates of the sites in the downstream polylinker cannot be given. This polylinker contains the same sites as the upstream polylinker but in the opposite orientation.

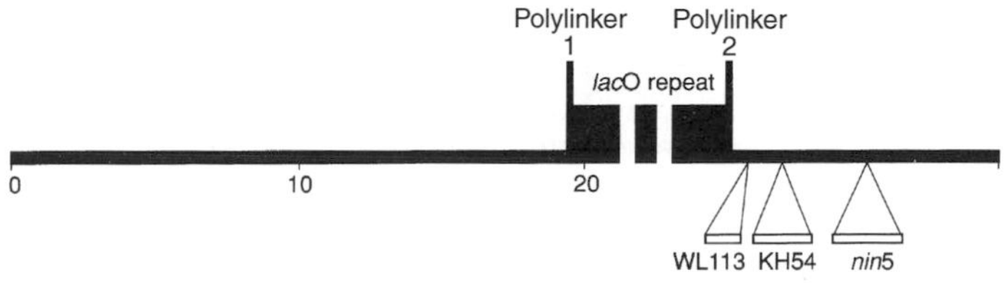

Type	Replacement vector
Capacity	9–22 kb
Left arm	20.0 kb
Right arm	8.8 kb
Properties of recombinants	cI^- Red^- Gam^-
Recognition of recombinants	Spi
Genotype	λ*sbh*Iλ1° *b*189 <T3 promoter-polycloning site *sr*Iλ3° *nin*L44 *bio* polycloning site–T7 promoter> KH54 *chi*C *sr*Iλ4° *nin*5 *shnd*III6° *sr*Iλ5°

Cloning sites and coordinates

Polylinker 1:	*Xba*I	19990
	*Sal*I	20000
	*Not*I	20000
	(T3 promoter)	
	*Eco*RI	20020
	*Bam*HI	20020
	*Hin*dIII	20030
	*Sac*I	20030
	*Xho*I	20030
	*Xba*I	20040
Polylinker 2:	*Xba*I	32770
	*Xho*I	32770
	*Sac*I	32780
	*Hin*dIII	32780
	*Bam*HI	32780
	*Eco*RI	32790
	(T7 promoter)	
	*Not*I	32810
	*Sal*I	32810
	*Xba*I	32810

Notes

1. λDASH II and λFIX II are identical except that the T3 and T7 promoters are interchanged and the polylinker sequences are slightly different.
2. λDASH is an older version that lacks *Not*I sites in the polylinkers.

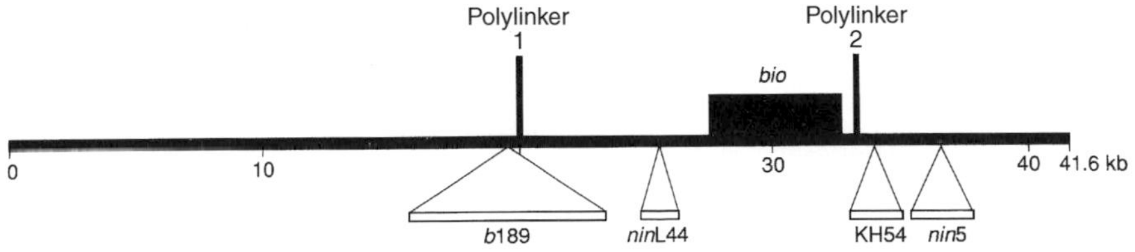

λEMBL3
λEMBL4

Type	Replacement vector
Capacity	7–20 kb
Left arm **Right arm**	19.9 kb 8.8 kb
Properties of recombinants	cI^- Red^- Gam^-
Recognition of recombinants	Spi
Genotype	λ*sbh*Iλ1° *b*189 <polycloning site *int*29 *nin*L44 *trp*E polycloning site> KH54 *chi*C *sr*Iλ4° *nin*5 *sr*Iλ5°

Cloning sites and coordinates

λEMBL3

Polylinker 1:	*Sal*I	19940
	*Bam*HI	19940
	*Eco*RI	19950
Polylinker 2:	*Eco*RI	33110
	*Bam*HI	33120
	*Sal*I	33120

λEMBL4

Polylinker 1:	*Eco*RI	19940
	*Bam*HI	19940
	*Sal*I	19950
Polylinker 2:	*Sal*I	33110
	*Bam*HI	33120
	*Eco*RI	33120

Reference 68

Note λEMBL3A is the same as λEMBL3 but carries the *A*am2 and *B*am1 mutations.

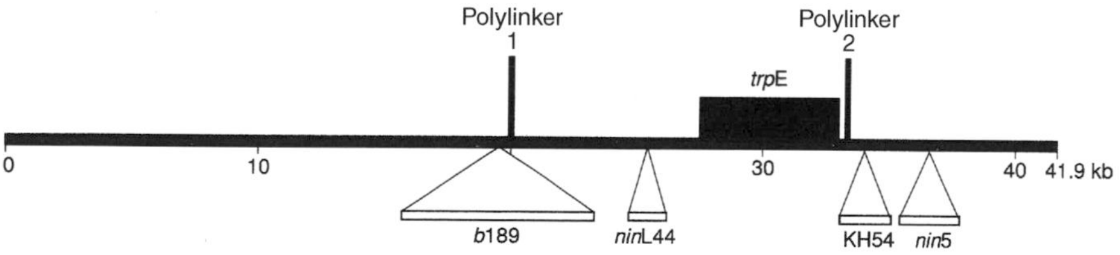

Cloning Vectors

Type: Replacement vector

Capacity: 9–22 kb

Left arm: 20.0 kb
Right arm: 8.8 kb

Properties of recombinants: cI^- Red^- Gam^-

Recognition of recombinants: Spi

Genotype: λ*sbh*Iλ1° *b*189 <T7 promoter-polycloning site *sr*Iλ3° *nin*L44 *bio* polycloning site–T3 promoter> KH54 *chi*C *sr*Iλ4° *nin*5 *shn*dIII6° *sr*Iλ5°

Cloning sites and coordinates

Polylinker 1:

*Xba*I	19990
*Sac*I	20000
*Not*I	20000
*Sac*I	20010
(T7 promoter)	
*Sal*I	20020
*Xho*I	20020
*Eco*RI	20030
*Xba*I	20030

Polylinker 2:

*Xba*I	32760
*Eco*RI	32760
*Xho*I	32760
*Sal*I	32770
(T3 promoter)	
*Sac*I	32780
*Not*I	32790
*Sac*I	32790
*Xba*I	32790

Notes

1. λDASH II and λFIX II are identical except that the T3 and T7 promoters are interchanged and the polylinker sequences are slightly different.
2. λFIX is an older version that lacks *Not*I sites in the polylinkers.

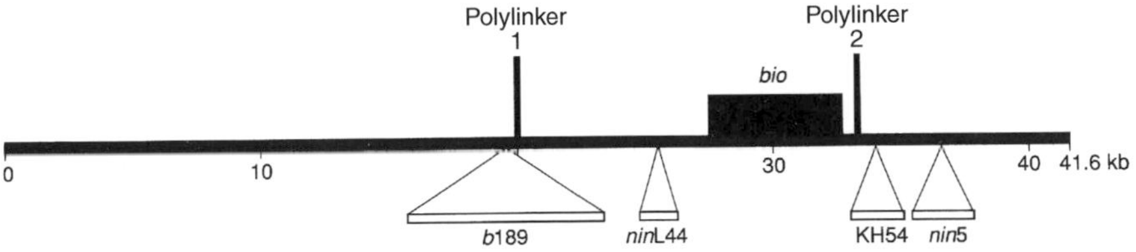

λGEM-11
λGEM-12

Marketed by Promega Corporation

Type Replacement vector

Capacity 9–23 kb

Left arm 19.9 kb
Right arm 8.8 kb

Properties of recombinants cI^- Red^- Gam^-

Recognition of recombinants Spi

Genotype λ*sbh*Iλ1° *b*189 <polycloning site *sr*Iλ3° *nin*L44 *bio* polycloning site> KH54 *chi*C *sr*Iλ4° *nin*5 *sr*Iλ5°

Cloning sites

λGEM-11		λGEM-12	
Polylinker 1		Polylinker 1	
*Sfi*I	19930	*Sfi*I	19930
(T7 promoter)		(T7 promoter)	
*Sac*I	19970	*Sac*I	19960
*Xho*I	19970	*Not*I	19970
*Bam*HI	19970	*Bam*HI	19980
*Avr*II	19980	*Eco*RI	19990
*Eco*RI	19980	*Xho*I	19990
*Xba*I	20000	*Xba*I	20010
Polylinker 2		Polylinker 2	
*Xba*I	32740	*Xba*I	32750
*Eco*RI	32760	*Xho*I	32770
*Avr*II	32760	*Eco*RI	32780
*Bam*HI	32760	*Bam*HI	32780
*Xho*I	32770	*Not*I	32790
*Sac*I	32780	*Sac*I	32790
(SP6 promoter)		(SP6 promoter)	
*Sfi*I	32800	*Sfi*I	32820

Note The arms of these two vectors are derived from λEMBL3 and the stuffer fragment from λ2001.

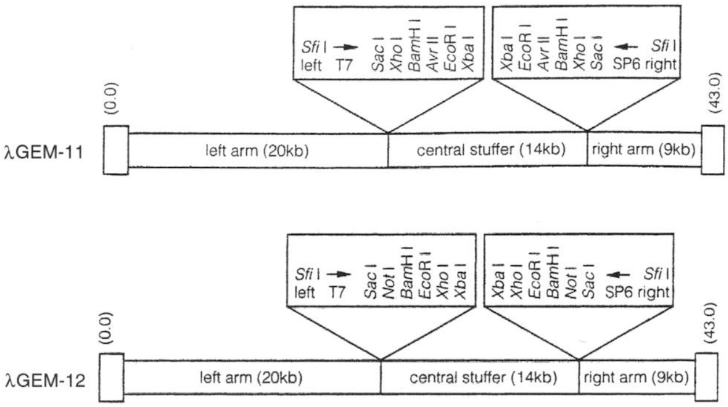

Cloning Vectors

λgt10

Type	Insertion vector
Capacity	0–6 kb
Left arm	32.7 kb
Right arm	10.6 kb
Properties of recombinants	cI^- Red^+ Gam^+
Recognition of recombinants	cI^-
Genotype	λ*sr*Iλ1° *b*527 *sr*Iλ3° *imm*434 (*sr*I434$^+$) *sr*Iλ4° *sr*Iλ5°
Cloning site and coordinates	*Eco*RI 32710
Reference	69

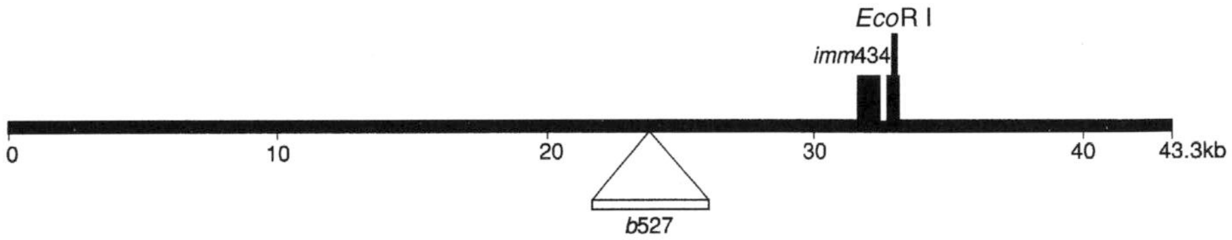

λgt11

Type	Insertion vector
Capacity	0–7.2 kb
Left arm **Right arm**	19.6 kb 24.1 kb
Properties of recombinants	cIts Red^+ Gam^+
Recognition of recombinants	Lac^-
Genotype	λ*lac*5 *shnd*IIIλ2-3 *sr*Iλ3° c*Its*857 *sr*Iλ4° *nin*5 *sr*Iλ5° *S*am100
Cloning site and coordinates	*Eco*RI 19600
Reference	70
Note	Promega Corporation market a version with *Not*I and *Sfi*I sites flanking the *Eco*RI site.

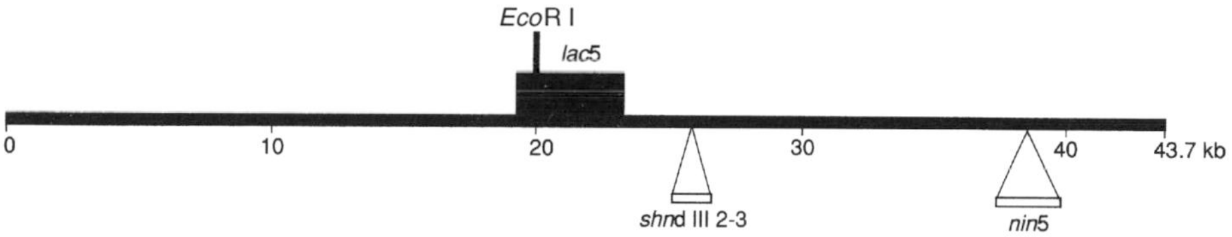

Cloning Vectors

λgt18
λgt19

Type	Insertion vector
Capacity	0–7.7 kb
Left arm	19.6 kb
Right arm	23.6 kb
Properties of recombinants	cIts Red^- Gam^+
Recognition of recombinants	Lac^-
Genotype	λ*lac*5 *shnd*IIIλ2-3 *sr*Iλ3° *ssl*Iλ1-2 c*Its*857 *sr*Iλ4° *nin*5 *sr*Iλ5° *S*am100
Cloning sites and coordinates	λgt18
	Polylinker: *Sal*I 19600
	*Eco*RI 19630
	λgt19
	Polylinker: *Eco*RI 19590
	*Sal*I 19620
Reference	71

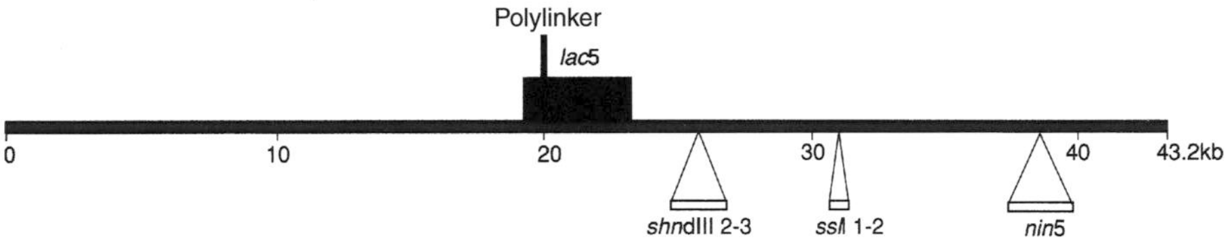

λgt20
λgt21

Type	Insertion vector
Capacity	0–8.2 kb
Left arm **Right arm**	19.6 kb 23.1 kb
Properties of recombinants	cIts Red⁻ Gam⁻
Recognition of recombinants	Lac⁻
Genotype	λ*lac*5 *sx*Iλ1° *chi shn*dIIIλ2-3 *sr*Iλ3° *ssl*Iλ1-2 c*Its*857 *sr*Iλ4° *nin*5 *sr*Iλ5° *S*am100
Cloning sites and coordinates	λgt20 Polylinker: *Sal*I 19600; *Xba*I 19620; *Eco*RI 19630 λgt21 Polylinker: *Eco*RI 19590; *Xba*I 19600; *Sal*I 19620
Reference	72
Note	λgt20/21 and λgt22/23 are identical except for different polycloning sites.

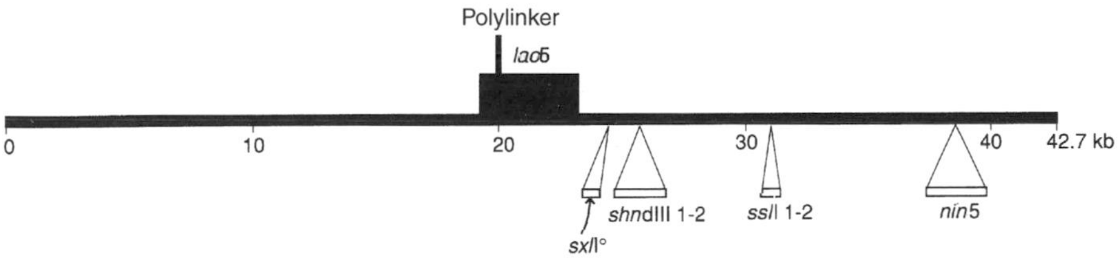

Cloning Vectors

λgt22
λgt23

Type	Insertion vector
Capacity	0–8.2 kb
Left arm	19.6 kb
Right arm	23.1 kb
Properties of recombinants	cIts Red⁻ Gam⁻
Recognition of recombinants	Lac⁻
Genotype	λ*lac*5 *sx*Iλ1° *chi shn*dIIIλ2-3 *sr*Iλ3° *ssl*Iλ1-2 *cIts*857 *sr*Iλ4° *nin*5 *sr*Iλ5° *S*am100

Cloning sites and coordinates

λgt22

Polylinker:		
	*Not*I	19600
	*Xba*I	19610
	*Sac*I	19610
	*Sal*I	19620
	*Eco*RI	19630

λgt23

Polylinker:		
	*Eco*RI	19600
	*Sal*I	19610
	*Sac*I	19610
	*Xba*I	19620
	*Not*I	19630

References 73, 74

Notes

1. λgt20/21 and λgt22/23 are identical except for different polycloning sites.
2. λgt22A is the same as λgt22 but with the *Sac*I site replaced by *Spe*I.

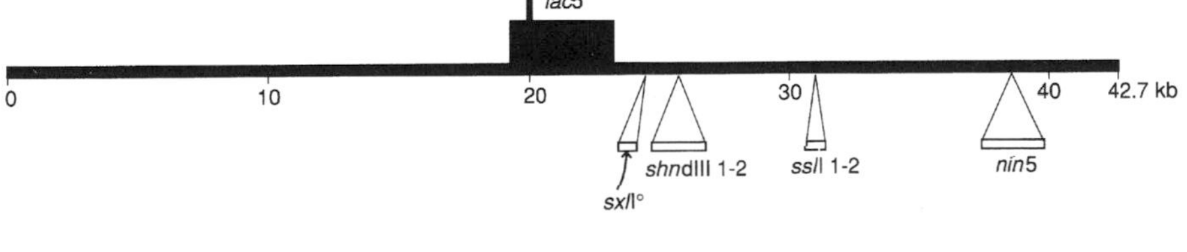

λgtWES.λB′

Type	Replacement vector
Capacity	2–13 kb
Left arm **Right arm**	21.2 kb 13.7 kb
Properties of recombinants	cIts Red$^-$ Gam$^+$
Recognition of recombinants	None
Genotype	λ*W*am403 *E*am1100 *inv* (*sr*Iλ1–*sr*Iλ2) Δ(*sr*Iλ2–*sr*Iλ3) *c*I*ts*857 *sr*Iλ4° *nin*5 *sr*Iλ5° *S*am100
Cloning site and coordinates	*Eco*RI 21230, 26100
Reference	75
Notes	1. The insertion in λgtWES.λB′ is *Eco*RI fragment 2 (coordinates 21230–26100) from the wild-type λ genome, in reverse orientation. This region therefore occurs twice in the vector. 2. *Sac*I (coordinates 21450 and 22550; S1 and S2 on the map below) can also be used for cloning. Capacity is unchanged, left arm 21.5 kb, right arm 17.2 kb. Recombinants have the same properties.

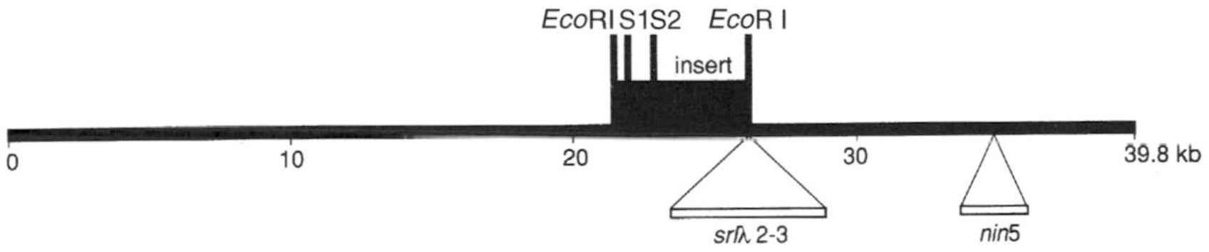

Type	Insertion vector
Capacity	0–9 kb
Left arm	22.1 kb
Right arm	18.7 kb
Properties of recombinants	cIts
Recognition of recombinants	Lac^-
Genotype	Uncertain: λ<T pSurfscript I> *cIts*857 *nin*5 + others
Cloning sites and coordinates	(T3 promoter) *Eco*RI 22090 *Not*I 22150 *Sfi*I 22180 *Spe*I 22190 *Xba*I 22850 (T7 promoter)

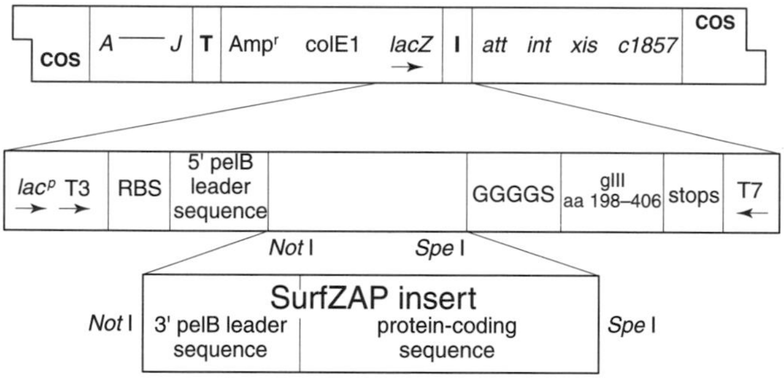

Type	Insertion vector
Capacity	0–10 kb
Left arm	22.0 kb
Right arm	18.8 kb
Properties of recombinants	cIts Red^+ Gam^+
Recognition of recombinants	Lac^-
Genotype	λ*sbh*Iλ1° *chi*A131 <T pBluescript SK- I> *sr*Iλ3° c*Its*857 *sr*Iλ4° *nin*5 *sr*Iλ5°
Cloning sites and coordinates	(T3 promoter) *Sac*I 22030 *Not*I 22050 *Xba*I 22050 *Spe*I 22060 *Eco*RI 22090 *Xho*I 22120 (T7 promoter)

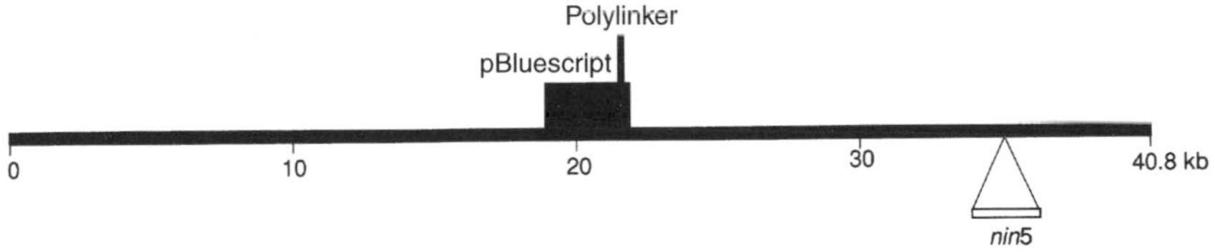

Type Insertion vector

Capacity 0–12 kb

Left arm 28.3 kb
Right arm 10.5 kb

Properties of recombinants cI^- Red^- Gam^+

Recognition of recombinants Lac^-

Genotype Unclear: λ*bio* <T pBK-CMV I>*red* WL113 KH34 *nin*5 + others

Cloning sites and coordinates

(T3 promoter)

*Sac*I	28330
*Sal*I	28350
*Spe*I	28360
*Bam*HI	28370
*Eco*RI	28380
*Hin*dIII	28380
*Xho*I	28390
*Xba*I	28400
*Not*I	28410
*Apa*LI	28410
*Sma*I	28420
*Kpn*I	28430

(T7 promoter)

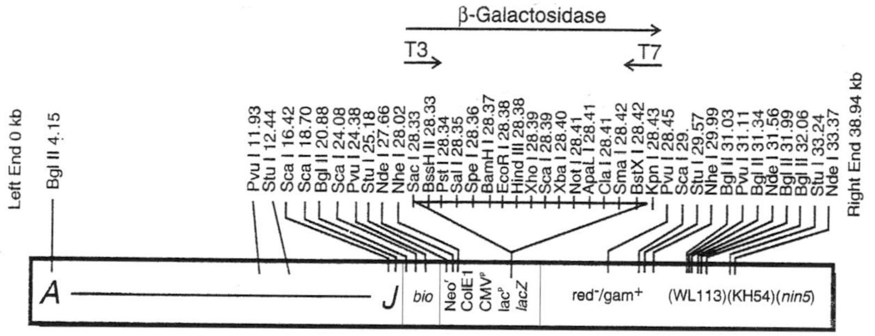

3. M13 CLONING VECTORS

The M13mp series of cloning vectors were developed specifically for the production of single-stranded DNA versions of cloned genes for use in chain termination sequencing (13, 52, 75, 76). Each vector is approximately 7.2 kb in size and carries the *lacZ′* gene. The sequences and restriction sites of the polylinkers carried by the various M13mp vectors are shown in *Figure 2*.

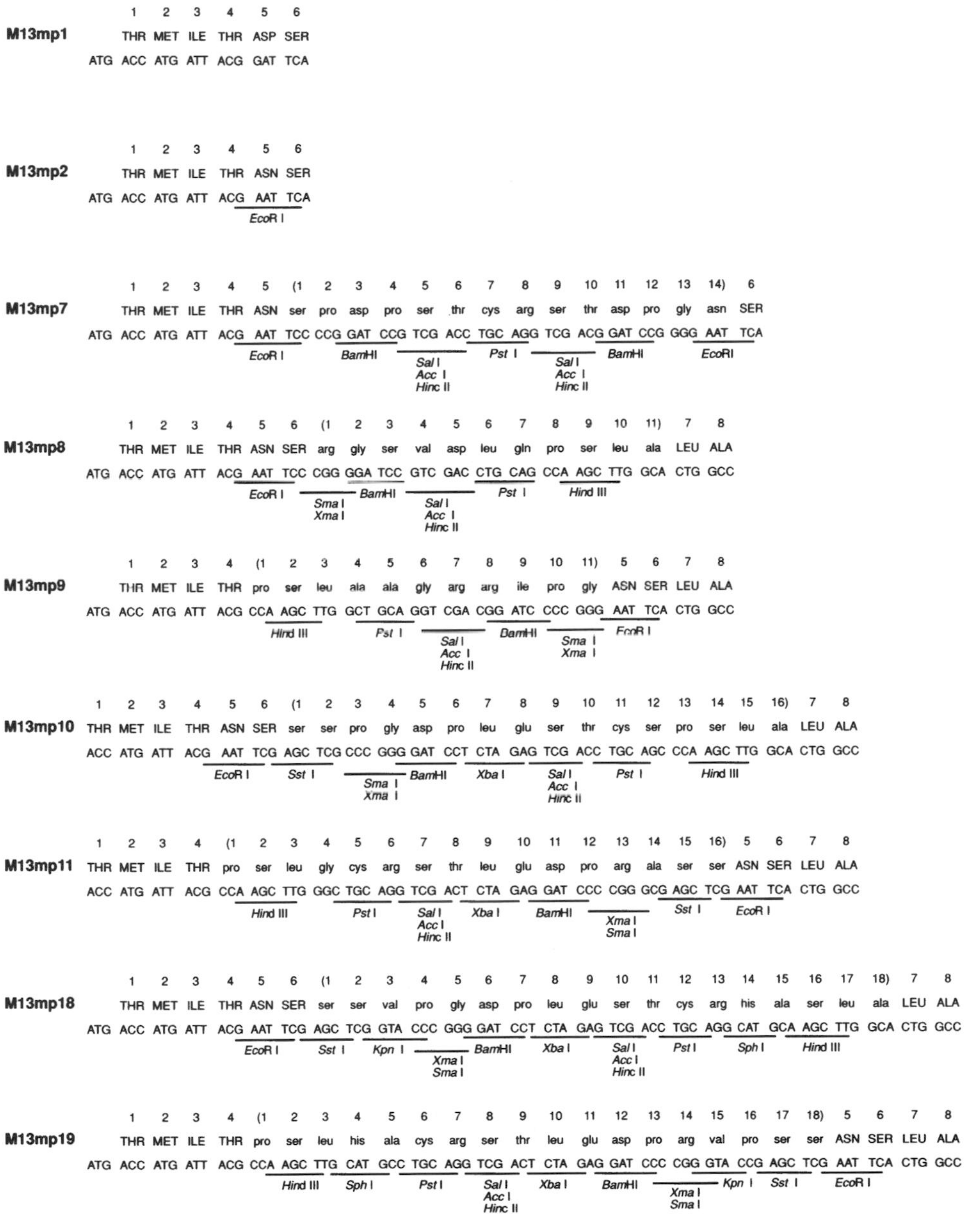

Figure 2. Polylinkers in M13mp vectors.

4. REFERENCES

1. Pouwels, P.H., Enger-Valk, B.E. and Brammar, W.J. (1985) *Cloning Vectors*. Elsevier, Amsterdam.

2. Kahn, M., Kolter, R., Thomas, C., Figurski, D., Meyer, R., Remaut, E. and Helinski, D.R. (1979) *Meth. Enzymol.*, **68**, 268.

3. Zinder, N.D. and Boeke, J.D. (1982) *Gene*, **19**, 1.

4. Chang, A.C.Y. and Cohen, S.N. (1978) *J. Bacteriol.*, **134**, 1141.

5. Bolivar, F., Rodriguez, R.L., Greene, P.J., Betlach, M.C., Heyneker, H.L., Boyer, H.W., Crosa, J.H. and Falkow, S. (1977) *Gene*, **2**, 95.

6. Clewell, D.B. (1972) *J. Bacteriol.*, **110**, 667.

7. Dotto, G.P. and Zinder, N.D. (1984) *Nature*, **311**, 279.

8. Bates, P.F. and Swift, R.A. (1983) *Gene*, **26**, 137.

9. Short, J.M., Fernandez, J.M., Sorge, J.A. and Huse, W.D. (1988) *Nucl. Acids Res.*, **16**, 7583.

10. Sutcliffe, J.G. (1978) *Nucl. Acids Res.*, **5**, 2721.

11. Bolivar, F. (1978) *Gene*, **4**, 121.

12. Soberon, X., Covarrubias, L. and Bolivar, F. (1980) *Gene*, **9**, 287.

13. Yanisch-Perron, C., Vieira, J. and Messing, J. (1985) *Gene*, **33**, 103.

14. Melton, D.A., Krieg, P.A., Rebegliati, M.R., Maniatis, T., Zinn, K. and Green, M.R. (1984) *Nucl. Acids Res.*, **12**, 7035.

15. Poustka, A., Rackwitz, H.-R., Frischauf, A.-M., Hohn, B. and Lehrach, H. (1984). *Proc. Natl. Acad. Sci. USA*, **81**, 4129.

16. Weiner, M.P.(1993) *Biotechniques*, **15**, 502.

17. Aslanidis, C. and de Jong, P.J. (1990) *Nucl. Acids Res.*, **18**, 6069.

18. Huan, R.S., Serventi, I.M. and Moss, J. (1992) *Biotechniques*, **13**, 515.

19. Dente, L., Cesarini, G. and Cortese, R. (1983) *Nucl. Acids Res.*, **11**, 1645.

20. Studier, F.W. and Moffatt, B.A. (1986) *J. Mol. Biol.*, **189**, 113.

21. Studier, F.W., Rosenberg, A.H., Dunn, J.J. and Dubendorff, J.W. (1990) *Meth. Enzymol.*, **185**, 60.

22. Rosenberg, A.H., Lade, B.N., Chui, D.S., Lin, S.W., Dunn, J.J. and Studier, F.W. (1987) *Gene*, **56**, 125.

23. Löwenadler, B., Jansson, B., Paleus, S., Holmgren, E., Nilsson, B., Moks, T., Palm, G., Josephson, S., Philipson, L. and Uhlen, M. (1987) *Gene*, **58**, 87.

24. Nilsson, B., Moks, T., Jansson, B., Abrahmsen, L., Elmblad, A., Holmgren, E., Henrichson, C., Jones, T.A. and Uhlen, M. (1987) *Protein Eng.*, **1**, 107.

25. Nilsson, B., Forsberg, G. and Hartmanis, M. (1991) *Meth. Enzymol.*, **198**, 3.

26. Churchward, G., Belin, D. and Nagamine, Y. (1984) *Gene*, **31**, 165.

27. Smith, D.B. and Johnson, K.S. (1988) *Gene*, **67**, 31.

28. Kaelin, W.G., Krek, W., Sellars, W.R., Decaprio, J.A., Ajchenbaum, F., Fuchs, C.S., Chittenden, T., Li, Y., Farnham, P.J., Blanar, M.A., Livingston, D.M. and Flemington, E.K. (1992) *Cell*, **70**, 351.

29. Ish-Horowicz, D. and Burke, J.F. (1981) *Nucl. Acids Res.*, **9**, 2989.

30. Brosius, J. (1984) *Gene*, **27**, 151.

31. Brosius, J., Ullroch, A., Raker, M.A., Gray, A., Dull, T.J., Gutell, R.R. and Noller, H.F. (1981) *Plasmid*, **6**, 112.

32. Brosius, J. and Holy, A. (1984) *Proc. Natl. Acad. Sci. USA*, **81**, 6929.

33. Gentz, R., Langner, A., Chang, A.C.Y., Cohen, S.N. and Bujard, H. (1981) *Proc. Natl. Acad. Sci. USA*, **78**, 4936.

34. Brosius, J. (1984) *Gene*, **27**, 161.

35. Amann, E. and Brosius, J. (1985) *Gene*, **40**, 183.

36. Shimizu, Y., Takabayashi, E., Yano, S., Shimizu, N., Yamada, K. and Gushima, H., (1988) *Gene*, **65**, 141.

37. Straus, D. and Gilbert, W. (1985) *Proc. Natl. Acad. Sci. USA*, **82**, 2014.

38. Brosius, J. (1988) in *Vector: A Survey of Molecular Cloning Vectors and Their Uses* (R.L. Rodriguez and D.T. Denhardt, eds). Butterworth, Boston, p. 205.

39. Talmadge, K., Stahl, S. and Gilbert, W. (1980) *Proc. Natl. Acad. Sci. USA*, **77**, 3369.

40. Talmadge, H. and Gilbert, W. (1980) *Gene*, **12**, 235.

41. Evans, P.D., Cook, S.N., Riggs, P.D. and Noren, C.J. (1995) *Biotechniques*, **19**, 130.

42. Riggs, P. (1994) in *Current Protocols in Molecular Biology* (F.M. Ausubel, R. Brent, R.E. Kingston, D.D. Moore, J.G. Seidman, J.A. Smith and K. Struhl, eds). John Wiley, New York. p. 16.6.14–16.6.14.

43. Shapira, S.K., Chou, J. Richard, F.V. and Casadaban, M.J. (1983) *Gene*, **25**, 71.

44. Casadaban, M.J., Arias, A.M., Shapira, S.K. and Chou, J. (1983) *Meth. Enzymol.*, **100**, 293.

45. Dean, D. (1981) *Gene*, **15**, 99.

46. Nilsson, B., Abrahmsen, L. and Uhlen, M. (1985) *EMBO J.*, **4**, 1075.

47. Brosius, J. (1989) *DNA*, **8**, 759.

48. Hough, D., Andrews, D.W., Petre, M. and Vaillancourt, P. (1994) *Strategies*, **7**, 15.

49. Tabor, S. and Richardson, C.C. (1985) *Proc. Natl. Acad. Sci. USA*, **82**, 1074.

50. Mead, D.A., Szczesna-Skorupa, E. and Kemper, B. (1986) *Protein Eng.*, **1**, 67.

51. Vieira, J. and Messing, J. (1982) *Gene*, **19**, 259.

52. Messing, J. (1983) *Meth. Enzymol.*, **101**, 20.

53. Norrander, J., Kempe, T. and Messing, J. (1983) *Gene*, **26**, 101.

54. Lobet, Y., Peacock, M. and Cieplak, W. (1989) *Nucl. Acids, Res.*, **17**, 4897.

55. Vieira, J. and Messing, J. (1987) *Meth. Enzymol.*, **153**, 3.

56. Wahl, G.M., Lewis, K.A., Ruiz, J.C., Rothenberg, B., Zhao, J. and Evans, G.A. (1987) *Proc. Natl. Acad. Sci. USA*, **84**, 2160.

57. Lau, Y.-F. and Kan, Y.W. (1983) *Proc. Natl. Acad. Sci. USA*, **80**, 5225.

58. Rossi, J.J., Soberon, X., Marumoto, Y., McMahon, J. and Itakura, K. (1983) *Proc. Natl. Acad. Sci. USA*, **80**, 3203.

59. Sambrook, J., Fritsch, E.F. and Maniatis, T. (1989) *Molecular Cloning: A Laboratory Manual (2nd edn)*. Cold Spring Harbor Laboratory Press, New York.

60. Karn, J., Brenner, S., Barnett, L. and Cesareni, G. 91980) *Proc. Natl. Acad. Sci. USA*, **77**, 5172.

61. Karn, J., Matthes, H.W.D., Gait, M.J. and Brenner, S. (1984) *Gene*, **32**, 217.

62. Blattner, F.R., Williams, B.G., Blechl, A.E., Thompson, K.D., Faber, H.E., Furlong, L.-A., Grunwald, D.J., Kiefer, D.O., Moore, D.D., Schumm, J.W., Sheldon, E.L. and Smithies, O. (1977) *Science*, **196**, 161.

63. Williams, B.G. and Blattner, F.R. (1979) *J. Virol.*, **29**, 555.

64. de Wet, J.R., Daniels, D.L., Schroeder, J.L., Williams, B.G., Thompson, K.D., Moore, D.D. and Blattner, F.R. (1980) *J. Virol.*, **33**, 401.

65. Nordstrom, K., Molin, S. and Hansen, H.A. (1980) *Plasmid*, **4**, 215.

66. Loenen, W.A.M. and Blattner, F.R. (1983) *Gene*, **26**, 171.

67. Dunn, I.S. and Blattner, F.R. (1987) *Nucl. Acids Res.*, **15**, 2677.

68. Frischauf, A.-M., Lehrach, H., Poustka, A. and Murray, N. (1983) *J. Mol. Biol.*, **170**, 827.

69. Huynh, T.V., Young, R.A. and Davis, R.W. (1985) in *DNA Cloning: A Practical Approach* (D.M. Glover, ed.). IRL Press, Oxford, Vol. 1, p. 49.

70. Young, R.A. and Davis, R.W. (1983) *Proc. Natl. Acad. Sci. USA*, **80**, 1194.

71. Han, J.H., Stratowa, C. and Rutter, W.J. (1987) *Biochemistry*, **26**, 1617.

72. Han, J.H. and Rutter, W.J. (1988) in *Genetic Engineering: Principles and Methods* (J.K. Setlow, ed.). Plenum, New York, Vol. 10, p. 195.

73. Han, J.H. and Rutter, W.J. (1987) *Nucl. Acids Res.*, **15**, 6304.

74. Leder, P., Tiemeier, D. and Enquist, L. (1977) *Science*, **196**, 175.

75. Messing, J., Gronenborn, B., Muller-Hill, B. and Hofschneider, P.H. (1977) *Proc. Natl. Acad. Sci, USA*, **74**, 3642.

76. Yanisch-Perron, C., Vieira, J. and Messing, J. (1985) *Gene*, **33**, 103.

INDEX